V The Particle Structure of the Weakly Coupled $P(\varphi)_2$ Model and Other Applications of High Temperature Expansions: Part I. Physics of Quantum Field Models

Contemporary Physicists

Gian-Carlo Rota
David Sharp

Editors

Birkhäuser
Boston · Basel · Stuttgart

James Glimm
Arthur Jaffe

Collected Papers
Volume 1

Quantum Field Theory and
Statistical Mechanics

Expositions

Birkhäuser
Boston · Basel · Stuttgart
1985

James Glimm
Courant Institute
New York, N.Y.
USA

Arthur Jaffe
Harvard University
Cambridge, Mass.
USA

Library of Congress Cataloging in Publication Data

Glimm, James.
Quantum field theory and statistical mechanics.

(Contemporary physicists)
Reprint of articles originally published 1969–1977.
Includes bibliographies.
1. Quantum field theory—Addresses, essays, lectures.
2. Statistical mechanics—Addresses, essays, lectures.
I. Jaffe, Arthur, 1937– . II. Title. III. Series.
QC174.46.G585 1985b 530.1'43 84–21524
ISBN-13: 978-1-4612-5417-1 e-ISBN-13: 978-1-4612-5415-7
DOI: 10.1007/978-1-4612-5415-7

CIP-Kurztitelaufnahme der Deutschen Bibliothek

Glimm, James:
Collected papers / James Glimm ; Arthur Jaffe.
Boston ; Basel ; Stuttgart : Birkhäuser

(Contemporary physicists)

NE: Jaffe, Arthur:; Glimm, James: [Sammlung];
Jaffe, Arthur: [Sammlung]

Vol. 1. Quantum field theory
and statistical mechanics : expositions. – 1985.

Softcover reprint of the hardcover 1st Edition 1984

James Glimm/Arthur Jaffe

Collected Papers – Volume 1

Collected Papers – Volume 2

I Construction of $P(\varphi)_2$

II Phase Cell Localization and $\varphi_3{}^4$ Stability

III Phase Transitions Exist

IV Phase Transitions and Critical Behavior

V Particle Structure

VI Bounds on Coupling Constants

VII Confinement and Instantons

VIII Reflection Positivity

Introduction

This volume contains a selection of expository articles on quantum field theory and statistical mechanics by James Glimm and Arthur Jaffe. They include a solution of the original interacting quantum field equations and a description of the physics which these equations contain. Quantum fields were proposed in the late 1920s as the natural framework which combines quantum theory with relativity. They have survived ever since.

The mathematical description for quantum theory starts with a Hilbert space H of state vectors. Quantum fields are linear operators on this space, which satisfy nonlinear wave equations of fundamental physics, including coupled Dirac, Maxwell and Yang-Mills equations. The field operators are restricted to satisfy a "locality" requirement that they commute (or anti-commute in the case of fermions) at space-like separated points. This condition is compatible with finite propagation speed, and hence with special relativity. Asymptotically, these fields converge for large time to linear fields describing free particles. Using these ideas a scattering theory had been developed, based on the existence of local quantum fields.

Whenever physicists had attempted to find solutions to such equations for quantum fields, they ran into formal difficulties. The requirement of Lorentz invariance forced the quantum fields to exhibit local singularities which appeared incompatible with the possibility to interpret nonlinear functions of the fields and therefore with nonlinear equations of motion. In practical terms, calculations led to infinite answers, which physicists learned to avoid only by adhering to a set of rules known as "renormalization." One interpretation of renormalization is that certain constants in the quantum mechanical Hamiltonian must be infinite, as is demonstrated in the paper **An infinite renormalization of the Hamiltonian is necessary.**

The rules to perform calculations were developed in the 1940s for the coupled equations of Maxwell and Dirac, i.e., for electrodynamics, by Bethe, Dyson, Feynman, Schwinger, Tomonaga, Weisskopf, and others. They gave a procedure for calculating two important observable deviations from Dirac's equations: the Lamb shift in the hydrogen spectrum and the "anomalous" magnetic moment of the electron and hence led physicists to conclude that quantum field theory was a correct interpretation of nature. The tenth of one percent correction to the magnetic moment predicted by Dirac can now be measured to one part in 10^7 and provides the most accurate comparison in physics between experiment and calculation. For this reason, physicists accept quantum field theory and apply these ideas to understand weak and strong forces as well as electromagnetic and gravitational ones. The quest for appropriate equations to describe elementary particles led to a unified framework for weak and electromagnetic interactions, namely a quantum theory with $SU(2) \times U(1)$ gauge fields. One speculates that SU(5) may also unify the strong forces and even further that a quantum theory including gravity can be formulated.

In spite of the success of renormalization in electrodynamics, and its extension to other quantum field interactions, there was no picture of how to fit quantum fields into a traditional mathematical framework. In the mid 1950s, a number of physicists including Haag, Jost, Lehmann, Symanzik, Wightman, and Zimmermann set up a general scheme into which a putative field theory could fit, while Hepp, Ruelle, and others investigated consequences of their proposal. Within this structure one could derive general relations such as the connection between spin and statistics, the PCT theorem, and scattering theory. These features are independent of the detailed equations of motion. However no non-trivial examples were known. This gap led to an effort over the last twenty years to demonstrate that quantum fields do exist and to analyze their structure starting from classical Lagrangians. This program has become known as *Constructive Quantum Field Theory,* or CQFT.

Ultimately CQFT has opened new directions both in physics and in mathematics. The two sets of lecture notes, **Field Theory Models and Boson Quantum Field Models,** develop the basic theme of CQFT from the Hamiltonian point of view. They appeared shortly after the first success of constructing two dimensional quantum field models and were intended as a first draft of a comprehensive monograph explaining the subject to both physicists and mathematicians; as such, they contain a comprehensive exposition and review of the literature. They were presented at summer schools in Les Houches (1970) and in London (1971) respectively, but the monograph was never completed.

The basic idea which led to understanding CQFT was *phase cell localization.* Phase space is the product of position and momentum spacer. The uncertainty principle (for instance in the form that a function and its Fourier transform cannot simultaneously have bounded support) places limits on allowed phase cell localization. One looked at parts of quantum fields, or degrees of freedom, approximately localized in a bounded region of phase space, namely in phase cell volumes of size O(1). One generically finds a finite number of degrees of freedom per unit volume. This guiding principle of analyzing degrees of freedom in geometrically increasing or geometrically decreasing momentum ranges, but O(1) phase space volume, was implemented throughout CQFT. It led to expansion methods for the study both of infrared problems and of ultraviolet problems. Related ideas underlie the renormalization group approach.

In 1972 the subject turned in a different direction by incorporating certain decade-old ideas of K. Symanzik relating Euclidean field theory to quantum mechanics via function space integration. While function space integration had been the major tool in the proof of estimates for the Hamiltonian approach above, and entered through the Feynman-Kac representation of the heat kerne!, the explicit Euclidean symmetry in Symanzik's approach came to play a central role. Work focused on studying the existence and properties of Euclidean field; using the Osterwalder-Schrader reconstruction theorem one could infer the analytic continuation from Euclidean fields to quantum fields and consequences for the study of the Hamiltonian and its spectrum. Work from this point of view was developed by Nelson, Guerra, Rosen, Simon, and Fröhlich, as well as the authors.

The 1973 Erice lectures, **The particle structure of the weakly coupled $P(\phi)_2$ model and other applications of high temperature expansions,** reflect the point of view adopted by the authors that a combination of both methods can provide deep insights. **Part 1: physics of quantum field models** deals with physics which can be studied in the laboratory provided by mathematically exact quantum field models, as well as how to extend the constructive approach to dimension three. Questions of particle structure – e.e., the spectrum of the quantum mechanical Hamiltonian H, its bound states and scattering, and questions of phase transitions, symmetry breaking, and critical phenomena had begun to complement renormalization as central issues of interest. Later developments led to work by Spencer, Zirilli, Dimock, Eckmann and others on two particle scattering via an exact Bethe-Salpeter equation. In **Particles and bound states and progress toward unitarity and scaling,** the authors propose a program to study n-particle scattering using kernels they define which are n-particle irreducible in a given channel. Their application to the spectral properties of scattering has only been partially understood.

The second portion of the Erice lectures, **The cluster expansion,** presents one version of a technical tool to investigate some of these problems. This expansion technique has come to be known as the "cluster expansion," because it invokes a decoupling strategy similar to high temperature expansions in statistical physics also known by that name. The cluster expansion also adopts in detail the phase cell localization ideas discussed above. Originally, the cluster expansion was developed to study large-distance effects in correlations, so called infrared-phenomena associated with the spectrum of H. It now turns out that these ideas can be used in the analysis of with certain short-distance questions of local regularity, or ultraviolet phenomena.

It is within the Euclidean framework that the relation between statistical mechanics and quantum theory became evident. The study of Euclidean fields boils down to the study of the statistical mechanics of classical field theory (classical Grassmanian fields in the case of Fermions). The probability distribution is a Gibbs distribution with the exponential weight exp(-S), where the functional of the fields S is the classical (Euclidean) action functional. For this reason, it is not surprising that the phase transitions and critical phenomena of statistical physics appear in quantum field theory. This connection between field theory and statistical mechanics led to the first existence proof for degenerate ground states in a continuum quantum field theory (i.e. the existence of multiple vacua and symmetry breaking), through an analysis of the equilibrium states of the corresponding statistical mechanics system in Euclidean field theory. One aspect of this joint work of the authors and Thomas Spencer is described in **Existence of phase transitions for ϕ_2^4 quantum fields,** a paper presented at a June 1975 conference in Marseilles. The detailed proof, as well as the unification of these ideals with the cluster expansion methods, is published in journal articles. Once again the theme of phase cell localization enters, with a decomposition of the field into high and low momentum parts. The low momentum degrees of freedom in each bounded space-time volume are analyzed according to a Peierls-type argument from statistical physics, while the high momentum degrees of freedom are shown to be fluctuations which have only a small influence on the structure of the phase diagram. Extension of these ideas and methods have led to other developments in the theory of phase transitions in statistical physics as well as in quantum field theory.

The physics of the particle structure and its connection with phase transitions through the approach to the critical point is discussed further in the articles, **Critical problems in quantum field theory, Critical exponents and renormalization in the ϕ^4 scaling limit,** and **A tutorial course in constructive field theory.**

In summary it is fitting to publish together here expositions of the two alternative approaches to CQFT: the Hamiltonian approach based on Hilbert space methods and the Euclidean approach based on functional integrals and probability theory. The text *Quantum Physics* by the authors concentrates on the latter. In many ways the technical estimates for scalar fields are more straightforward within the probabilistic framework. In the case of fermion and gauge theories, the question is still debatable. Physicists and mathematicians now are interested again in understanding the operator formulation of quantum field theory, as well as the Euclidean formulation, since it may have advantages for the study of certain open problems. Thus the authors hope that making these papers accessible will provide a valuable working reference.

Bibliography

An infinite renormalization of the Hamiltonian is necessary, J. Math. Phys. 10, 2213–2214 (1969).

Field theory models, in *Statistical Mechanics and Quantum Field Theory,* C. De Witt and R. Stora, Editors, Gordon and Breach, New York 1971.

Boson quantum field models, in *Mathematics of Contemporary Physics,* R. Streater, Editor, Academic Press, London, 1972. (References at the end of the volume.)

The particle structure of the weakly coupled $P(\phi)_2$ model and other applications of high temperature expansions, Part I: physics of quantum field models, and Part II: the cluster expansion, with T. Spencer, in *Constructive Quantum Field Theory,* G. Velo and A. S. Wightman, Editors, Lecture Notes in Physics 25, Springer, New York 1973.

Particles and bound states and and progress toward unitarity and scaling, in *International Symposium on Mathematical Problems in Theoretical Physics,* H. Araki, Editor, Lecture Notes in Physics 39, Springer, New York 1975.

Critical problems in quantum fields, in *Les Méthods Mathématiques de la Théorie Quantique des Champs,* F. Guerra, D. Robinson, and R. Stora, Editors, CNRS, Paris 1976.

Existence of phase transitions for ϕ_2^4 quantum fields, with T. Spencer, in *Les Méthods Mathématiques de la Théorie Quantique des Champs,* F. Guerra, D. Robinson, and R. Stora, Editors, CNRS, Paris 1976.

Critical exponents and renormalization in the ϕ^4 scaling limit, in *Quantum Dynamics: Models and Mathematics,* L. Streit, Editor, Springer, Vienna 1976.

A tutorial course in constructive field theory, in *New Developments in Quantum Field Theory and Statistical Mechanics, Cargèse 1976,* P. Mitter and M. Levy, Editors, Plenum, New York 1977.

I Infinite Renormalization of the Hamiltonian Is Necessary

Infinite Renormalization of the Hamiltonian Is Necessary

JAMES GLIMM*
Courant Institute of Mathematical Sciences, New York University, New York, New York

AND

ARTHUR JAFFE†
Lyman Laboratory of Physics, Harvard University, Cambridge, Massachusetts

(Received 28 April 1969)

We show that the unrenormalized Hamiltonian in quantum field theory is unbounded from below whenever lowest-order perturbation theory indicates that this is true. We conclude that perturbation theory is an accurate guide to the divergence of the vacuum energy in quantum field theory.

In this paper we show that the unrenormalized Hamiltonian is unbounded from below when lowest-order perturbation theory predicts that this is true.

The proof is a simple calculation. The unrenormalized Hamiltonian H is a densely defined bilinear form on Fock space.[1] We choose a sequence Ω_κ of unit vectors in the domain of H. As $\kappa \to \infty$, the expectation values of the Hamiltonian in the vectors Ω_κ tends to $-\infty$. The ground state of H given by first-order perturbation theory motivates our choice of the vectors Ω_κ.

We concentrate on the boson self-interaction Φ^{2n} in $(s+1)$-dimensional space–time. The second-order vacuum energy has a momentum divergence for $s \geq 2$, $n \geq 2$ and for $s \geq 3$, $n = 1$, and these cases will be treated first. Afterwards, the remaining cases (which have a volume divergence) will be treated.

The Hamiltonian that we study is

$$H = H_0 + \lambda \int :\Phi(x)^{2n}: dx = H_0 + \lambda H_I. \tag{1}$$

The methods and the results hold equally for the spatially cut-off Hamiltonian:

$$H(g) = H_0 + \lambda \int :\Phi(x)^{2n}: g(x)\, dx, \quad 0 \leq g(x) \leq 1. \tag{2}$$

We work on the Fock space; H_0 is the free-field Hamiltonian for mass $m > 0$, and Φ is the standard boson field.

We study first the cases $s \geq 2$, $n \geq 2$ and $s \geq 4$, $n \geq 1$. The vectors Ω_κ are defined by

$$\Omega_\kappa = c(\psi_0 - \lambda \psi_{2n}),$$

where ψ_0 is the Fock no-particle vector and ψ_{2n} is a $2n$-particle vector. The constant c is chosen so that $\|\Omega_\kappa\| = 1$, and

$$\psi_{2n}(k_1, \cdots, k_{2n}) = \left(\sum_{i=1}^{2n} \mu(k_i)\right)^{-1} ((2n)!)^{\frac{1}{2}} h\left(\sum_{i=1}^{2n} k_i\right) \prod_{i=1}^{2n} [\mu(k_i)^{\tau - \frac{1}{2}} \chi_\kappa(k_i)]. \tag{3}$$

Here $\mu(k) = (k^2 + m^2)^{\frac{1}{2}}$ and h is a smooth, positive, rapidly decreasing function. The function $\chi_\kappa(k)$ equals unity if $|k| \leq \kappa$; it equals zero otherwise. We choose

$$\tau = -(s-1)\left(1 - \frac{1}{2n}\right) + \frac{\epsilon + 2}{2n},$$

* Supported by the New York Science and Technology Foundation, Grant SSF-(8)-8.
† Alfred P. Sloan Foundation Fellow. Supported in part by the Air Force Office of Scientific Research, AF49(638)-1380.

[1] A. Galindo Proc. Natl. Acad. Sci. (U.S.) **48**, 1128 (1962).

where ϵ is in the interval $0 < \epsilon < \frac{1}{2}$. We remark that the choices $h = \delta$, $\kappa = \infty$, $\tau = 0$ would give the ground state in first-order perturbation theory. With the above restrictions on n, s, and ϵ, we have $\tau < -\frac{1}{8}$. Furthermore,

$$\epsilon = 2n\tau + (2n - 1)s - (2n + 1).$$

Theorem 1:

$$\lim_{\kappa \to \infty} (\Omega_\kappa, H\Omega_\kappa) = -\infty.$$

Proof: We compute the inner product as

$$(\Omega_\kappa, H\Omega_\kappa) = c^2\{P_0 + \lambda P_1 + \lambda^2 P_2 + \lambda^3 P_3\}.$$

In this expansion it is easy to see that

$$P_0 = P_1 = 0$$

and that

$$P_2 = (\psi_{2n}, H_0\psi_{2n}) - (\psi_0, H_I\psi_{2n}) - (\psi_{2n}, H_I\psi_0).$$

The proof is completed by showing that, for large κ,

$$c^{-2} = 1 + \lambda^2 \|\psi_{2n}\|^2 = O(1),$$
$$(\psi_{2n}, H_0\psi_{2n}) = O(1),$$
$$P_3 = O(1),$$

and that for some positive constant D,

$$D\kappa^\epsilon \leq (\psi_0, H_I\psi_{2n}) = (\psi_{2n}, H_I\psi_0).$$

These orders of growth are established by standard power-counting arguments.[2] This completes the proof. The remaining case $n = 1, s = 3$ is handled by similar methods, modified to deal with a logarithmic divergence.

In the cases not covered by Theorem 1, perturbation theory predicts no momentum divergence. Thus, when g has compact support, perturbation theory predicts that $H(g)$ is bounded from below. This lower bound has been proved rigorously.[3] Perturbation theory predicts that H has a vacuum-energy divergence which is linear in the volume, and thus it predicts a lower bound for $H(g)$ which is linear in the volume. (The "volume" here is the area of the support of g.)

It is known that the true bounds are no worse than this prediction. Thus, for $s = 1$ or $s = 2$, $n = 1$, the lower bound diverges no faster than a constant times the volume.[3] We now show that, for $s = 1$, the H defined in Eq. (1) is unbounded from below. The same proof shows that the lower bound on $H(g)$ in Eq. (2) tends to $-\infty$ as $g \to 1$, and similar results hold for the case $s = 2$, $n = 1$.

Let

$$h_V(k) = Vh(kV).$$

In the definition (3) for ψ_{2n}, we substitute h_V for h and set $\tau = 0$, $\kappa = \infty$. We define

$$\Omega_V = \psi_0 - V^{-\frac{1}{2}}\lambda\psi_{2n}.$$

As before, one proves the following theorem:

Theorem 2:

$$\lim_{V \to \infty} (\Omega_V, H\Omega_V) = -\infty,$$

$$\limsup_{V \to \infty} \|\Omega_V\|^2 < \infty.$$

We conclude that perturbation theory is an accurate guide to the divergence of the vacuum energy in quantum field theory.

[2] S. Weinberg, Phys. Rev. **118**, 839 (1960).

[3] For the case $s = 1$, any n, see E. Nelson in *Mathematical Theory of Elementary Particles* (M.I.T. Press, Cambridge, Mass., 1966); J. Glimm, Commun. Math. Phys. **8**, 12 (1968); J. Glimm and A. Jaffe (to be published). The case $s = 2$, $n = 1$ can be computed explicitly or estimated.

II Quantum Field Theory Models: Part I. The φ_2^{2n} Model

* Supported in part by the U.S. Air Force Office of Scientific Research, Contract AF-49(638)-1719.

† Alfred P. Sloan Foundation Fellow. Supported in part by the U.S. Air Force Office of Scientific Research, Contract F 44 620-70-C-0030.

Introduction

Quantum fields are believed to provide a correct description of the interactions between particles. In the case of quantum electrodynamics, calculations based on perturbation theory provide exact agreement between theory and experiment within the limits of experimental accuracy. These calculations rely on infinite renormalizations and therefore reveal that quantum fields are highly singular. In these lectures, we will be concerned with the qualitative structure of quantum fields. Given a classical Lagrangian, we seek a mathematically complete construction of the corresponding quantum fields. Because of the difficulty of this problem, we have restricted ourselves to two space time dimensions. With this restriction, the program has been successful. Quantum fields have been constructed for the ϕ^{2n} and Yukawa interactions in two dimensions. Many properties of the fields have been verified for these models, and the ϕ^{2n} theory obeys all the Haag-Kastler axioms. Two major unanswered questions are:

a) Can the existence of local quantum fields be extended to three space time dimensions?

b) For the two dimensional models, can detailed properties of the fields be established in order to describe elementary particles and their bound states? Is standard perturbation theory asymptotic to the exact solution of the models? Do standard dispersion relations hold for the cross sections?

In quantum field theory there are four types of divergences, the infra red, ultraviolet, infinite volume and particle number divergences. The infra red divergence is associated with the divergence of integrals such as

$$\int_{|k|\leq 1} |k|^{-1}\, dk.$$

This divergence occurs at low frequencies (i.e. small momenta) and it occurs only when at least one particle has zero mass. This divergence is more singular in lower dimensions. Thus there is no necessity to analyse the infra red difficulties in two dimensions where they are worse than in the physical world of four dimensions, and we take all masses to be positive.

The ultraviolet divergences are associated with integrals such as

$$\int (k^2 + m^2)^{-1/2}\, dk$$

and these integrals become more divergent in higher dimensions. The ultraviolet divergences arise in the definition of the nonlinear terms in the field equations. For the ϕ^{2n} model, the classical field equation is

$$\phi_{tt} - \phi_{xx} + m_0^2\phi + 2n\phi^{2n-1} = 0.$$

The ultraviolet divergences have a rather trivial nature in the corresponding quantum field equation, because of the restriction to two dimensions. The Yukawa model is more singular and there is an infinite shift in the boson mass (and in the vacuum energy) due to the ultraviolet divergences. These divergences are predicted by second order perturbation theory. The mass divergence is cancelled by a renormalization of the field equation. We substitute

$$m^2 - \delta(m^2) = m_0^2$$

for m_0^2 in the field equation. Here m_0 and $-\delta m^2$ are infinite for the Yukawa model but finite for the ϕ^{2n} model. The infinite part of δm^2 is chosen to cancel other infinities in the theory. This leaves δm^2 uniquely fixed, modulo some finite contribution. According to conventional ideas, the latter is determined by a requirement placed on the solution ϕ of the field equations. Namely, ϕ should describe an elementary particle of (physical) mass $m = m_{\text{phys.}}$. Since we do not have satisfactory control over the finite part of the mass renormalization in our models, we do not discuss this point further.

The infinite volume divergence arises from the translation invariance of the theory, i.e. from conservation of momentum. In calculations, this divergence occurs as integrals of the form

$$\int_R dx = \int_{\hat{R}} \delta(k)^2\, dk.$$

In a similar but less obvious fashion, the ultraviolet divergences are forced by the invariance of the theory under Lorentz rotations.

The divergence for large particle numbers has a different character from the ultraviolet or volume divergences. This divergence is not cancelled by some other infinity. Rather, it remains in the theory and reflects the fact that the Taylor's expansion of the S-matrix,

$$S = S(\lambda) \sim \sum \frac{\lambda^n}{n!}\left[\left(\frac{d}{d\lambda}\right)^n S(\lambda)\Big|_{\lambda=0}\right],$$

appears to diverge. Here λ is the coupling constant, and is inserted into the field equation as a coefficient multiplying the nonlinear term. The Taylor's series is simply the perturbation expansion, and because it presumably diverges, we must use nonperturbative methods in an essential manner.

We now discuss the methods used to control the ultraviolet and volume divergences; we then discuss the particle number divergence separately. We first introduce an approximation into the problem to eliminate the divergences. For the ultraviolet divergences we use a momentum cutoff.

For example we may replace ϕ by $\phi_\varkappa = h_\varkappa * \phi$, where $h_\varkappa(x) = \varkappa h(\varkappa x)$ is a suitable function with total integral one. For the infinite volume divergence, we may use either (or occasionally both) of two approximations. We may destroy translation invariance by introducing a space cutoff into the nonlinear terms of the field equations, e.g.

$$\phi^{2n-1}(x, t) \to g(x)\, \phi^{2n-1}(x, t)$$

where the cutoff function g is nonnegative and has compact support. Alternatively, we may replace R by a circle (quantization in a box with periodic boundary conditions). For the cutoff theories there are no infinite divergences, but there are cancellations between large finite quantities. We perform the cancellations first, and then estimate the remainders in order to take the limits $\varkappa \to \infty$, $V \to \infty$ and $g \to 1$. In this program, the major difficulty is to obtain estimates which are valid uniformly as the cutoffs are removed.

In order to avoid the particle number divergence, we use nonperturbative methods. We use a Hamiltonian, which serves to reduce the nonlinear field equation to an equivalent linear problem. Let H be the total energy operator and let $H(g)$ and $H(g, \varkappa)$ be the total energy operators associated with the cutoff fields and cutoff field equations. The main point is to realize H (or $H(g)$) as a self adjoint operator, because a simple calculation shows that

$$\phi(x, t) = e^{itH(g)}\, \phi(x, 0)\, e^{-itH(g)}$$

solves the g-cutoff field equations. Furthermore, using finite propagation speed (i.e. the hyperbolic character of the field equations), we can piece together different g-cutoff fields for different g's to obtain a single field ϕ independent of g. The study of $H(g)$ is the linear problem which must be approached by nonperturbative methods. In fact,

$$H(g) = \lim_{\varkappa \to \infty} H(g, \varkappa) \quad \text{and} \quad H = \lim_{g \to 1} H(g)$$

are defined as limits, and most of the difficulty in our proofs occurs in the estimates on operators that we need to control these limits.

The Cauchy data $\phi(x, 0)$ and $\dot{\phi}(x, 0) = \pi(x, 0)$ in the equation above are time zero free fields. We begin our construction with $\phi(x, 0)$ given in the standard fashion, acting on Fock space. However, a complete treatment of the limit $g \to 1$ and the corresponding infinite volume renormalization will force us to modify this hypothesis. We must choose a non-Fock representation of $\phi(x, 0)$ and $\pi(x, 0)$, and the particular representation is related intimately to the interaction and the field equations we are solving.

In summary, our construction of the ϕ_2^{2n} model follows the outline of Figure 1. The Yukawa$_2$ model follows the same outline, but is at present completed only through the construction of the field operators and the verification of the field equations. A similar outline should hold for models in three space time dimensions.

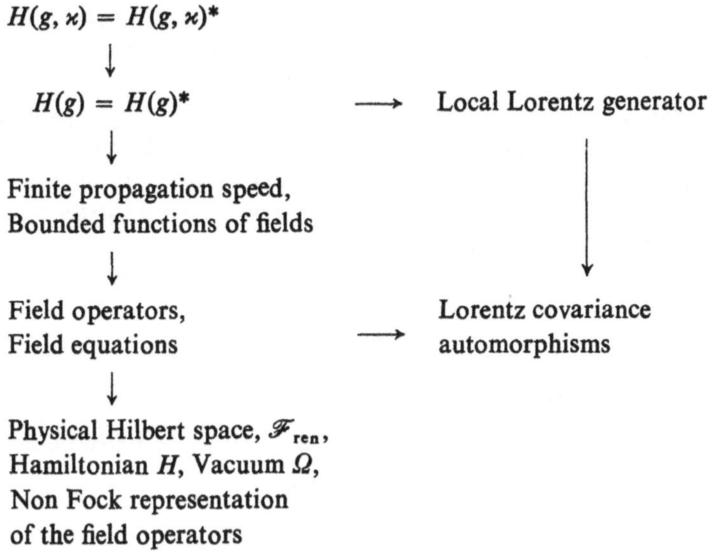

Figure 1. Construction of the ϕ_2^{2n} quantum field theory.

Our final field theory should satisfy the usual requirements of physics: For the ϕ_2^{2n} model we show that the Hamiltonian is positive, $0 \leqq H$, and that H has a ground state Ω, $H\Omega = 0$. On the physical Hilbert space $\mathscr{F}_{\text{ren}}$, $\phi(x, t) = e^{itH} \phi(x, 0) e^{-itH}$. The field is local, so that a boson field satisfies $[\phi(x, t), \phi(x', t')] = 0$ if $|x - x'| > |t - t'|$. The theory is Lorentz covariant; the Lorentz transformations should be unitarily implemented, and so we expect that

$$U(a, \Lambda)\, \phi(x, t)\, U(a, \Lambda)^* = \phi(\{a, \Lambda\}\, (x, t)).$$

Physical particles should occur in the theory as discrete eigenvectors for the operator $(H^2 - P^2)^{1/2}$, where P is the momentum operator; see Glimm-Jaffe[9] for a more detailed discussion. The scattering of particles, in principle, is completely described by the fields, see Jost[1].

Part I The ϕ_2^{2n} Model

The ϕ_2^4 theory is the least singular and most highly developed of our quantum field theory models. The $P(\phi)_2$ theory is essentially as complete, with P a positive polynomial. Wightman[2] proposed the construction of local field theories as a limit of cutoff models. Two early results were the proof of semiboundedness of the Hamiltonian by Nelson[3] and the construction of space and momentum cutoff models by Jaffe(2,3). Subsequent development of the theory was made by Glimm(3), by Glimm and Jaffe(2,3,4,5) and by Cannon and Jaffe(1). These results were extended to the $P(\phi)_2$ theory by Rosen(1,2,3). Simplifications of some of the proofs have been given by Segal(2,3) and by Hoegh-Krohn and Simon(1). Segal(2) announced results for $P(\phi)_2$ which are independent of (and somewhat weaker than) Rosen's(1). Other contributions to the $P(\phi)_2$ program are due to Lanford (see Wightman(1), Glimm and Jaffe(2)), Guenin(1), Segal(1), Jaffe and Powers(1), Federbush(1), Glimm and Jaffe(6,10–11), Jaffe, Lanford and Wightman(1), Hoegh-Krohn(1,2) and Simon(1). For a further discussion of bibliographical points, see the cited articles.

1 Fock Space

1.1 Annihilation-creation forms

The Fock space $\mathscr{F}$ for our $P(\phi)_2$ model is the Hilbert space completion of the symmetric tensor algebra over $L_2(R)$,

$$\mathscr{F} = \mathfrak{S}(L_2(R)) = \bigoplus_{n=0}^{\infty} \mathscr{F}_n, \tag{1.1.1}$$

where $\mathscr{F}_n$ is the space of n noninteracting particles:

$$\mathscr{F}_n = L_2(R) \otimes_s L_2(R) \otimes_s \dots \otimes_s L_2(R) \qquad (n \text{ factors}). \tag{1.1.2}$$

The variable $k \in R$ denotes momentum. For $\psi = \{\psi_0, \psi_1, \dots\} \in \mathscr{F} = \mathscr{F}_0 \oplus \mathscr{F}_1 \oplus \dots$, we have $\|\psi\|^2 = \sum_{n=0}^{\infty} \|\psi_n\|^2$. The no particle space $\mathscr{F}_0$ is the complex numbers and $\Omega_0 = \{1, 0, \dots\} \in \mathscr{F}$ is the (bare) vacuum or (bare) no particle vector. We define the fractional order energy operator N_τ by

$$(N_\tau \psi)_n (k_1, \dots, k_n) = \sum_{j=1}^{n} \mu(k_j)^\tau \psi_n(k_1, \dots, k_n) \tag{1.1.3}$$

where $\mu(k) = (k^2 + m^2)^{1/2}$. Thus $N_0 = N$ is the number operator and $N_1 = H_0$ is the free energy operator. The annihilation operator $a(k)$ is defined by the formula

$$(a(k)\psi)_{n-1} (k_1, \dots, k_{n-1}) = n^{1/2} \psi_n(k, k_1, \dots, k_{n-1}). \tag{1.1.4}$$

We take as a domain for $a(k)$ the set $\mathscr{D}$ of vectors $\psi = \{\psi_n\}$ with a finite number of particles and with wave functions ψ_n in the Schwartz space $\mathscr{S}(R^n)$. Any monomial $A(k)^* A(k')$, where

$$A(k') = a(k_1') \dots a(k_s'),$$

$$A(k)^* = a(k_1)^* \dots a(k_r)^*,$$

is a bilinear form on $\mathscr{D} \times \mathscr{D}$, and for $\theta \in \mathscr{D}$,

$$\langle \theta, A(k)^* A(k') \theta \rangle \in \mathscr{S}(R^{r+s}).$$

Thus we define the weak integral of bilinear forms

$$W = \int A^*(k) w(k, k') A(k') \, dk \, dk' \tag{1.1.5}$$

for any distribution w in $\mathscr{S}'(R^{r+s})$. On $\mathscr{D} \times \mathscr{D}$, we compute that if $w(k, k') = \mu(k)^\tau \delta(k - k')$,

$$\begin{aligned} N_\tau &= \int a^*(k) \mu(k)^\tau \delta(k - k') a(k') \, dk \, dk' \\ &= \int a^*(k) \mu(k)^\tau a(k) \, dk. \end{aligned} \tag{1.1.6}$$

For the Y_2 model of Part II, our Fock space will be

$$\mathscr{F} = \mathscr{F}_b \otimes \mathscr{F}_f = \bigoplus_0^\infty \mathscr{F}_{n_b n_f} \tag{1.1.7}$$

where $\mathscr{F}_b$ is given by (1.1.1) and $\mathscr{F}_f$ is the Hilbert space completion of the antisymmetric tensor algebra

$$\begin{aligned} \mathscr{F}_f &= \textcircled{a}(L_2(R) \oplus L_2(R)) \\ &= \bigoplus_{n=0}^{\infty} (L_2(R) \oplus L_2(R)) \otimes_a \dots \otimes_a (L_2(R) \oplus L_2(R)). \end{aligned} \tag{1.1.8}$$

We have fermion particle and antiparticle annihilation operators

$$b(p) = b(p, 1), \qquad b'(p) = b(p, -1)$$

satisfying the commutation relations

$$b(p_1, \varepsilon_1) b(p_2, \varepsilon_2)^* + b(p_2, \varepsilon_2)^* b(p_1, \varepsilon_1) = \delta(p_1 - p_2) \delta_{\varepsilon_1, \varepsilon_2}$$

and also

$$a(k_1) a(k_2)^* - a(k_2)^* a(k_1) = \delta(k_1 - k_2).$$

The notation

$$a(k) = b(k, 0)$$

will also be convenient. On the Fock space (1.1.7),

$$N_\tau = \sum_{\varepsilon = 0, \pm 1} \int b(p, \varepsilon)^* \mu(p, \varepsilon)^\tau b(p, \varepsilon) \, dp$$

where $\mu(p, \varepsilon) = (p^2 + m(\varepsilon)^2)^{1/2}$ and $m(\varepsilon)$ is the rest mass of the corresponding particle. We take the domain $\mathscr{D}$ as before, and then W defined by

$$W = W_{\varepsilon, \varepsilon'} = \int A(k, \varepsilon)^* w(k, k', \varepsilon, \varepsilon') A(k', \varepsilon') \, dk \, dk' \tag{1.1.9}$$

is a bilinear form on $\mathscr{D} \times \mathscr{D}$ and a weak integral of bilinear forms where

$$A(k, \varepsilon)^* = b(k_1, \varepsilon_1)^* \dots b(k_r, \varepsilon_r)^*.$$

We find it convenient to introduce a periodic approximation in configuration space. Under this approximation, the Fourier transform or momentum space variable $k \in R$ is replaced by a discrete variable $k \in Z_V$, where $Z_V = (2\pi/V)\, Z$ is the lattice of multiples (by $2\pi/V$) of integers. Thus we define $\mathscr{F}_V$, the boson Fock space for volume V, as

$$\mathscr{F}_V = \mathfrak{S}(l_2(Z_V)) = C \oplus l_2(Z_V) \oplus \left[l_2(Z_V) \underset{s}{\otimes} l_2(Z_V)\right] \oplus \ldots. \quad (1.1.10)$$

We identify $\mathscr{F}_V$ with the subspace of $\mathscr{F}$ consisting of piecewise constant functions which are constant on each cube of volume $(2\pi/V)^j$ centered about a lattice point

$$\{k_1, \ldots, k_j\} \in Z_V \times \ldots \times Z_V = Z_V^j.$$

The periodic annihilation and creation operators $a_V(k)$ and $a_V^*(k)$ can be extended from $\mathscr{F}_V$ to $\mathscr{F}$ through the formulas

$$a_V^{\#}(k) = \left(\frac{V}{2\pi}\right)^{1/2} \int_{-\pi/V}^{\pi/V} a^{\#}(k + l)\, dl, \quad (1.1.11)$$

where $a^{\#}(k) = a(k)^*$ or $a(-k)$. We define

$$H_{0,V} = \int a(k)^* \mu(k_V)\, a(k)\, dk \quad (1.1.12)$$

with k_V a lattice point close to k,

$$k_V \in Z_V, \qquad |k - k_V| \leqq \pi/V. \quad (1.1.13)$$

On $\mathscr{F}_V$, this definition agrees with the standard formula

$$\sum_{k \in Z_V} a_V^*(k)\, \mu(k)\, a_V(k).$$

There is a similar periodic approximation to the Yukawa Fock space (1.1.7) which we will use without further comment. On (1.1.7) we have the periodic approximation

$$N_{\tau,V} = \sum_{\varepsilon=0,\pm 1} \int b(k, \varepsilon)^* \mu(k_V, \varepsilon)^\tau\, b(k, \varepsilon)\, dk \quad (1.1.14)$$

to N_τ. On the boson Fock space, only the term $\varepsilon = 0$ occurs in N_τ or $N_{\tau,V}$.

1.2 N_τ estimates

The important operators of field theory have simple expressions in terms of the annihilation creation forms (1.1.5) and (1.1.9). These forms are dominated by powers of the operator N_τ. We call such bounds N_τ estimates.

Vectors in the tensor product $L_2(R^s) \otimes \mathscr{F}$ describe s labeled particles and an indefinite number of unlabeled particles, just as vectors in $\mathscr{F}$ describe an indefinite number of particles. Corresponding to the smooth domain $\mathscr{D} \subset \mathscr{F}$ defined in Section 1.1, we introduce the domain

$$\mathscr{S}(R^s) \otimes \mathscr{D} \subset L_2(R^s) \otimes \mathscr{F}$$

consisting of vectors with a finite number of particles (at least s particles; the first s variables are not symmetrized) and wave functions in $\mathscr{S}$. The annihilation operator $A(k, \varepsilon) = b(k_1, \varepsilon_1) \dots b(k_s, \varepsilon_s)$ maps $\mathscr{D}$ into $\mathscr{D}$, and depends on the parameter k. If we regard $k \in R^s$ as a variable and not a parameter, then

$$A(., \varepsilon) : \mathscr{D} \to \mathscr{S}(R^s) \otimes \mathscr{D} \subset L_2(R^s) \otimes \mathscr{F}. \tag{1.2.1}$$

The operator

$$B(., \varepsilon) = A(., \varepsilon) \prod_{i=1}^{s} (\mu(k_i, \varepsilon_i)^{\tau_i/2} N_{\tau_i}^{-1/2}) \tag{1.2.2}$$

also maps $\mathscr{D}$ into $\mathscr{S}(R^s) \otimes \mathscr{D}$. Here $N_\tau^{-1/2}$ denotes the operator which gives zero on Ω_0 and is the inverse to $N_\tau^{1/2}$ on $\Omega_0^\perp$.

PROPOSITION 1.2.1 $\qquad \|B(., \varepsilon)\| \leqq 1$.

Proof We express B as a product of operators mapping $\mathscr{S}(R^j) \otimes \mathscr{D} \to \mathscr{S}(R^{j+1}) \otimes \mathscr{D}$. It is sufficient to bound each of these operators; this amounts to proving the proposition in the case $s = 1$. To rearrange B into such a product we note that

$$b(k, \varepsilon)\, N_\tau^{-1/2} = (N_\tau + \mu(k, \varepsilon)^\tau)^{-1/2}\, b(k, \varepsilon).$$

Let $s = 1$, $\theta \in \mathscr{D}$. Then

$$\|\mu(., \varepsilon)^{\tau/2}\, b(., \varepsilon)\, \theta\|^2 = \int \langle \theta, b(k, \varepsilon)^* \mu(k, \varepsilon)^\tau b(k, \varepsilon)\, \theta \rangle\, dk$$

$$\leqq \langle \theta, N_\tau \theta \rangle = \|N_\tau^{1/2} \theta\|^2.$$

If $\theta \perp \Omega_0$, we substitute $\psi = (N_\tau + aI)^{1/2}\, \theta$, $0 \leqq a$, to complete the proof.

Let $\prod_{i=1}^{r+s} \mu(k_i)^{-\tau_i/2}\, w$ be a bounded operator from $L_2(R^s) \otimes \mathscr{F} \to L_2(R^r) \otimes \mathscr{F}$. Then the integral (1.1.9) defines W as a bilinear form on

$$\mathscr{D}\left(\prod_{i=1}^{r} N_{\tau_i}^{1/2}\right) \times \mathscr{D}\left(\prod_{i=r+1}^{s} N_{\tau_i}^{1/2}\right)$$

and we have

PROPOSITION 1.2.2

$$\left\|\left(\prod_{i=1}^{r} N_{\tau_i}^{-1/2}\right) W \prod_{i=r+1}^{s} N_{\tau_i}^{-1/2}\right\| \leq \left\|\prod_{i=1}^{r+s} \mu(k_i)^{-\tau_i/2}\, w\right\|.$$

Proof Both norms are operator norms. We apply Proposition 1.2.1.

We see that $N_\tau^{1/2}$ is sufficient to dominate each annihilation or creation operator. In the case of fermions, fewer powers of N_τ are needed provided the kernel $w(k)$ of w is sufficiently well behaved and is a scalar function. Let

$$\{1, \ldots, r + s\} = A \cup B \cup C$$

be a disjoint union with

$$C \subset \{1, \ldots, r\}, \qquad A \subset \{r + 1, \ldots, r + s\}$$

and let $|A|$, $|B|$ and $|C|$ denote the number of elements of A, B, C.

We now let $w(k, k') \in \mathscr{S}'(R^{r+s})$, so that w is a densely defined bilinear form on $\mathscr{S}(R^r) \times \mathscr{S}(R^s)$. We denote by $\|w\|$ the norm of the operator w from $L_2(R^s)$ to $L^2(R^r)$ given by the kernel $w(k, k')$, $k = k_1, \ldots, k_r$; $k' = k_{r+1}, \ldots, k_{r+s}$. Let W be given by (1.1.9).

PROPOSITION 1.2.3 (*a*) *There is a constant less than* $0(|r - s| + 1|)^{|B|/2}$ *such that*

$$\|N_\tau^{-|C|/2} W N_\tau^{-|A|/2} (N + I)^{-|B|/2}\| \leqq \text{const.} \left\| \prod_{l \in A \cup C} \mu(k_l)^{-\tau/2} w \right\| .$$

(*b*) *Suppose that B contains at least one fermion variable.*

$$\|N_\tau^{-|C|/2} W N_\tau^{-|A|/2} (N + I)^{-(|B|-1)/2}\| \leqq \text{const.} \left\| \prod_{l \in A \cup C} \mu(k_l)^{-\tau/2} w \right\|_2 .$$

(*c*) *Let W contain exactly j boson variables. Then*

$$\| W(N + I)^{-j/2} \| \leqq |w|_s$$

where $|w|_s$ *is an appropriate Schwartz space norm.*

(*d*) *Let* $D \subset B$ *be a set of fermion variables* $p_1, \ldots, p_j$. *We write* $w = w(k, p)$ *to distinguish the variables p in D, and we assume* $w(k, p) = w(k, p_V)$. *Then*

$$\|N_\tau^{-|C|/2} W N_\tau^{-|A|/2} (N + I)^{-(|B|-|D|)/2}\| \leqq |w|_{1,2},$$

where

$$|w|_{1,2} = 0(V^{|D|/2}) \left\| \prod_{l \in A \cup C} \mu(k_l)^{-\tau/2} \int |w(\,.\,, p)|\, dp \right\|_2 .$$

Proof (a) We use Proposition 1.2.2 with $\tau_l = \tau$ or $\tau_l = 0$, but replace factors $N^{-1/2}$ by factors $(N + I)^{-1/2}$, using $\|(N + I)^{1/2} N^{-1/2}\| \leqq 2$. These factors can be placed on either side of W because w is a scalar kernel and

$$W(N + I)^{-1/2} = (N + I)^{-1/2} W(N + (r - s + 1) I)^{1/2} (N + I)^{-1/2}.$$

Since $\|(N + (r - s + 1) I)^{1/2} (N + I)^{-1/2}\| \leqq (|r - s| + 1)^{1/2}$, part (*a*) is established.

(b) Let k_j be a fermion variable in B. The cannonical anticommutation relations yield

$$\{b(f, \varepsilon), b(f, \varepsilon)^*\} = \|f\|^2$$

and $\|b(f, \varepsilon)\| = \|f\|$, since bb^* and b^*b have orthogonal ranges. Hence

$$\left\| \int w(k, k')\, b(k_j, \varepsilon_j)\, dk_j \right\|^2 = \int |w|^2\, dk_j$$

and the same bound holds if k_j is a creating variable. We apply Proposition 1.2.2 to the operator valued kernel

$$v(k_1, \ldots, k_{j-1}, k_{j+1}, \ldots, k_{r+s}) = \int dk_j\, w(k, k')\, b(k_j, \varepsilon_j)$$

with $\tau_i = \tau$ or 0. The operator norm of v satisfies

$$\begin{aligned}\left\| \prod_{i \in A \cup C} \mu(k_i)^{-\tau/2}\, v \right\| &\leq \left(\int \prod_{i \in A \cup C} \mu(k_i)^{-\tau}\, \|v(k)\|^2\, dk_1 \ldots dk_{j-1}\, dk_{j+1} \ldots dk_{r+s} \right)^{1/2} \\ &= \left(\int \prod_{i \in A \cup C} \mu(k_i)^{-\tau}\, |w(k, k')|^2\, dk\, dk' \right)^{1/2} \\ &= \left\| \prod_{i \in A \cup C} \mu(k_i)^{-\tau/2}\, w \right\|_2 .\end{aligned}$$

Since w is a scalar kernel, the proof of part (a) above yields (b).

(c) We treat the fermion variables first. As in the proof of (b), the boson variables are handled by an application of Proposition 1.2.2, with an operator valued kernel. Thus we may suppose $j = 0$. Let $\{e_i\}$ be an orthonormal basis for $L_2(R)$ consisting of eigenfunctions of the Hermite operator

$$-\frac{d^2}{dk^2} + k^2 = H(k)$$

with eigenvalues $\lambda_i = 2i + 1$. We may suppose $w \in L_2$, and then

$$w = \Sigma \alpha_{i_1 \ldots i_{r+s}}\, e_{i_1} \otimes \ldots \otimes e_{i_{r+s}}$$

with $\Sigma |\alpha|^2 = \|w\|_2^2 < \infty$. Since

$$\left\| \int e_{i_\nu}(k_\nu)\, b(k_\nu, \varepsilon_\nu)\, dk_\nu \right\| = 1,$$

it is sufficient to show that $\Sigma|\alpha| \leq |w|_s$ for some Schwartz norm $|.|_s$. However,

$$\begin{aligned}\Sigma|\alpha| &= \Sigma |\alpha_{i_1 \ldots i_{r+s}}| \prod_\nu (2i_\nu + 1) \prod_\nu (2i_\nu + 1)^{-1} \\ &\leq \left(\Sigma_{i_1 \ldots i_{r+s}} \prod_\nu (2i_\nu + 1)^{-2} \right)^{1/2} \left(\Sigma_{i_1 \ldots i_{r+s}} |\alpha|^2 \prod_\nu (2i_\nu + 1)^2 \right)^{1/2} \\ &= \text{const.} \left\| \prod_\nu H(k_\nu)\, w \right\|_2 = |w|_s .\end{aligned}$$

We remark that for $j = 0$, the norm $|w|_s$ may be replaced by

$$\inf \Sigma |\alpha|,$$

where the infinum ranges over the choice of an orthonormal basis for $L^2(R)$.

(d) We write W in the form

$$W = \int A(k_1)^* A(p_1)^* w(k, p) A(p_2) A(k_2) \, dk \, dp$$

where $\{p\} = \{p_1, p_2\}$, $\{k\} = \{k_1, k_2\}$. We first estimate

$$X(k) = \int A(p_1)^* w(k, p) A(p_2) \, dp$$

$$= \left(\frac{2\pi}{V}\right)^{|D|/2} \sum_{p \in Z_V \times \ldots \times Z_V} A_V(p_1)^* w(k, p) A_V(p_2).$$

Since $\|b_V(k, \varepsilon)\| = 1$ for $\varepsilon = \pm 1$, we have

$$\|X(k)\| \leqq \left(\frac{2\pi}{V}\right)^{|D|/2} \sum_{p \in Z_V \times \ldots \times Z_V} |w(k, p)|$$

$$\leqq 0(|V|^{|D|/2}) \int |w(k, p)| \, dp.$$

By applying Proposition 1.2.2 to $\int A(k_1)^* X(k) A(k_2) \, dk$, we obtain, by the methods of part (a),

$$\|N_\tau^{-|C|/2} W N_\tau^{-|A|/2} (N + I)^{-(|B|-|D|)/2}\|$$

$$\leqq \text{const.} \left\| \prod_{l \in A \cup C} \mu(k_l)^{-\tau/2} \|X(k)\| \right\|_2$$

$$\leqq |w|_{1,2}.$$

To simplify the discussion of the Yukawa$_2$ interaction in part II, we give the following corollary to Proposition 1.2.3.

Corollary 1.2.4 *Let W_i have scalar kernels w_i and equal*

$$W_1 = \int a(k)^* b(p_1)^* b'(p_2)^* w_1(k, p_1, p_2) \, dk \, dp_1 \, dp_2,$$

$$W_2 = \int b(p_1)^* b'(p_2)^* a(-k) w_2(k, p_1, p_2) \, dk \, dp_1 \, dp_2,$$

$$W_3 = \int a(k)^* b(p_1, \varepsilon)^* b(-p_2, \varepsilon) w_3(k, p_1, p_2) \, dk \, dp_1 \, dp_2, \quad \varepsilon = \pm 1,$$

$$W_4 = \int b(p_1) b(p_2) w_4(p_1, p_2) \, dp_1 \, dp_2.$$

Then

$$\|(N_\tau + I)^{-1/2} W_i (N_\tau + I)^{-1/2}\| \leqq \text{const.} \, M_i,$$

and

$$\| W_4(N_\tau + I)^{-1/2} \| \leqq \text{const. } M_4,$$

where

$$M_1 = \|(\mu + \omega_1 + \omega_2)^{-\tau/2} w_1\|_2, \quad M_2 = \|(\omega_1 + \omega_2)^{-\tau/2} \mu^{-\tau/2} w_2\|_2$$

$$M_3 = \|\mu^{-\tau/2} \omega_2^{-\tau/2} w_3\|_2, \quad M_4 = \|(\omega_1 + \omega_2)^{-\tau/2} w_4\|_2.$$

Proof We prove the estimate on W_1; the other cases are similar. Write $W_1 = R_1 + R_2 + R_3$ where the kernel $r_j = wE_j$ and E_j is the characteristic function of the subset of R^3 in which the jth variable is largest in magnitude. Apply Proposition 1.2.3 (b) to each R_j. Since $\mu^{-\tau/2}E_1 \leqq \text{const.} (\mu + \omega_1 + \omega_2)^{-\tau/2} E_1$, $\omega_1^{-\tau/2}E_2 \leqq \text{const.} (\mu + \omega_1 + \omega_2)^{-\tau/2} E_2$, $w_2^{-\tau/2}E_3 \leqq \text{const.} (\mu + \omega_1 + \omega_2)^{-\tau/2} E_3$, the bound

$$\|(N_\tau + I)^{-1/2} R_j(N_\tau + I)^{-1/2}\| \leqq \text{const. } \|(\mu + \omega_1 + \omega_2)^{-\tau/2} r_j\|_2$$

follows.

We now apply the N_τ estimates to estimate Wick polynomials in the boson field operators. Since

$$[\mu(k)/\mu(k_V)]^{\pm 1} \leqq \text{const.}$$

we have

$$[N_\tau/N_{\tau,V}]^{\pm 1} \leqq \text{const.}$$

and so the estimates could be given in terms of the operators $N_{\tau,V}$ instead.

Our time zero fields coincide with free fields,

$$\phi(x) = (4\pi)^{-1/2} \int e^{-ikx} \mu(k)^{-1/2} \{a^*(k) + a(-k)\} \, dk \tag{1.2.3}$$

$$\pi(x) = i(4\pi)^{-1/2} \int e^{-ikx} \mu(k)^{1/2} \{a^*(k) - a(-k)\} \, dk. \tag{1.2.4}$$

Similarly we define periodic fields $\phi_V(x)$ and $\pi_V(x)$ by replacing $\mu(k)$ above by $\mu(k_V)$. The canonical commutation relations

$$[\phi(x), \pi(y)] = i\,\delta(x - y) \tag{1.2.5}$$

hold in exponentiated form. The Wick powers of ϕ are given by the formula

$$:\phi(x)^n: = \sum_{j=0}^{n} \binom{n}{j} w_x(k)\, a^*(k_1) \dots a^*(k_j)\, a(-k_{j+1}) \dots a(-k_n) \tag{1.2.6}$$

where

$$w_x(k) = (4\pi)^{-n/2} \{\mu(k_1) \dots \mu(k_n)\}^{-1/2} e^{-i(\Sigma_i k_i)x}. \tag{1.2.7}$$

The space averaged Wick powers,

$$:\phi^n(g): = \int :\phi(x)^n: g(x)\, dx, \tag{1.2.8}$$

$$:\phi_V^n: = \int_{-V/2}^{V/2} :\phi_V(x)^n: dx, \tag{1.2.9}$$

have kernels

$$w_g(k) = (2\pi)^{1/2}(4\pi)^{-n/2}\{\mu(k_1)\ldots\mu(k_n)\}^{-1/2}\tilde{g}(\Sigma k_i) \quad (1.2.10)$$

$$w_V(k) = V(4\pi)^{-n/2}\{\mu(k_{1,V})\ldots\mu(k_{n,V})\}^{-1/2}\delta_{0,\Sigma k_{i,V}} \quad (1.2.11)$$

PROPOSITION 1.2.5 *Let* $g \in L_2$ *and let* $0 < \varepsilon$. *Then*

$$\sup_i \mu(k_i)^{-\varepsilon+1/2} w_g \in L_2(R^n)$$

$$\sup_i \mu(k_i)^{-\varepsilon+1/2} w_V \in l_2(Z_V^n)$$

and the L_2 *norms are dominated by* const. $\|g\|_2$.

We omit the elementary proof. Let $P = \Sigma c_j \xi^j$ be a real polynomial, and (henceforth) let g be real. Then the operators

$$:P(\phi, g): = \Sigma c_j \int :\phi(x)^j: g(x)\,dx,$$

$$:P(\phi_V): = \Sigma c_j \int_{-V/2}^{V/2} :\phi_V(x)^j:\,dx$$

are symmetric and defined on the domain $\mathscr{D}(N^{n/2})$, where n is the degree of P, by Propositions 1.2.3-4. These operators are also essentially self adjoint on this domain, as we will see later.

PROPOSITION 1.2.6 *Let* $(-\Delta + 1)^{-1/4} f \in L_2$ *and let* $(-\Delta + 1)^{1/4} h \in L_2$ *with* f *and* h *real. Then* $\phi(f)$ *and* $\pi(h)$ *are essentially self adjoint on* $\mathscr{D}$.

Proof *For* $\theta \in \mathscr{D}$, we compute

$$\|\phi(f)^n \theta\| \leqq 0(c^n n!^{1/2}),$$

using Proposition 1.2.3 (a). Hence θ is an analytic vector for $\phi(f)$ and since $\mathscr{D}$ is dense, $\phi(f)$ is essentially self adjoint on $\mathscr{D}$, by a theorem of Nelson[1]. The same proof works for $\pi(h)$; the essential point is that the kernel lies in L_2.

The introduction of Q space in Chapter 2 is based on the following proposition. Let $\mathfrak{M}$ be the strongly closed operator algebra (von Neumann algebra) generated by the spectral projections of the time zero fields $\phi(f)$ $f \in C_0^\infty$.

PROPOSITION 1.2.7 $\mathfrak{M}\Omega_0$ *is dense in* $\mathscr{F}$.

Proof Let $\theta \perp \mathfrak{M}\Omega_0$. Since Ω_0 is an analytic vector for $\phi(f)$, the Taylor series for $\exp\left(i \sum_{\mu=1}^{\nu} t_\mu \phi(f_\mu)\right)\Omega_0$ converges strongly and we have

$$0 = \partial_{t_1} \ldots \partial_{t_\nu} \left\langle \theta, \exp\left(i \sum_{\mu=1}^{\nu} t_\mu \phi(f_\mu)\right)\Omega_0 \right\rangle\bigg|_{t=0}$$

$$= i^\nu \langle \theta, \phi(f_1) \ldots \phi(f_\nu)\Omega_0 \rangle.$$

Let θ_j be the j particle component of θ. We prove by induction that $\theta_j = 0$. If $\nu = 0$, we have from the above equation $\theta_0 = 0$. Suppose $\theta_i = 0$ for all i less than j. Then

$$0 = \langle \theta, \phi(f_1) \ldots \phi(f_j) \Omega_0 \rangle = \langle \theta_j, a^*(f_1) \ldots a^*(f_j) \Omega_0 \rangle$$

and $\theta_j = 0$ also, since the vectors $a^*(f_1) \ldots a^*(f_j) \Omega_0$ are total in $\mathscr{F}_j$. Hence $\theta = 0$ and the proof is complete.

The Hamiltonians for bosons with a $P(\phi)$ interaction are

$$H(g) = H_0 + :P(\phi, g):, \tag{1.2.12}$$

$$H(V) = H_0 + :P(\phi_V):. \tag{1.2.13}$$

Later we will add a constant to renormalize these operators, and much later we will take the limit $g \to 1$. We also consider

$$H(g, V) = H_{0,V} + :P(\phi_V, g): \tag{1.2.14}$$

and the momentum cutoff Hamiltonians

$$H(g, \varkappa), \quad H(V, \varkappa) \quad \text{and} \quad H(g, V, \varkappa).$$

In these operators, the field ϕ or ϕ_V in the interaction term P is replaced by the momentum cutoff field $\phi_\varkappa$ or $\phi_{V,\varkappa}$:

$$\phi_\varkappa = (4\pi)^{-1/2} \int_{|k| \leqq \varkappa} e^{-ikx} \mu(k)^{-1/2} \{a(k)^* + a(-k)\}\, dk$$

and for $\phi_{V,\varkappa}$ we replace $\mu(k)$ by $\mu(k_V)$. We define

$$Z_{V,\varkappa} = Z_V \cap \{k : |k| \leqq \varkappa\}.$$

1.3 Direct sums and tensor products

The operators H_0 and $H_0 + H_I$ are partial differential operators in a continuum of variables. The only way we are able to make use of this fact is to approximate these operators by partial differential operators in a finite number of variables. This section is devoted to the Fock space aspects of this approximation. Let $\mathscr{H}$ be a closed linear subspace of $\mathscr{F}_1 = L_2(R)$. We will be interested in the case $\mathscr{H} = l_2(Z_{V,\varkappa}) \subset L_2(R)$. Let

$$\mathscr{H}^\perp = \mathscr{F}_1 \ominus \mathscr{H} = L_2(R) \ominus \mathscr{H}.$$

The Fock spaces $\mathfrak{S}(\mathscr{H})$ and $\mathfrak{S}(\mathscr{H}^\perp)$ are both subspaces of $\mathscr{F}$. Let S be the symmetrization projection which maps each space $L_2(R^n)$ onto $\mathscr{F}_n$ and let $\mathfrak{S}_i(\mathscr{H})$ be the i particle subspace of $\mathfrak{S}(\mathscr{H})$.

LEMMA 1.3.1 *The mapping*

$$S \binom{i+j}{i}^{1/2} : \mathfrak{S}_i(\mathscr{H}) \otimes \mathfrak{S}_j(\mathscr{H}^\perp) \to \mathscr{F}_{i+j} \tag{1.3.1}$$

is isometric.

Proof The symmetrization projection onto $\mathscr{F}_n$ is a sum over the permutation group, acting by interchange of variables, divided by $n!$. Acting on a function $f \in \mathfrak{S}_i(\mathscr{H}) \otimes \mathfrak{S}_j(\mathscr{H}^\perp)$, the $(i+j)!$ permutations yield $(i+j)!/i!\,j! = \binom{i+j}{i}$ distinct (and orthogonal) permuted functions, and each such function occurs $i!\,j!$ times. It follows that

$$\|Sf\|^2 = \binom{i+j}{i}\frac{i!^2\,j!^2}{(i+j)!^2}\|f\|^2 = \binom{i+j}{i}^{-1}\|f\|^2$$

and the proof is complete.

Let $N(\mathscr{H})$ be the number operator on $\mathfrak{S}(\mathscr{H})$. We can write the isometry above as

$$S\left(\frac{N(\mathscr{H})\otimes I + I \otimes N(\mathscr{H}^\perp)}{N(\mathscr{H})\otimes I}\right)^{1/2} = U. \tag{1.3.2}$$

PROPOSITION 1.3.2 *(1.3.2) defines a unitary mapping of* $\mathfrak{S}(\mathscr{H}) \otimes \mathfrak{S}(\mathscr{H}^\perp)$ *onto* $\mathscr{F}$.

Proof For distinct i and j, the ranges of the maps (1.3.1) are orthogonal. This is easily seen for functions

$$f = \{S(f_1 \otimes \dots \otimes f_i)\} \otimes \{S(f_{i+1} \otimes \dots \otimes f_{i+j})\} \tag{1.3.3}$$

which factor, and such functions span the spaces $\mathfrak{S}_i(\mathscr{H}) \otimes \mathfrak{S}_j(\mathscr{H}^\perp)$. To complete the proof we show that the range of (1.3.2), restricted to

$$\sum_{i+j=n} \mathfrak{S}_i(\mathscr{H}) \otimes \mathfrak{S}_j(\mathscr{H}^\perp)$$

equals $\mathscr{F}_n$. If $h \in \mathscr{F}_n$ is orthogonal to the range, then $h \perp f_1 \otimes \dots \otimes f_n$ where each $f_l \in \mathscr{H}$ or $\mathscr{H}^\perp$, using the fact that h is symmetric. However, the $f = f_1 \otimes \dots \otimes f_n$ which factor span $L_2(R^n)$ and so $h = 0$. This completes the proof.

For $h \in L_2(R)$, we define $a(h) = \int a(k)\,h(k)^-\,dk$, and if $h \in \mathscr{H}$ or $\mathscr{H}^\perp$, we define the annihilation operator $a_1(h)$ or $a_2(h)$ acting on $\mathfrak{S}(\mathscr{H})$ or $\mathfrak{S}(\mathscr{H}^\perp)$.

PROPOSITION 1.3.3 *For* $h \in \mathscr{H}$,

$$a(h)\,U = U(a_1(h) \otimes I).$$

Proof It is enough to establish the identity for vectors (1.3.3) in $\mathfrak{S}_i(\mathscr{H}) \otimes \mathfrak{S}_j(\mathscr{H}^\perp)$ which factor. For such f the left and right members of this identity are

$$\binom{i+j}{i}^{1/2}(i+j)^{-1/2}\sum_{l=1}^{i+j}\left(\int hf\,dk\right) S(f_1 \otimes \dots \otimes f_{l-1} \otimes f_{l+1} \otimes \dots \otimes f_{i+j})$$

and

$$\binom{i+j-1}{i-1}^{1/2} i^{-1/2}\sum_{l=1}^{i}\left(\int hf_l\,dk\right) S(f_1 \otimes \dots \otimes f_{l-1} \otimes f_{l+1} \otimes \dots \otimes f_{i+j}).$$

Since $h \perp f_l$ for $l > i$, the summations in the two expressions above are equal. The factorials are also equal and the proposition is proved.

THEOREM 1.3.4 *Let $\mathscr{H} = l_2(Z_{V,\varkappa})$, regarded as a subspace of $L_2(R)$ by conventions of section 1.1. The unitary operator U transforms $H_{0,V}$ and $H_I(g, V, \varkappa)$ as follows:*

$$H_{0,V} \to (H_{0,V} \restriction \mathfrak{S}(\mathscr{H})) \otimes I + I \otimes (H_{0,V} \restriction \mathfrak{S}(\mathscr{H})^{\perp}), \tag{1.3.3}$$

$$H_I(g, V, \varkappa) \to (H_I(g, V, \varkappa) \restriction \mathfrak{S}(\mathscr{H})) \otimes I \tag{1.3.4}$$

Proof (1.3.4) follows from Proposition 1.3.3 and the fact that the kernels of $H_I(g, V, \varkappa)$ belong to $\mathscr{H} \otimes \ldots \otimes \mathscr{H}$. (1.3.3) is most easily seen by a direct calculation, and is based on the fact that $\mathscr{H}$ is an invariant subspace for the multiplication operator $\mu(k_V)$, acting on $L_2(R)$.

2 Q space

To say that $\phi(x)$ is a field operator means that $\langle \theta, \phi(x)\, \theta \rangle$ is the expected value of the field strength at the space point x if the field is in the state θ. We diagonalize the $\phi(x)$'s simultaneously so that $\mathscr{F}$ is represented as an L_2 space,

$$\mathscr{F} \approx L_2(Q)$$

and under this isomorphism, each $\phi(f)$ goes into a multiplication operator. For each point $q \in Q$, the field strength at x (or averaged near x) takes on a definite value and so Q may be regarded as the configuration space of the classical field. The state $\theta(q) \in L_2(Q)$ determines a probability distribution, $|\theta|^2\, dq$, which gives the probability that the quantum field will be in some region in Q space. The interaction Hamiltonian $:P(\phi, g):$ is a multiplication operator in the Q space representation; this is the principal merit of Q space. Fock space was set up in order to diagonalize H_0. On $L_2(Q)$, H_0 is a Hermite operator and the particle structure (1.1.1) in $\mathscr{F}$ results from an expansion of $L_2(Q)$ in Hermite functions.

2.1 $:\phi^n:$ as a multiplication operator

PROPOSITION 2.1.1 *There is a measure space Q with a measure dq of total mass 1, and a unitary operator $W: \mathscr{F} \to L_2(Q)$ such that*

$$W \mathfrak{M} W^* = L_\infty(Q), \tag{2.1.1}$$

$$W\Omega_0 = 1. \tag{2.1.2}$$

Q is the spectrum of $\mathfrak{M}$ or of a weakly dense subalgebra of $\mathfrak{M}$.

The existence of Q and W follows from one form of the spectral theorem for a family of commuting normal operators (see for instance Kunze and Segal[1]) and from the fact that $\mathfrak{M}$ has a cyclic vector Ω_0, Proposition 1.2.7.

Proposition 2.1.2 *Let* $g = \bar{g} \in L_2$. *Then* $:P(\phi, g):$ *is essentially self adjoint on the domain* $\mathscr{D}$, *and* $W :P(\phi, g):^- W^* \in L_p(Q)$ *for any* $p < \infty$.

Proof Let $T = (:P(\phi, g): \upharpoonright \mathscr{D})^-$. We assert that $\mathfrak{M}\Omega_0 \in \mathscr{D}(T)$ and that T commutes with $\mathfrak{M}$ on the domain $\mathfrak{M}\Omega_0$. We have

$$\lim_n \sum_{j \leq n} :P(\phi, g): (i\phi(f))^j \,\Omega_0/j!$$

$$= \lim_n \sum_{j \leq n} (i\phi(f))^j :P(\phi, g): \Omega_0/j!$$

$$= e^{i\varphi(f)} T\Omega_0.$$

Thus $\exp(i\phi(f))\,\Omega_0 \in \mathscr{D}(T)$ and

$$T \exp(i\phi(f))\,\Omega_0 = \exp(i\phi(f))\, T\Omega_0.$$

The same argument applies to linear combinations in the exponentials $\exp(i\phi(f))$, and to polynomials in the exponentials (since polynomials in the exponentials can be written as linear combinations). An arbitrary element M of $\mathfrak{M}$ is the strong limit of polynomials M_j in the operators $\exp(i\phi(f))$. Hence $M\Omega_0 = \lim M_j\Omega_0$ and

$$\lim TM_j\Omega_0 = \lim M_j T\Omega_0 = MT\Omega_0,$$

which proves our assertion.

On Q space, our assertion states that $L_\infty \subset \mathscr{D}(WTW^*)$ and that WTW^* commutes with bounded multiplication operators. Let $t = WT\Omega_0 \in L_2(Q)$. Then WTW^*, on the domain L_∞, is multiplication by t. Since multiplication by a real L_2 function is essentially self adjoint on the domain L_∞, WTW^* and T are extensions of self adjoint operators. Since they are symmetric, they are self adjoint.

On Fock space, Proposition 1.2.3 (a) shows that $\Omega_0 \in \mathscr{D}(T^j)$ for any integer j. Thus 1 is in the domain of multiplication by t^j and $t \in L_p$ for any $p < \infty$. This completes the proof. In the following we let $:P(\phi, g): = T$.

Let B be an open subset of R and let $\mathfrak{M}(B)$ be the von Neumann algebra generated by the operators of the form $\exp(i\phi(f))$, where $f = \bar{f} \in C_0^\infty$ and $\operatorname{supp} f \subset B$.

Proposition 2.1.3 *Let* $g \in L_2$ *and let* $\operatorname{supp} g \subset B^-$. *Then* $\exp(it :P(\phi, g):) \in \mathfrak{M}(B)$ *for all real* t.

Proof If $T_\varkappa \to T$ strongly on $\mathscr{D}$ and if each $T_\varkappa$ is self adjoint, then $\exp(iT_\varkappa t) \underset{s}{\to} \exp(iTt)$ since $\mathscr{D}$ is a core for T. See Kato [1, page 502]. We approximate T by $T_\varkappa = :P(\phi_\varkappa, g_\varkappa):$. Here the $g_\varkappa$ are chosen so that $g_\varkappa \to g$ in L_2, and the distance from $\operatorname{supp} g_\varkappa$ to $\sim B$ is greater than $\varkappa^{-1}$. We let $\phi_\varkappa(x) = (h_\varkappa * \phi)(x)$, where $h_\varkappa(x) = \varkappa h(\varkappa x)$, $\int h(x)\,dx = 1$ and $h \in C_0^\infty$ with support in the interval $[-1/2, 1/2]$. We note that the Fourier

2*

transform of $h_\varkappa$ is $\tilde{h}(k/\varkappa)$, which converges pointwise to 1 as $\varkappa \to \infty$. By Proposition 1.2.5, the kernels of $T_\varkappa$ converge in L_2 to the kernels of T, so by Proposition 1.2.3 (a)

$$\|(T_\varkappa - T)(N + I)^{-n/2}\| \to 0.$$

We note that by Wick's theorem (as in the proof of Lemma 2.1.5 below) there is a polynominal P_1 such that

$$:P(\phi_\varkappa, g_\varkappa): = P_1(\phi_\varkappa, g_\varkappa).$$

By definition, the bounded functions of $\phi_\varkappa(x), x \in \operatorname{supp} g_\varkappa$, belong to $\mathfrak{M}(B)$, and hence so do the bounded functions of $P_1(\phi_\varkappa, g_\varkappa)$. By limits, the same is true for $:P(\phi, g):$ and the proof is complete.

We saw that $:P(\phi, g):$, as a function on Q, is in L_p for any $p < \infty$; it is thus "almost bounded". When $0 \leqq P, 0 \leqq g$, we have $:P(\phi, g):$ almost bounded from below in a much more restricted sense.

THEOREM 2.1.4 *Let* $0 \leqq P$ *and* $0 \leqq g$. *Then as a function on* Q, $\exp(-:P(\phi, g):) \in L_p$ *for all* $p < \infty$.

We write

$$:P(\phi, g): = :P(\phi_\varkappa, g): + (:P(\phi, g): - :P(\phi_\varkappa, g):) \qquad (2.1.3)$$

$$= P_{g,\varkappa} + P'_{g,\varkappa}$$

and we bound the two terms above separately in the following lemmas.

LEMMA 2.1.5 *Let P have degree n. Then*

$$-\int g(x)\, dx\, 0 (\log \varkappa)^{n/2} \leqq P_{g,\varkappa}.$$

Proof Let $c_\varkappa = \|\phi_\varkappa(x)\, \Omega_0\|^2$. By (1.2.15), $c_\varkappa = 0(\log \varkappa)$. Wick's theorem asserts that

$$:\phi_\varkappa(x)^j: = \sum_{i=0}^{[j/2]} (-1)^i \frac{j!}{(j-2i)!\, i!} c_\varkappa^i \phi_\varkappa(x)^{j-2i}.$$

This formula follows from the commutation relations and a combinatorial argument. The factorials give the number of ways of selecting i pairs from j objects. Since $0 \leqq P$, it follows that

$$-0(c_\varkappa^{n/2}) \leqq :P(\phi_\varkappa(x)):,$$

which yields the lemma.

LEMMA 2.1.6 *Regard* $P'_{g,\varkappa}$ *as a function on* Q. *Let* $\varepsilon > 0$. *With a constant independent of* $\varkappa$ *and* j *(but depending on* g*)*

$$\int |P'_{g,\varkappa}(q)|^{2j}\, dq \leqq j!^n (\text{const. } \varkappa^{-1/2+\varepsilon})^{2j}$$

Proof We use the identity

$$\int |P'_{g,\kappa}(q)|^{2j}\, dq = \|(P'_{g,\kappa})^j \Omega_0\|^2$$

to estimate the integral. If $P'_{g,\kappa}$ is written as a sum of monomials (1.1.5) in the annihilation creation operators, then the kernels w_κ which occur satisfy the bound

$$\|w_\kappa\|_2 \leqq 0(\kappa^{-1/2+\varepsilon})$$

by Proposition 1.2.5. Since $(P'_{g,\kappa})^l \Omega_0$ is a state with at most ln particles, we have

$$\|(P'_{g,\kappa})^l \Omega_0\| \leqq \|P'_{g,\kappa}(N+I)^{-n/2}\| \; \|(N+I)^{n/2}(P'_{g,\kappa})^{l-1}\Omega_0\|$$

$$0(\kappa^{-(1/2)+\varepsilon})\,(nl+1)^{n/2}\, \|(P'_{g,\kappa})^{l-1}\Omega_0\|,$$

using Proposition 1.2.3(a). An induction completes the proof.

Proof of Theorem 2.1.4 Let $\kappa_\nu = 2^\nu$ and let

$$Q_\nu = \{q : |P'_{g,\kappa_\nu}(q)| \leqq 1,\ |P'_{g,\kappa_{\nu-1}}(q)| > 1\}$$

for $\nu = 2, 3, \ldots,$

$$Q_1 = \{q : |P'_{g\kappa_1}(q)| \leqq 1\}.$$

On Q_ν, $-0(\log \kappa_\nu)^{n/2} \leqq :P(\phi, g):$ by Lemma 2.1.5.
Thus

$$\int \exp(-t :P(\phi, g):)\, dq \leqq \sum_{\nu=1}^{\infty} \exp(0(\log \kappa_\nu)^{n/2}) \int_{Q_\nu} dq$$

and the proof is completed by the estimate

$$\int_{Q_\nu} dq \leqq \exp(-0(\kappa_\nu^{-(1-2\varepsilon)/n})). \tag{2.1.4}$$

To derive this bound on the measure of Q_ν, we use Lemma 2.1.6, being careful to make an optimal choice for j. For $\nu \geqq 2$,

$$\int_{Q_\nu} dq \leqq \int |P'_{g,\kappa_{\nu-1}}|^{2j}\, dq$$

$$\leqq j!^n\, 0(\kappa_\nu^{-1/2+\varepsilon})^{2j}.$$

We use Sterling's formula to bound $j!$; the choice $j = \kappa_\nu^{-(1-2\varepsilon)/n}$ then completes the proof.

The Theorem 2.1.4 is sufficient for the study of $H(g)$, but to control the limit $g \to 1$, we need an estimate independent of g, given in the following theorem. Since the proof is lengthy, we only indicate the main ideas of the proof. For details, see Glimm-Jaffe[5], where a similar result is proved.

THEOREM 2.1.7 *Let $0 \leqq P$, $0 \leqq g$ and let $|g(x)| \leqq M$. With a constant independent of g,*

$$\int \exp(-t{:}P(\phi, g){:})\, dq \leqq e^{\text{const.}\, V},$$

where V is the length of the set of points within distance one of suppt g.

Proof We decompose g,

$$g = \sum_{\alpha=-\infty}^{\infty} g_\alpha$$

so that suppt $g_\alpha \subset [\alpha - \frac{1}{2}, \alpha + \frac{1}{2}]$. There are V nonzero terms in the sum. Thus

$${:}P(\phi, g){:} = \Sigma_\alpha P(\phi, g_\alpha) = \Sigma P_\alpha$$

also decomposes, and as in (2.1.3), each P_α is split into a low momentum portion and a tail. The idea behind the proof is that the functions P_α and P_β are nearly independent, for $|\alpha - \beta|$ large. If they were exactly independent for $\alpha \neq \beta$, then we would have

$$\int \exp(-t{:}P(\phi,g){:})dq = \Pi_\alpha \int \exp(-t\, P_\alpha)\, dq = e^{\text{const.}\, V},$$

using Theorem 2.1.4 to bound each factor. Although the integral does not factor as above, we obtain bounds for the integral which do factor and this gives the upper bound $\Pi_\alpha e^{\text{const.}} = e^{\text{const.}\, V}$ for the integral.

Let $\nu = \{\nu(\alpha)\}$ be a multi-index and let

$$Q_\nu = \{q : |P'_{g_\alpha, \varkappa_{\nu(\alpha)}}(q)| \leqq 1 \text{ for all } \alpha \text{ and } |P'_{g_\alpha, \varkappa_{\nu(\alpha)-1}}(q)| > 1 \text{ if } \nu(\alpha) > 1\}.$$

Then

$$\int \exp(-t{:}P(\phi, g){:})\, dq$$
$$\leqq \Sigma_\nu \exp(0\, \Sigma_\alpha (\log \varkappa_\alpha)^{n/2}) \int_{Q_\nu} dq.$$

The next lemma completes the proof; on comparison with (2.1.4) we see that this is the bound which factors. The proof is as in Glimm-Jaffe[5].

LEMMA 2.1.8 *With the above notation,*

$$\int_{Q_\nu} dq \leqq \exp(-0(\Sigma_\alpha \varkappa_{\nu(\alpha)}^{-(1-2\varepsilon)/n})).$$

Remark on the proof Let $j = \{j(\alpha)\}$ be a multi-index. Then

$$\int_{Q_\nu} dq \leqq \int \Pi_\alpha |P'_{g_\alpha, \varkappa_{\nu(\alpha)-1}}|^{2j(\alpha)}\, dq;$$

we omit from the product α's with $g_\alpha = 0$ or with $\varkappa_{\nu(\alpha)} = 1$. This integral is equal to

$$\|\Pi_\alpha(P'_{g_\alpha,\varkappa_{\nu(\alpha)}-1})^{j(\alpha)}\,\Omega_0\|^2.$$

The detailed estimates take advantage of the fact that P_α and P_β or $P'_{g_\alpha,\varkappa_{\nu(\alpha)}-1}$ and $P'_{g_\beta,\varkappa_{\nu(\beta)}-1}$ create and annihilate particles with nearly orthogonal wave functions for $|\alpha - \beta|$ large.

2.2 H_0 as a Hermite operator

The semigroup $\exp(-tH_0)$ is smoothing; it maps L_p into L_q for $1 < p$, $q < \infty$ and t sufficiently large.

PROPOSITION 2.2.1 *Let f_1 and f_2 be nonnegative functions in $L_2(Q)$. Then $0 \leqq \langle f_1, \exp(-tH_0)\, f_2\rangle$, i.e. $\exp(-tH_0)$ has a nonnegative kernel.*

Proof We approximate H_0 strongly on $\mathscr{D}$ by operators of the form $A = A_1 + \ldots + A_j$ where

$$A_l = \mu(k_l)\, a^*(e_l)\, a(e_l),$$

$$a^\#(e_l) = a_V^\#(k_l), \qquad k_l \in Z_V.$$

Since $\mathscr{D}$ is a core for H_0, strong convergence of the operators A implies strong convergence of the resolvents (see Kato [1, page 429]) and strong convergence of the semigroups, Kato [1, page 502]. Strong convergence is simple to verify, so it is sufficient to prove that e^{-tA} has a nonnegative kernel.

Let $\mathscr{H}$ be the span of the e_l and let $\mathfrak{M}(\mathscr{H})$ be the von Neumann algebra generated by the spectral projections of the $\phi(e)$, $e \in \mathscr{H}$. Let $\mathfrak{M}^+$ and $\mathfrak{M}(\mathscr{H})^+$ be the positive cones in $\mathfrak{M}$ and $\mathfrak{M}(\mathscr{H})$. We postpone the proof of the following lemma.

LEMMA 2.2.2

$$\cup_{\mathscr{K}}\{\mathfrak{M}(\mathscr{K})^+\,\Omega_0 : \mathscr{H} \subset \mathscr{K} \subset L_2(R), \dim \mathscr{K} < \infty\}$$

is dense in $\mathfrak{M}^+\Omega_0$.

Using the lemma, we may suppose that f_1 and f_2 are in $\mathfrak{M}(\mathscr{K})^+$. Thus we are reduced to a problem with a finite number of degrees of freedom.
Let

$$\mathfrak{S}(\mathscr{K}) = C \oplus \mathscr{K} \oplus (\mathscr{K} \otimes_s \mathscr{K}) \oplus \ldots \subset \mathscr{F}$$

be the Fock space over $\mathscr{K}$. Let $e_{j+1}, \ldots, e_l$ be an orthonormal basis for $\mathscr{K} \ominus \mathscr{H}$ and let

$$q_l = 2^{-1/2}(a^*(e_l) + a(e_l)),$$

$$p_l = 2^{-1/2}\, i(a^*(e_l) - a(e_l)).$$

Then

$$A = \sum_{i=1}^{j} \frac{\mu_i}{2} (p_i^2 + q_i^2 - 1)$$

and $[q_\alpha, p_\beta] = i\,\delta_{\alpha\beta}$. By the von Neumann uniqueness theorem for irreducible representations of the commutation relations (in a finite number of degrees of freedom) or by explicit calculations using Hermite functions (see Bargmann[1]) we have a unitary equivalence of $\mathfrak{S}(\mathscr{K})$ with $L_2(R^l, \pi^{-l/2} e^{-|q|^2} dq)$. Furthermore this unitary equivalence carries q_α into multiplication by the αth coordinate, q_α, and it carries p_α into

$$-ie^{-|q|^2/2} \frac{\partial}{\partial q_\alpha} e^{|q|^2/2}.$$

The equivalence carries Ω_0 into the function 1 and $\mathfrak{M}(\mathscr{K})^+ \Omega_0$ into positive functions, and so it is sufficient to show that e^{-tA} is carried into an operator with a positive kernel. In fact, this kernel is given by the formula

$$\prod_{\alpha=1}^{j} (1 - e^{-2\mu_\alpha t})^{-1/2} \exp\left[-\frac{(q'_\alpha - e^{-\mu_\alpha t} q_\alpha)^2}{1 - e^{-2\mu_\alpha t}} + (q'_\alpha)^2 \right] \tag{2.2.1}$$
$$\times \prod_{\beta=j+1}^{l} \delta(q_\beta - q'_\beta).$$

The quickest derivation of this formula is to check that the kernel and (2.2.1) satisfy the same parabolic equation with the same Cauchy data, because then by a uniqueness theorem, they must coincide.

Proof of Lemma 2.2.2 Let $h \in \mathfrak{M}^+$. Since

$$\cup\{\mathfrak{S}(\mathscr{K}) : \mathscr{H} \subset \mathscr{K} \subset L_2(R), \dim \mathscr{K} < \infty\}$$

is dense in $\mathscr{F}$, we may approximate $h\Omega_0$ by vectors $h_\mathscr{K}\Omega_0, h_\mathscr{K} \in \mathfrak{M}(\mathscr{K})$. Thus as functions on Q, the $\mathfrak{M}(\mathscr{K})$ are dense in $\mathfrak{M}$ in the L_2 norm. By the triangle inequality,

$$-|h - h_\mathscr{K}| \leqq |h| - |h_\mathscr{K}| \leqq |h - h_\mathscr{K}|,$$

and it follows that the $\mathfrak{M}(\mathscr{K})^+$ are dense in $\mathfrak{M}^+$ in the L_2 norm, which proves the lemma.

PROPOSITION 2.2.3 *(Glimm[3]) e^{-tH_0} is a contraction on $L_1(Q)$ for $t \geqq 0$ and for sufficiently large t, e^{-tH_0} is a constraction from $L_2(Q)$ to $L_4(Q)$.*

Remark A semigroup e^{-tH_0} acting on a measure space Q, dq, with a positive self adjoint generator H_0 is called *hypercontractive* (Hoegh Krohn-Simon[1]) if it is a contraction on L_1 for all t and a bounded operator from L_2 to L_4 for some $t > 0$. As we will see, these hypotheses give strong conclusions, by a simple application of the Riesz-Thorin and Stein convexity theorems.

Proof $\|e^{-tH_0} f\|_1 \leqq \|e^{-tH_0}|f|\|_1$

$$= \langle \Omega_0, e^{-tH_0}|f|\Omega_0\rangle = \langle \Omega_0, |f|\Omega_0\rangle = \|f\|_1,$$

and so e^{-tH_0} is a contraction on L_1.

To prove that e^{-tH_0} is a contraction from L_2 to L_4, we may replace the semigroup by strongly approximating semigroups, namely by $e^{-tH_{0,V}}$. Furthermore, we may replace $f \in L_2$ by vectors in a dense subset of L_2. We choose the approximating f's to lie in subspaces of the form $\mathfrak{S}(\mathscr{H})$ where $\mathscr{H}$ is a finite dimensional subspace of $L_2(R)$ with the orthonormal basis $e_1, \ldots, e_j$ given above. For such an f, $e^{-tH_0} f = e^{-tA} f$, with A as in Proposition 2.2.1. Thus we are reduced to the case of a finite number of degrees of freedom, and the kernel of e^{-tA} is given by (2.2.1) with $j = l$. The following lemma reduces the problem to a single degree of freedom.

LEMMA 2.2.4 *Let K_1 and K_2 be operators on σ finite measure spaces Q_1 and Q_2, acting as contractions from $L_2(Q_i)$ to $L_4(Q_i)$. Then $K_1 \otimes K_2$ is a contraction from $L_2(Q_1 \times Q_2)$ to $L_4(Q_1 \times Q_2)$.*

Proof We identify K_1 and $K_1 \otimes I$, K_2 and $I \otimes K_2$. Then

$$\begin{aligned}
\|K_1 \otimes K_2 f\|_4^4 &= \iint |K_1(K_2 f)(q_1, q_2)|^4 \, dq_1 \, dq_2 \\
&\leqq \int \Big(\int |(K_2 f)(q_1, q_2)|^2 \, dq_1 \Big)^2 dq_2 \\
&= \iint |(K_2 f)(q_1, q_2)|^2 \, dq_1 \int |(K_2 f)(p_1, q_2)|^2 \, dp_1 \, dq_2 \\
&= \iiint |(K_2 f)(q_1, q_2)|^2 \, |(K_2 f)(p_1, q_2)|^2 \, dq_2 \, dp_1 \, dq_1 \\
&\leqq \iint \Big(\int |(K_2 f)(q_1, q_2)|^4 \, dq_2 \Big)^{1/2} \\
&\qquad \times \Big(\int |(K_2 f)(p_1, p_2)|^4 \, dp_2 \Big)^{1/2} dp_1 \, dq_1 \\
&\leqq \iint \Big(\int |f(q_1, q_2)|^2 \, dq_2 \Big) \Big(\int |f(p_1, p_2)|^2 \, dp_2 \Big) dp_1 \, dq_1 \\
&= \|f\|_2^4.
\end{aligned}$$

We now take $A = A_1$ and $j = 1$ in (2.2.1). Let

$$\begin{aligned}
a^2(q) &= \pi^{-1} \int (1 - e^{-2\mu t})^{-1} \exp\Big(-2 \frac{(q' - e^{-\mu t} q)^2}{1 - e^{-2\mu t}} + 2(q')^2 \Big) e^{-(q')^2} \, dq' \\
&= \pi^{-1} \int (1 - e^{-2\mu t})^{-1} \exp\Big(-2 \frac{(q' - e^{-\mu t} q)^2}{1 - e^{-2\mu t}} + (q')^2 \Big) dq'.
\end{aligned}$$

By the Schwarz inequality

$$|(e^{-tA} f)(q)| \leqq a(q) \, \|f\|_2,$$

and so the norm of e^{-tA}, as a mapping from L_2 to L_4, is bounded by $\|a\|_4$. The exponent in the integrand defining a^2 is proportional to

$$-2(q' - e^{-\mu t} q)^2 + (1 - e^{-2\mu t})(q')^2$$
$$= -\left[(1 + e^{-2\mu t})^{1/2} q' - \frac{e^{-\mu t}}{(1 + e^{-2\mu t})^{1/2}} q\right]^2 + \left[\frac{e^{-\mu t}}{1 + e^{-2\mu t}} - 2e^{-2\mu t}\right] q^2.$$

The first term above makes the integral defining a^2 converge. Thus

$$a^2(q) \leqq \text{const.} \exp\left(\left[\frac{e^{-\mu t}}{1 + e^{-2\mu t}} - 2e^{-2\mu t}\right] q^2\right)$$

and since the coefficient of q^2 is small for large t, $a^2 \in L_4$ for large t, and e^{-tA} is a bounded map from L_2 to L_4.

Let $\mathscr{H}' = 1^{\perp}$ in $L_2(R, \pi^{-1/2} e^{-q^2} dq)$. Since the norm of $e^{-tA} : \mathscr{H}' \to \mathscr{H}'$ tends to zero as $t \to \infty$, so does the norm of $e^{-tA} : \mathscr{H}' \to L_4$. Let $f' \in \mathscr{H}'$ and let $f = 1 + f'$. Then

$$(e^{-tA} f)^4 = \sum \binom{4}{j} (e^{-tA} f')^j$$

and for large t,

$$\|e^{-tA} f\|_4^4 \leqq \sum_{j=0,2,3,4} \binom{4}{j} \|e^{-tA} f'\|_j^j$$
$$\leqq \sum_{j=0,2,3,4} \binom{4}{j} \|e^{-tA} f'\|_4^j$$
$$\leqq 1 + O(1)(\|e^{-tA} f'\|_4^2 + \cdots + \|e^{-tA} f'\|_4^4)$$
$$\leqq 1 + o(1)(\|f'\|_2^2 + \|f'\|_2^4) \leqq \|f\|.$$

The $j = 1$ term drops out because $\int e^{-tA} f' = \langle 1, e^{-tA} f' \rangle = \langle 1, f' \rangle = 0$. We used the inequalities $\|h\|_2 \leqq \|h\|_3 \leqq \|h\|_4$, valid because the total measure is 1.

THEOREM 2.2.5 *For some* $M > 0, e^{-tH_0}$ *is a contraction from* L_p *to* L_q *if*

$$\frac{1}{p} \leqq (1 + Mt)\frac{1}{q}$$

and

$$\left(1 - \frac{1}{q}\right) \leqq (1 + Mt)\left(1 - \frac{1}{p}\right).$$

Proof By duality ($H_0 = H_0^*$) the $L_p \to L_q$ norm equals the $L_{q'} \to L_{p'}$ norm if $L_{p'} = L_p^*, L_{q'} = L_q^*$. By the Riesz-Thorin convexity theorem (Dunford Schwartz[1]), the $L_p \to L_q$ norm is bounded by one in a convex region in the square $0 \leqq 1/p \leqq 1$, $0 \leqq 1/q \leqq 1$. Combining these facts with

Proposition 2.2.3, we see that e^{-tH_0} is a contraction from L_p to L_p for $0 \leqq t$ and $1 \leqq p \leqq \infty$. Since the identity operator is a contraction from L_p to L_q for $q > p$, so is e^{-tH_0}. For large t, we again use duality, convexity and Proposition 2.2.3. We find that e^{-tH_0} is a contraction from L_4 to L_8, and from L_{2^n} to $L_{2^{n+1}}$. Thus for $t = Mn$ it is a contraction from L_2 to L_{2^n} and (by duality) from $L_{1/(1-2^{-n})}$ to L_2. For large t, the theorem follows.

For small t, we use the Stein interpolation theorem, Stein[1,2], following Segal[3]. Since $0 \leqq H_0$, e^{-tH_0} is complex analytic in the strip $0 \leqq \operatorname{Re} t \leqq T$. On the boundary $\operatorname{Re} t = 0$, it is a contraction from L_2 to L_2 and on the boundary $\operatorname{Re} t = T$ it is a contraction from L_2 to L_4. Thus it is a contraction from L_2 to L_q, where $1/q$ is a convex combination of $\frac{1}{2}$ and $\frac{1}{4}$, namely,

$$\frac{(1 - \operatorname{Re} t/T)}{2} + \frac{(\operatorname{Re} t/T)}{4} = \frac{1}{q}\cdot$$

The theorem now follows from duality and the Riesz-Thorin convexity theorem.

We complete this section by analysing the Hamiltonian $H(g, V, \varkappa)$. After a trivial reduction, the Hamiltonian has only a finite number of degrees of freedom, and is transformed into the operator $-\Delta$ + potential.

THEOREM 2.2.6 (*Jaffe*[3,4]) *$H(g, V, \varkappa)^n$ is essentially self adjoint on* $C^\infty(H_0) = \bigcap_j \mathscr{D}(H_0^j)$.

Proof We use Theorem 1.3.4 to transform the Hamiltonian and write (1.3.3-4) as

$$H_0 \to A_1 \otimes I + I \otimes A_2 = A_1^\sim + A_2^\sim, \tag{2.2.2}$$

$$H_I(g, V, \varkappa) \to B \otimes I = B^\sim. \tag{2.2.3}$$

LEMMA 2.2.7 *Let A_1 and A_2 be self adjoint operators in Hilbert spaces $\mathscr{H}_1$ and $\mathscr{H}_2$ respectively. If for some $n \geqq 1$, A_1^n and A_2^n are essentially self adjoint on domains $\mathscr{D}_1$ and $\mathscr{D}_2$ respectively, then $C_j = (A_1 \otimes I + I \otimes A_2)^j$ is essentially self adjoint on $\mathscr{D}_1 \otimes \mathscr{D}_2$ for $1 \leqq j \leqq n$. Here $\mathscr{D}_1 \otimes \mathscr{D}_2$ is the algebraic tensor product.*

Proof We first note that

$$\mathscr{D}(A_1^{n}) \otimes \mathscr{D}(A_2^{n}) \subset \mathscr{D}(C_j^-),$$

which is a consequence of the inequality

$$\|A_i^j x_i\| \leqq \|A_i^n x_i\| + \|x_i\|$$

for $x_i \in \mathscr{D}_i$, $i = 1,2$. Hence if $y_\nu = (x_{1\nu} \otimes x_{2\nu}) \to y$ and $A_i^n x_{i\nu} \to A_i^n x_i$, $i = 1,2$, we have

$$C_j y_\nu \to C_j^- y, \qquad j = 1, \ldots, n.$$

Furthermore

$$C_j^-(x_1 \otimes x_2) = (A_1 \otimes I + I \otimes A_2)^j (x_1 \otimes x_2).$$

Let x_i have compact support with respect to the spectral measure of A_i. Then x_i is an analytic vector for A_i^n, and $x_1 \otimes x_2$ is an analytic vector for C_j^-. The linear span of such vectors is dense, and so C_j^- is self adjoint by Nelson's theorem.

We apply the lemma to $A_1 + B$ and A_2 of (2.2.2-3). A_2^n is essentially self adjoint on $C^\infty(A_2)$. We are now reduced to considering $(A_1 + B)^n$, a problem with a finite number of degrees of freedom. We use a Schrödinger representation, similar to that of Lemma 2.2.2,

$$\mathfrak{S}(\mathscr{H}) \approx L_2(R^l, dq),$$

$$A_1 + B \to H = -\Delta + W$$

where W is a semibounded polynomial. (The lower bound on B, used here and above follows from the proof of Lemma 2.1.5.) The above isomorphism carries $C^\infty(A_1)$ onto the Schwartz space $\mathscr{S}$. By a Kato perturbation argument, we may without loss of generality add a constant to W, so that $1 \leqq W$, and we must then prove the essential self adjointness of H^n on $\mathscr{S}$.

LEMMA 2.2.8 *Let $\chi \in \mathscr{S}'$ be a distribution solution of the equation $\Delta\chi = W\chi$. Then $\chi \in \mathscr{O}_M$, i.e. χ and all its derivatives are continuous and polynomially bounded.*

Proof The fundamental solution to the Laplace equation is const. r^{2-l}. Let $\phi \in C_0^\infty$ be chosen so that ϕ is a constant in a neighborhood of the origin and

$$\Delta E = \delta - \omega, \quad \omega \in C_0^\infty,$$

where $E = r^{2-l}\phi$. Note that E and $\partial_j E$ are integrable and have compact support. Thus convolution by E or $\partial_j E$ maps $\mathscr{S}$ into $\mathscr{S}$ and $\mathscr{S}'$ into $\mathscr{S}'$ and convolution by E takes a C^m polynomially bounded function into a C^{m+1} polynomially bounded function. Let $F\chi = E * (W\chi)$. Then

$$\chi - \omega * \chi = \Delta E * \chi = E * \Delta\chi = E * (W\chi) = F\chi$$

and iterating n times, we get

$$\chi = \sum_{j=0}^{n-1} F^j(\omega * \chi) + F^n\chi.$$

Since $\chi \in \mathscr{S}'$ and $\omega \in C_0^\infty \subset \mathscr{S}$, it follows that $\omega * \chi \in \mathscr{O}_M$, Schwartz[1], and so $F^j(\omega * \chi) \in \mathscr{O}_M$. Thus we need only study $F^n\chi$. We write $\chi = \partial^\nu\theta$ where ν is a multi-index and θ is continuous and polynomially bounded. We integrate the derivatives in $F^n\chi$ by parts to move them from θ to the W's or the E's, and for n sufficiently large a given E is differentiated at most once. There are at least $n - |\nu|$ undifferentiated E's and so $F^n\chi$ (and χ) is a $C^{n-|\nu|}$ polynomially bounded function. Thus $\chi \in \mathscr{O}_M$.

The n-fold application of the next lemma completes the proof of Theorem 2.2.6.

LEMMA 2.2.9 *Let $\mathscr{S}_0$ be a dense subspace of $\mathscr{S}$. Then $H\mathscr{S}_0 = (-\Delta + W)\mathscr{S}_0$ is dense in $\mathscr{S}$.*

Proof H maps $\mathscr{S}$ continuously into $\mathscr{S}$. If $H\mathscr{S}_0$ is not dense in $\mathscr{S}$, there is a $\chi \in \mathscr{S}'$ orthogonal to $H\mathscr{S}_0$ (and to $H\mathscr{S}$) and then χ is a distribution solution of the equation $\Delta\chi = W\chi$. By Lemma 2.2.8, $\chi \in \mathscr{O}_M$. Since W is real, we can suppose χ is real also. By Green's theorem and the fact that $1 \leqq W$, we have

$$\int_{r^2 \leqq a^2} \chi^2 r^{l-1}\, dr\, d\omega \leqq 2^{-1} r^{l-1} \int_{r=a} (\partial_r \chi)^2\, d\omega$$

where $r^{l-1}\, dr\, d\omega = dq$. If we define $F(a) = \int_{r=a} \chi^2\, d\Omega$ then F has at most polynomial growth, since $\chi \in \mathscr{O}_M$. But by Green's theorem above,

$$0 \leqq \int_0^a r^{l-1} F(r)\, dr \leqq a^{l-1} F'(a).$$

Thus $F(a)$ is monotone increasing and for $1 < a$,

$$\left(\frac{a-1}{a}\right)^{l-1} F(a-1) \leqq \int_{a-1}^{a} \left(\frac{r}{a}\right)^{l-1} F(r)\, dr$$
$$\leqq \int_0^a \left(\frac{r}{a}\right)^{l-1} F(r)\, dr \leqq F'(a).$$

Integrating from a to $a + 1$,

$$((a-1)/a)^{l-1} F(a-1) \leqq F(a+1) - F(a),$$
$$(1 + ((a-1)/a)^l) F(a-1) \leqq F(a+1)$$

and F grows exponentially, a contradiction. Thus $F = 0$ and the proof is complete.

3 The Hamiltonian $H(g)$

3.1 Positivity of the energy and higher order estimates

Let

$$E(g, \varkappa) = \inf \text{spectrum}\, (H_0 + H_I(g, \varkappa)), \tag{3.1.1}$$

$$H(g, \varkappa) = H_0 + H_I(g, \varkappa) - E(g, \varkappa). \tag{3.1.2}$$

These equations change the definition of $H(g, \varkappa)$ by including the finite vacuum energy renormalization. We prove that $E(g) = \lim_{\varkappa \to \infty} E(g, \varkappa)$ is finite, so that after the finite renormalization, $0 \leqq H(g) = H_0 + H_I(g) - E(g)$. The finiteness of the vacuum energy is a special feature of the space cutoff ϕ_2^{2n} model (Glimm-Jaffe[6]). However, a fundamental property, which should be true for all models, is that the shift in the vacuum energy under suitable perturbations A is finite. With $H(g, \varkappa)$ renormalized by (3.1.1-2), let

$$\delta E(g, \varkappa, A) = \inf \text{spectrum } (H(g, \varkappa) + A). \tag{3.1.3}$$

If $\delta E(\pm A)$ remains finite as $\varkappa \to \infty$ or $g \to 1$ then we have uniform estimates

$$\pm A \leqq \text{const.} (H(g, \varkappa) + I) \tag{3.1.4}$$

and

$$\|R^{1/2} A R^{1/2}\| \leqq \text{const.} \tag{3.1.5}$$

where $R = R(\zeta) = (H(g, V, \varkappa) - \zeta)^{-1}$. If $0 \leqq A$, we also have $\|A^{1/2} R^{1/2}\| \leqq \text{const.}$

The higher order estimates have the form

$$\pm A \leqq \text{const.} (H(g, \varkappa) + I)^n$$

and lead to the bound $\|R^{n/2} A R^{n/2}\| \leqq$ const. The A's which can be so dominated depend upon the model. They can be N_τ or related operators, or, for higher order estimates, N_τ^j. Combining these estimates with the N_τ estimates of Sec. 1.2, we can estimate $\|R^{n/2} A R^{m/2}\|$ where A is an annihilation creation form whose kernel has a finite L_2 or operator type norm.

THEOREM 3.1.1 (*Glimm-Jaffe*[5]) *With the hypothesis and assumptions of Theorem 2.1.7,*

$$0 \leqq -E(g, \varkappa) \leqq \text{const. diam. suppt. } g.$$

The constant is independent of g and $\varkappa$, for $\varkappa \leqq \infty$, $g \to 1$.

Remark Nelson[3] derived a bound uniform in $\varkappa$ but depending on a periodic volume V. The box was replaced by a fixed g cutoff by Glimm[3]. The passage from separate estimates on H_0 and H_I (cf. Theorem 2.1.3 or 2.1.6 and 2.2.5) to the bound on E has been simplified by Segal[2] and is valid in the hypercontractive setting, with Theorem 2.1.3 as hypothesis on the perturbing potential. See also Glimm-Jaffe[12].

Proof $E(g, \varkappa) \leqq 0$ since $\langle \Omega_0, (H_0 + H_I)\Omega_0 \rangle = 0$. Our proof of the lower bound on E requires essential self adjointness of the sum $H_0 + H_I$ on the domain $\mathscr{D}(H_0) \cap \mathscr{D}(H_I)$. Thus for the present we estimate only $E(g, V, \varkappa)$ (cf. Theorem 2.2.6) and E_J, the vacuum energy of

$$H_J = H_0 + H_I^{(J)} = H_0 + F_J(H_I(g)),$$

where

$$F_J(\lambda) = \begin{cases} \lambda & \text{for } |\lambda| \leqq j \end{cases}$$

Since $H_I^{(J)}$ is bounded, $H_0 + H_I^{(J)}$ is self adjoint. The proof for $E(g, \varkappa)$ will follow from essential self adjointness proved in section 3.2.

Let c = const. diameter support g. For t fixed but sufficiently large, and H_I equal to either $H_I(g, V, \varkappa)$ or $H_I^{(J)}$,

$$\|e^{-tH_0/2}\, e^{-tH_I}\, e^{-tH_0/2}\| \leqq e^c,$$

since Theorems 2.1.7 and 2.2.5 bound the three factors above (from right to left) as operators from $L_2 \to L_4$, $L_4 \to L_2$ and $L_2 \to L_2$. In fact, by Hölder's inequality

$$\|e^{-tH_I}\psi\|_2 \leqq \|e^{-tH_I}\|_4\, \|\psi\|_4$$

so $e^c = \|e^{-tH_I}\|_4$. Thus $e^{-tH_I} \leqq e^c\, e^{tH_0}$, and by monotonicity of the square root, see Kato [1, p. 292],

$$e^{-(t/n)H_I} \leqq e^{c/n}\, e^{(t/n)H_0},$$

or

$$e^{-(t/2n)H_0}\, e^{-(t/n)H_I}\, e^{-(t/2n)H_0} \leqq e^{c/n}.$$

We use the Trotter product formula

$$e^{tC} = \operatorname*{st\,lim}_{n\to\infty} (e^{tA/n}\, e^{tB/n})^n,$$

valid when A, B and $C = (A + B)^-$ generate contraction semigroups, see Nelson[(4)]. It follows that

$$\|e^{-t(H_0+H_I)}\| \leqq e^c = \|e^{-tH_I}\|_4.$$

Hence $-(c/t) \leqq (H_0 + H_I)$, and the proof for $E(g, V, \varkappa)$ or E_J is complete.

Corollary 3.1.2 *Let P_1 be a polynomial with degree less than or equal to the degree of P. Then with*

$$A = H_0/(\text{diam. suppt. } g)$$

or with $|f| \leqq$ const. g and

$$A = \int :P_1(\phi_\varkappa(x)):\, f(x)\, dx/(\text{diam. suppt. } g)$$

the estimates (3.1.4-5),

$$\pm A \leqq \text{const.}\,(H(g, V, \varkappa) + I),$$

$$\|R^{1/2} A R^{1/2}\| \leqq \text{const.},$$

are valid uniformly as $\varkappa \to \infty$ and $g \to 1$.

Proof $0 \leqq 2^{-1}H_0 + H_I(g, V, \varkappa) - 2^{-1}E(2g, V, \varkappa)$ and so

$$H_0 \leqq 2(2^{-1}H_0 + 2^{-1}H_0 + H_I(g, V, \varkappa) - 2^{-1}E(2g, V, \varkappa))$$

$$= 2(H_0 + H_I(g, V, \varkappa) - E(g, V, \varkappa)) + 2E(g, V, \varkappa) - E(2g, V, \varkappa).$$

The bound on E from Theorem 3.1.1 completes this case and the estimate in the second case is similar.

Remark The estimates (3.1.4-5), uniform in $\varkappa$ and g, for $A = H_0^{\text{loc}}$ or N_τ^{loc} would lead to estimates on the vacuum expectation values in the limit $\varkappa \to \infty$, $g \to 1$. (See chapter 4.)

We now turn to higher order estimates. These estimates are used in the study of the field equations, in the proof of Lorentz covariance and in the construction of in and out fields for $H(g)$. First we state the results.

THEOREM 3.1.3 (*Rosen*[2]) *Let i be a positive integer. For $\varepsilon > 0$, $i \geqq 3$ and $j = j(i, \deg. P, \varepsilon)$ sufficiently large, there is a constant such that*

$$H_0^{3-\varepsilon} N^{i+\varepsilon-3} \leqq \text{const.}\,(H(g) + \text{const.})^j.$$

If $\varepsilon > 2$, we may take $j = i$.

If degree $P \leqq 4$, we have (Glimm-Jaffe[2]*)*

$$H_0^2 \leqq \text{const.}\,(H(g)\ \text{const.})^2.$$

The theorem is proved first for $H(g, V, \varkappa)$ with constants independent of V and $\varkappa$ and then transferred to the limit, $H(g)$. (See Sec. 3.2.) The proof is based on the pull through formula

$$a(k)\, R(\zeta) = R(\zeta - \mu(k))\, a(k) - R(\zeta - \mu(k))\, [a(k), H_I]\, R(\zeta). \quad (3.1.6)$$

This formula has also been used in standard field theory calculations, see for instance Schweber [1, p. 359].

PROPOSITION 3.1.4 *Let R be the resolvent of $H(g, V, \varkappa)$. With a constant independent on V and $\varkappa$,*

$$\|R^{1/2}[a(k), H_I]\, R^{1/2}\| \leqq \text{const.}\, \mu(k)^{-1/2}.$$

Proof We apply Corollary 3.1.2 with $P_1 = \mu(k)^{1/2}\,[a(k), H_I]$. P_1 has k and x dependent coefficients but they are bounded uniformly and $\deg P_1 = \deg P - 1$. The factor $\mu^{1/2}$ is permitted because of the factor $\mu^{-1/2}$ in the field occuring in H_I.

To illustrate the method, we give a formal derivation of the weak second order estimate

$$H_0 N_\tau \leqq \text{const.}\,(H(g, V, \varkappa) + \text{const.})^2$$

for $\tau < 0$. We have

$$H_0 N_\tau = N_{\tau+1} + \int a^*(k)\, H_0 \mu(k)^\tau\, a(k)\, dk,$$

and the first term is dominated by the first order estimate, Corollary 3.1.2, with $A = N_{\tau+1} \leqq \text{const.}\, H_0$. To bound the second term, we show

$$\int \|H_0^{1/2}\, a(k)\, R(\zeta)\, \theta\|^2\, \mu(k)^\tau\, dk \leqq \text{const.}\, \|\theta\|^2. \quad (3.1.8)$$

By (3.1.6) and the triangle inequality, the left member is bounded by

$$2\int \|H_0^{1/2}R(\zeta - \mu(k))\, a(k)\,\theta\|^2\, \mu(k)^\tau\, dk$$

$$+\, 2\int \|H_0^{1/2}\, R(\zeta - \mu(k))\, [a(k), H_I]\, R(\zeta)\,\theta\|^2\, \mu(k)^\tau\, dk. \qquad (3.1.9)$$

We use Corollary 3.1.2 to bound $\|H_0 R^{1/2}\|$. The first term in (3.1.9) is bounded by

$$\text{const.} \int \|R(\zeta - \mu(k))^{1/2}\, a(k)\,\theta\|^2\, \mu(k)^\tau\, dk$$

$$\leqq \text{const.} \int \|(N_\tau + I + \mu(k)^\tau)^{-1/2}\, a(k)\,\theta\|^2\, \mu(k)^\tau\, dk$$

$$= \text{const.} \int \|a(k)\,(N_\tau + I)^{-1/2}\,\theta\|^2\, \mu(k)^\tau\, dk$$

$$= \text{const.}\ \|N_\tau^{1/2}(N_\tau + I)^{-1/2}\,\theta\|^2 \leqq \text{const.}\ \|\theta\|^2$$

and the second term is bounded by

$$\text{const.} \int \|R(\zeta - \mu(k))^{1/2}\, [a(k), H_I]\, R(\zeta)\,\theta\|^2\, \mu(k)^\tau\, dk$$

$$\leqq \text{const.} \int \mu(k)^{-1+\tau}\, dk\, \|\theta\|^2 \leqq \text{const.}\ \|\theta\|^2.$$

3.2 Domains of essential self adjointness

THEOREM 3.2.1 *The Hamiltonian $H(g)$ is essentially self adjoint on $\mathscr{D}(H_0) \cap \mathscr{D}(H_I(g))$ and on the subdomains*

$$C^\infty(H_0) = \bigcap_n \mathscr{D}(H_0^n), \qquad (3.2.1)$$

$$\mathscr{D}(H_0) \cap L_p(Q), \qquad 2 < p < \infty. \qquad (3.2.2)$$

The two domains result from distinct proofs of essential self adjointness. The domain (3.2.1) is associated with the Fock space proof while (3.2.2) comes from the Q space proof. For the ϕ^4 theory, $H(g)$ is self adjoint on $\mathscr{D}(H_0) \cap \mathscr{D}(H_I(g))$.

The Fock space proof uses N_τ estimates and Rosen's higher order estimates, and it generalizes the original ϕ_2^4 proof of Glimm and Jaffe(2). The first ϕ_2^{2n} proof of essential self adjointness was given on path space (Rosen(1)). This path space proof was simplified and placed in Q space by Segal(3); see also Høegh-Krohn and Simon(1).

Proof of (3.2.1) in Fock space We will show that the resolvents

$$R(g, V, \varkappa) = (H_{0,V} + H_I(g, V, \varkappa) + \text{const.})^{-1}$$

converge in norm as $V \to \infty$, $\varkappa \to \infty$. Furthermore, the Hamiltonians converge strongly on $\mathscr{D}(H_0 N^i)$, provided $\deg P \leqq 2(i+1)$, by a calculation based on Proposition 1.2.3:

$$\|\{(H_0 + H_I(g)) - (H_{0,V} + H_I(g, V, \varkappa))\}(N+I)^{-i}(H_0 + I)^{-1}\| \to 0.$$

Hence $H_{0,V} + H_I(g, V, \varkappa)$ has a strong graph limit (see Sec. 6.2) and it follows that

$$R(g) = \lim_{V,\varkappa} R(g, V, \varkappa)$$

is the resolvent of a self adjoint operator T, and

$$\{H_0 + H_I(g) + \text{const.}\} \restriction D(H_0 N^i) \subset T.$$

The operator T is essentially self adjoint on the range of $R(g)^j$, for any j. We next show that

$$\text{Range } R(g)^j \subset D(H_0 N^i), \tag{3.2.3}$$

if j is large enough. The essential self adjointness on $\mathscr{D}(H_0 N^i)$ follows, and hence the essential self adjointness on (3.2.1) follows from an N_τ estimate.

Let A be a self adjoint operator and $\psi_j \in \mathscr{D}(A)$. If

$$\psi_n \to \psi, \qquad \|A\psi_n\| \leqq \text{const.}, \tag{3.2.4}$$

then $\psi \in \mathscr{D}(A)$. To prove this we note that for $\psi \in \mathscr{D}(A)$,

$$|\langle A\chi, \psi\rangle| = \left|\lim_n \langle A\chi, \psi_n\rangle\right| = \left|\lim_n \langle \chi, A\psi_n\rangle\right| \leqq \|\chi\|\,\|A\psi_n\|$$
$$\leqq \text{const.}\ \|\chi\|.$$

Hence $\psi \in \mathscr{D}(A^*) = \mathscr{D}(A)$.

By Theorem 3.1.3, for $\psi_{\varkappa,V} = R(g, V, \varkappa)^j \theta$, $\theta \in \mathscr{F}$, we have

$$\|H_0 N^i \psi_{\varkappa,V}\| \leqq \text{const.},$$

uniformly in $\varkappa$, V. Hence resolvent convergence and (3.2.4) prove (3.2.3).

To prove norm convergence of the resolvents, it is sufficient to prove the norm convergence of their j^{th} power. If $\delta R = R_1 - R_2$ and $\delta H = H_1 - H_2$ denote differences in the resolvents and the Hamiltonians arising from different cutoffs, then calculating as bilinear forms on the domain $C^\infty(H_1) \times C^\infty(H_2)$ we have by Theorem 3.1.3,

$$\|R_1^j - R_2^j\| \leqq \sum_{\alpha=0}^{j-1} \|R_1^\alpha\, \delta R\, R_2^{j-\alpha-1}\|$$
$$\leqq \sum_{\alpha=0}^{j-1} \|R_1^{\alpha+1}\, \delta H\, R_2^{j-\alpha}\|$$
$$\leqq \text{const.} \sum_{\alpha=0}^{j-1} \|(N+I)^{-(\alpha+1)}\, \delta H (N+I)^{-(j-\alpha)}\|.$$

Hence, we choose $j \geqq \frac{1}{2} \deg. P$.

Each term in the sum over α above converges to zero, by Proposition 1.2.3. This completes the proof, and furthermore shows that Theorem 3.1.3 is valid for $H(g)$, in addition to $H_{0,V} + H_I(g, V, \varkappa)$ as indicated in Sec. 3.1.

The Q space proof of (3.2.1) extends to the hypercontractive setting under the hypothesis that $H_I(g)$ and $\exp(-H_I(g))$ belong to L_p for all $p < \infty$, as in Proposition 2.1.2 and Theorem 2.1.4. Let H_j be the Hamiltonian (3.1.1).

LEMMA 3.2.2 *Let* $1 < p, q < \infty$. *The semigroups* $e^{-tH_j}: L_p \to L_p$ *are strongly continuous in* $t \geqq 0$ *and exponentially bounded with a bound uniform in* j. *If* t *is sufficiently large, the same is true for* $e^{-tH_j}: L_p \to L_q$.

Proof We use the Trotter product formula,

$$e^{-tH_j} f = L_2 \lim_n \left(e^{-tH_0/n} e^{-tH_I^{(j)}/n}\right)^n f,$$

valid since $H_I^{(j)}$ is a bounded perturbation of the positive and self adjoint operator H_0. Let $f \in L_\infty$. Each factor $e^{-tH_0/n}$ maps $L_r \to L_{r+Mt/n}$ with norm 1, by Theorem 2.2.5 and each factor $\exp(-tH_I^{(j)}/n)$ maps $L_r \to L_{r-Mt/2n}$ with norm $e^{O(t/n)}$ by Theorem 2.1.4 and the Hölder inequality, for r bounded away from 1 and ∞. Here $t > 0$. Thus the Trotter approximants to $e^{-tH_j}f$ have the required exponential bounds. The approximants converge strongly in L_2 and they are weakly compact in L_p (and in L_q for large t). Thus they converge weakly in L_p or L_q and $e^{-tH_j} f$ has the required exponential bound:

$$\|e^{-tH_j} f\|_p \leqq e^{O(t)} \|f\|_p, \qquad 0 \leqq t$$

$$\|e^{-tH_j} f\|_q \leqq e^{O(t)} \|f\|_p, \qquad \text{large } t.$$

For $p = 2$, strong continuity results from self adjointness. For $1 < p < 2$, it results from the denseness of L_2 in L_p and the continuity for $p = 2$. For $2 < p < \infty$, we use duality and reduce to the case $1 < p < 2$.

LEMMA 3.2.3 *Let* $1 < p, q < \infty$. *The semigroups* $e^{-tH_j}: L_p \to L_p$ *converge strongly as* $j \to \infty$ *to continuous exponentially bounded semigroups, and for large* t, *the maps* $e^{-tH_j}: L_p \to L_q$ *converge in norm.*

Proof In the Duhamel formula

$$e^{-tH_l} f - e^{-tH_j} f = \int_0^t e^{-(t-u)H_l} (H_j - H_l) e^{-uH_j} f \, du$$

we take $f \in L_\infty$. Then $e^{-uH_j} f$ is bounded in L_r with a bound uniform j and u, $0 \leqq u \leqq t$. We have

$$H_j - H_l = H_I^{(j)} - H_I^{(l)} \to 0 \text{ in } L_q,$$

for all $q \to \infty$. Strong convergence follows from Hölder's inequality; Lemma 3.2.2 controls the factor $e^{-(t-u)H_l}$. The convergence is uniform in bounded t, and consequently the limiting semigroups are continuous. For

3*

large t, we take $f \in L_p$ and we decompose the integration into two regions: $\int_0^t = \int_0^{t/2} + \int_{t/2}^t$. For $0 \leqq u \leqq t/2$, $e^{-uH_j}: L_p \to L_p$ and $e^{-(t-u)H_i}: L_{p-\varepsilon} \to L_q$ with bounds uniform in i and j. Also $H_i - H_j: L_p \to L_{p-\varepsilon}$ converges to zero as $i, j \to \infty$ by Hölder's inequality. Reversing the role of the exponential factors for $u > t/2$, the proof is complete.

Let T be the self adjoint generator of the limiting semigroup on L_2.

Proof of (3.2.2) Consider the domain $\mathscr{D}_t = e^{-tT} L_\infty(Q)$. Then $\mathscr{D}_t \subset L_p$, for all $p < \infty$. Let $f = e^{-tT} g \in \mathscr{D}_t, g \in L_\infty$, and let $f_j = e^{-tH_j} g$. Then $f_j \to f$ in L_p and $H_j f_j \to Tf$ in L_2 by the functional calculus for self adjoint operators. Moreover $f \in L_p \subset \mathscr{D}(H_I(g))$ and

$$H_I^{(j)} f_j \to H_I(g) f$$

in L_2 by Hölder's inequality. By subtraction,

$$H_0 f_j = H_j f_j - H_I^{(j)} f_j$$

converges in L_2. This implies that $f \in \mathscr{D}(H_0)$ and

$$H_0 f = \lim_j H_0 f_j = Tf - H_I(g) f.$$

Thus

$$H_0 + H_I(g) \restriction \mathscr{D}_t = T \restriction \mathscr{D}_t.$$

Since $H_0 + H_I(g)$ is symmetric on $\mathscr{D}_t$ and T is essentially self adjoint on $\mathscr{D}_t$, the theorem follows.

Remark The semigroups and resolvents, as operators on L_2, are norm continuous in the cutoff $\varkappa$ in the potential $H_I(g, \varkappa)$ and they are norm continuous in V and $\varkappa$ in $H_{0,V} + H_I(g, V, \varkappa)$. Using the formulas

$$\inf \operatorname{spectrum} A = t^{-1} \log \|e^{-tA}\| = \|(A + b)^{-1}\|^{-1} + b$$

we see that the vacuum energies $E(g, \varkappa)$ and $E(g, V, \varkappa)$ converge as $\varkappa, V \to \infty$. Henceforth, we renormalize all Hamiltonians as in (3.1.2), so that $H(g)$ now denotes

$$H(g) = H_0 + H_I(g) - E(g), \tag{3.2.5}$$

and

$$0 = \inf \operatorname{spectrum} H(g). \tag{3.2.6}$$

Since the E's converge, the semigroups and resolvents of the renormalized Hamiltonians still converge as before.

3.3 The spectrum of $H(g)$

Let m be the mass in H_0. After the vacuum energy renormalization (3.2.5) the spectrum of $H(g)$ has only separated discrete eigenvalues of finite multiplicity in the spectral interval $[0, m)$ (Glimm and Jaffe[4]). Høegh-Krohn[2] proved the existence of continuous spectrum for $H(g)$ in the interval $[m, \infty)$. The discrete eigenvalues in $(0, m)$ may result from a downward mass shift: particles localized in suppt g may have a physical mass less than m. We propose that such eigenvalues should be eliminated by the addition to $H(g)$ of a finite mass renormalization counter term,

$$-1/2\,\delta m^2 \int :\phi^2(x):\, g(x)^2\, dx.$$

For the box cutoff Hamiltonian $H(V)$ we lose the continuum, but since $H(V)$ and the periodic momentum operator P_V commute, we may discuss their joint spectrum.

THEOREM 3.3.1 *The joint spectrum of $H(V)$ and P_V is purely discrete and is contained in the forward light cone. The same is true for P_V and the momentum cutoff Hamiltonian $H(V, \varkappa)$.*

Proof We only show that $H(V)$ has discrete spectrum; see Glimm-Jaffe[10] for the remainder. We use the representation (2.2.2-3) for

$$H(V) = A_1^{\sim} + B^{\sim} + A_2^{\sim} - E(V) = (A_1 + B) \otimes I + I \otimes A_2 - E(V),$$

with $\varkappa = \infty$, $g = 1$, and then it is enough to show that $H_{0,V} + H_I(V)$, acting on $\mathfrak{S}(\mathscr{H}) = \mathfrak{S}(l_2(Z_V))$, has pure discrete spectrum. However $H_{0,V}$ has a compact resolvent on $\mathfrak{S}(\mathscr{H})$ and the first order estimate

$$H_{0,V} \leqq \text{const.}\,(H_{0,V} + H_I(V) + \text{const.})$$

implies that $H_{0,V} + H_I(V)$ has a compact resolvent on $\mathfrak{S}(\mathscr{H})$ also, so the proof is complete.

Remark The same proof shows that $H(g, V, \varkappa)$, as an operator on $\mathfrak{S}(l_2(Z_{V,\varkappa})) = \mathscr{F}_{\varkappa,V}$, has a compact resolvent. We now use this fact.

THEOREM 3.3.2 *In the spectral interval* $[0, m)$, *$H(g)$ has pure discrete spectrum with finite multiplicity and no accumulation points.*

Proof First we prove this result for $H(g, V, \varkappa)$, using the representation (1.3.3-4). Since

$$m \leqq A_2 = I \otimes (H_{0,V} \upharpoonright \mathfrak{S}(\mathscr{H}^{\perp})),$$

the spectrum and multiplicity of $H(g, V, \varkappa)$ in the interval $[0, m)$ coincides with that of $H(g, V, \varkappa) \upharpoonright \mathfrak{S}(l_2(Z_{V,\varkappa}))$. By the remark above, this proves the theorem for $H(g, \varkappa, V)$. The proof for $H(g)$ follows from the norm convergence of the resolvents and the following proposition.

PROPOSITION 3.3.3 *Let A_j be a sequence of self adjoint operators, and let the resolvents $R_j(\zeta) = (A_j - \zeta)^{-1}$ converge in norm to the resolvents of a self adjoint operator A. If each A_j has pure discrete spectrum with finite multiplicity and no accumulation points in some fixed spectral interval (a, b) then the same is true of A.*

Proof For any continuous function F, vanishing at infinity, $F(R_j(\zeta)) \to F(R(\zeta))$ in norm. If suppt $F \subset ((b-\zeta)^{-1}, (a-\zeta)^{-1})$ then $F(R_j(\zeta))$ is compact. Thus $F(R(\zeta))$ is compact and the proof is complete.

THEOREM 3.3.4 *$H(g)$ has a unique vacuum. That is, 0 is a simple eigenvalue for $H(g)$. The same statement is valid for the Hamiltonians $H(g,\varkappa)$, $H(g,V,\varkappa)$, $H(V)$ and $H(V) + \theta P_V$, $|\theta| < 1$.*

Let Ω_g be the vacuum for $H(g)$. Then $H(g)\Omega_g = 0$ and Ω_g is a limit of the $\Omega_{g,\varkappa}$ or $\Omega_{g,V,\varkappa}$. The existence of the vacuum follows from (3.2.6) and Theorem 3.3.2. The uniqueness is based on a general proposition.

DEFINITION 3.3.5 *Let A be a bounded operator on a measure space $L_2(Q, dq)$. A has a positive ergodic kernel provided $0 \leqq \langle f, Ah\rangle$ whenever f and h are nonnegative functions in L_2 and*

$$0 < \langle f, A^j h\rangle \tag{3.3.1}$$

whenever $0 \leqq f, h$, $0 \neq f, h$, for some j depending on f and h.

For positive kernels the ergodic hypothesis is trivially seen to be equivalent to the irreducibility of A and $L_\infty(Q)$ as operators on $L_2(Q)$.

PROPOSITION 3.3.6 *Let A have a positive ergodic kernel, and suppose that $\|A\|$ is an eigenvalue of A. Then $\|A\|$ is a simple eigenvalue and the corresponding eigenvector can be chosen to be a strictly positive function.*

Proof Since A maps positive functions into positive functions, it also maps real functions into real functions. If $f \in L_2$ satisfies $Af = \|A\|\, f$, then so do Re f and Im f. Therefore we may take f to be real. Since $\|A^j\| = \|A\|^j$ and $A^j f = \|A\|^j f$, we infer that

$$\|A^j\|\ \|f\|^2 = \langle f, A^j f\rangle \leqq \langle |f|, A^j|f|\rangle \leqq \|A^j\|\ \|f\|^2$$

and so $\langle f, A^j f\rangle = \langle |f|, A^j|f|\rangle$. Writing $f = f^+ - f^-$ where $f^\pm$ are the positive and negative parts of f,

$$\langle f^+, A^j f^+\rangle - \langle f^+, A^j f^-\rangle - \langle f^-, A^j f^+\rangle + \langle f^-, A^j f^-\rangle$$

$$= \langle f^+, A^j f^+\rangle + \langle f^+, A^j f^-\rangle + \langle f^-, A^j f^+\rangle + \langle f^-, A^j f^-\rangle,$$

or

$$\langle f^+, A^j f^-\rangle + \langle f^-, A^j f^+\rangle = 0.$$

Unless $f^+ = 0$ or $f^- = 0$, each of the two terms above is strictly positive, by a suitable choice of j. Thus $f^+ = 0$ or $f^- = 0$ and we may choose f to be nonnegative. If $0 \leqq h$, $0 \neq h$, then for some j, $0 < \langle h, A^j f\rangle = \|A\|^j \times \langle h, f\rangle$. Thus $hf \neq 0$ and f is positive almost everywhere. Finally if h is a second eigenvector for the eigenvalue $\|A\|$ then a scalar multiple of h is also positive almost everywhere, and h and f are not orthogonal. Hence $\|A\|$ must be a simple eigenvalue.

Proof of Theorem 3.3.4 Let $A = e^{-H(g)}$ in Proposition 3.3.6. The Trotter approximation to A shows that A has a positive kernel, by Proposition 2.2.1. The ergodic property (3.3.1) may be established directly (Glimm-Jaffe[4]) or the irreducibility of A and $L_\infty(Q)$ may be established (Høegh-Krohn and Simon[1]). Let B be a bounded operator commuting with A and $L_\infty(Q)$. We prove $B = \lambda I$. Since $H_0 + H_I(g) - H_I^{(j)}$ converges to H_0 on a core for H_0, we have

$$e^{-tH_0} = \text{st.} \lim_{j\to\infty} \lim_{n\to\infty} (e^{-(t/n)H(g)}\, e^{(t/n)H_I(j)})^n.$$

Hence B commutes with e^{-tH_0}, and $e^{-tH_0} B\Omega_0 = B\, e^{-tH_0}\Omega_0 = B\Omega_0$. Since zero is a simple eigenvalue of H_0, $B\Omega_0 = \lambda\Omega_0$, and since $\mathfrak{M}\Omega_0$ is dense, $B - \lambda I = 0$. This completes the proof.

Høegh-Krohn[2] has constructed the in and out annihilation and creation operators $a_\pm$ for the Hamiltonian $H(g)$. Let

$$a^{\#}(h) = \int a(k)\, h(k)^{-}\, dk \quad \text{or} \quad \int a^{*}(k)\, h(k)\, dk,$$

$h \in L_2(R)$ and let

$$a_t^{\#}(h) = e^{-itH(g)}\, e^{itH_0}\, a^{\#}(h)\, e^{-itH_0}\, e^{itH(g)}.$$

PROPOSITION 3.3.7 *On the domain $H(g)^{1/2}$, the following strong limits exist:*

$$a_\pm^{\#}(h) = \lim_{t\to\pm\infty} a_t^{\#}(h).$$

Sketch of proof For large j, one can evaluate the weak derivative

$$\frac{d}{dt} a_t^{\#}(h) = -e^{-itH(g)}\, [iH_I(g), a^{\#}(h_t)]\, e^{itH(g)}$$

on the domain $\mathscr{D}(H(g)^j) \times \mathscr{D}(H(g)^j)$, where $h_t = e^{it\mu(k)}\, h$. The proof uses the higher order estimate, Theorem 3.1.3. Thus with $n = \deg P$ and j large,

$$\|(a_{t_1}^{\#}(h) - a_{t_2}^{\#}(h))\, (H(g) + I)^{-j}\|$$

$$\leq \int_{t_1}^{t_2} \|[H_I(g), a^{\#}(h_s)]\, (H(g) + I)^{-j}\|\, ds$$

$$\int_{t_1}^{t_2} \|[H_I(g), a^{\#}(h_s)]\, (N + I)^{-n/2}\|\, ds.$$

The integrand is estimated by N_τ estimates of section 1.1.2 with $\tau = 0$. For $h \in C_0^\infty$, h vanishing in a neighborhood of the origin, the kernels in the commutator have L_2 norms bounded by $|s|^{-r}$ for any r, as $s \to \pm\infty$. This gives convergence for a restricted class of h's on a smaller domain, and the full result follows from the uniform estimate

$$\|a_\tau^\#(h)\,(H(g) + I)^{-1/2}\| \leqq \text{const. } \|h\|_2,$$

which is a consequence of Corollary 3.1.2 and an N_τ estimate.

Using the $a_\pm^\#$, we can realize $\mathscr{F}$ as a tensor product, $\mathscr{F} \approx \mathscr{F}_\pm \otimes \mathscr{V}_\pm$, where $\mathscr{V}_\pm$ is the space of simultaneous null vectors for the $a_\pm(h)$ and $F_\pm$ is the closure of the range of polynomials in the $a_\pm^*(h)$ applied to Ω_g.

THEOREM 3.3.8 (Høegh-Krohn[2]) *With the above tensor product decomposition, $H(g)$ is transformed into*

$$(H(g) \upharpoonright \mathscr{F}_\pm) \otimes I + I \otimes (H(g) \upharpoonright \mathscr{V}_\pm).$$

Moreover $H(g) \upharpoonright \mathscr{F}_\pm$ is unitarily equivalent to H_0. $H(g) \upharpoonright \mathscr{V}_\pm$ is a positive operator and in any interval $[0, m - \varepsilon]$, $\varepsilon > 0$, the spectrum is finite dimensional.

4 Removing the space cutoff

Having studied the cutoff Hamiltonian $H(g)$, we are now prepared to construct our ϕ_2^{2n} field theory model, by taking the limit $g \to 1$. The experimentally observed phenomena in quantum field theory are the particles and the scattering cross sections, which specify the measured interactions between the particles. According to conventional ideas in physics, particles are labeled by their masses, which are discrete eigenvalues of the mass operator $M = (H^2 - P^2)^{1/2}$. Here $H = \lim H(g)$ and P is the generator of space translations. The scattering cross sections are expressed in terms of the S matrix and via the Lehmann–Symanzik–Zimmermann reduction formulas (Schweber[1]), the S matrix is expressed in terms of the vacuum expectation values

$$\langle \Omega, \phi(x_1, t_1) \ldots \phi(x_n, t_n)\, \Omega \rangle \tag{4.1}$$

of the field operators. In this chapter we take the limit $g \to 1$ in $H(g)$, in its vacuum vector Ω_g and in the fields

$$\phi_g(x, t) = e^{itH(g)}\, \phi(x, 0)\, e^{-itH(g)}, \tag{4.2}$$

$$\pi_g(x, t) = e^{itH(g)}\, \pi(x, 0)\, e^{-itH(g)}. \tag{4.3}$$

Here $\phi(x, 0) = \phi(x)$ and $\pi(x, 0)$ are the time zero free fields (1.2.3-4). While we also expect that the limit $g \to 1$ exists for the vacuum expectation values (4.1), this has not been proved.

Field theory may also be expressed in terms of bounded operators and operator algebras, and we do this by verifying the Haag-Kastler axioms in Chapters 4 and 5. With sufficient control on the spectrum of M, the existence of the S matrix would follow from the Haag-Ruelle theory, see for instance Jost[1].

4.1 The limit g → 1 on Fock space

We begin by studying bounded operators and we remove the space cutoff g in the Heisenberg picture dynamics,

$$A \to A(t) = \sigma_t(A) = e^{itH(g)}\, A\, e^{-itH(g)}. \tag{4.1.1}$$

Using the finite propagation speed, we prove that σ_t is independent of g. Let B be a bounded open region of space and let $\mathfrak{A}(B)$ be the von Neumann algebra (weakly closed *-algebra) generated by the operators

$$e^{i\phi(h)},\quad e^{i\pi(h)} \tag{4.1.2}$$

where suppt $h \subset B$ and $\phi(h) = \int \phi(h)\, h(x)\, dx$ and $\pi(h)$ are time zero free fields. Let $\mathfrak{A}$ be the C^*-algebra (norm closed *-algebra) generated by the $\mathfrak{A}(B)$'s. $\mathfrak{A}$ is called the algebra of quasilocal observables, and an operator A in some $\mathfrak{A}(B)$ is a local observable. The topologies (weak vs. norm) in the definitions of $\mathfrak{A}(B)$ and $\mathfrak{A}$ are determined by the renormalizations required for ϕ_2^{2n}. When one studies distinct representations of a given algebra of operators, it is a general rule that von Neumann algebras and weak closures can be used whenever the represention (or quasi-equivalence class of the representation) does not change; otherwise C^*-algebras and norm closures are necessary. Thus the locally Fock property, that we establish for the algebras $\mathfrak{A}(B)$, justifies the use of the weak closure for $\mathfrak{A}(B)$ but not for $\mathfrak{A}$, see Sections 4.2-4.3. For the Yukawa$_2$ theory, we have a field algebra $\mathfrak{A}_f$, an algebra of observables $\mathfrak{A} \subset \mathfrak{A}_f$ and the corresponding local subalgebras $\mathfrak{A}_f(B)$ and $\mathfrak{A}(B)$. $\mathfrak{A}_f(B)$ is the von Neumann algebra generated by the operators (4.1.2) and the time zero fermion fields $\psi^\alpha(h)$, suppt $h \subset B$, while $\mathfrak{A}(B)$ is generated by (4.1.2) and the fermion currents $\psi^\alpha(h_1)^* \psi^\beta(h_2), \alpha, \beta = 1, 2$, denoting components of the fields. The quasilocal algebras $\mathfrak{A}$ and $\mathfrak{A}_f$ are defined as the C^*-algebras generated by the corresponding local subalgebras $\mathfrak{A}(B)$ or $\mathfrak{A}_f(B)$. For B and C disjoint the commutation relations yield

$$[\mathfrak{A}(B),\quad \mathfrak{A}(C)] = 0 \tag{4.1.3}$$

and in the fermion case

$$[\mathfrak{A}_f(B), \mathfrak{A}(C)] = 0. \tag{4.1.4}$$

Let B_t be the set of points within distance $|t|$ of B.

DEFINITION *Let σ be a one parameter automorphism group of A. The propagation speed $s(\sigma)$ is the infimum of the numbers v such that*

$$\sigma_t(\mathfrak{A}(B)) \subset \mathfrak{A}(B_{vt}) \tag{4.1.5}$$

for all t and all B. Let $\sigma_t^{(1)}$ and $\sigma_t^{(2)}$ be automorphism groups of $\mathfrak{A}$ having finite propagation speed. We define the product $\sigma^{(1)} \times \sigma^{(2)}$ of $\sigma^{(1)}$ and $\sigma^{(2)}$ if for each $A \in \cup_B \mathfrak{A}(B)$, the sequence $(\sigma_{t/n}^{(1)} \cdot \sigma_{t/n}^{(2)})^n (A)$ of Trotter approximants converges in the strong operator topology, as $n \to \infty$. The limit defines an automorphism group of $\mathfrak{A}$, which we denote by $\sigma^{(1)} \times \sigma^{(2)} = \sigma^{(3)}$.

For example, if $\sigma_t^{(j)}$, $j = 1, 2, 3$ are implemented by unitary groups with self-adjoint generators $H^{(j)}$, respectively,

$$\sigma_t^{(j)}(A) = \exp(iH^{(j)}t) A \exp(-iH^{(j)}t),$$

and if $H^{(3)} = (H^{(1)} + H^{(2)})^-$ then $\sigma^{(3)} = \sigma^{(1)} \times \sigma^{(2)}$ by Trotter's theorem.

PROPOSITION 4.1.1 *The propagation speed is subadditive: If* $\sigma^{(3)} = \sigma^{(1)} \times \sigma^{(2)}$,

$$s(\sigma^{(3)}) \leqq s(\sigma^{(1)}) + s(\sigma^{(2)}).$$

Proof Since the local algebras $\mathfrak{A}(B)$ are dense in $\mathfrak{A}$, the Trotter limit

$$\sigma_t^{(3)}(A) = \text{s.} \lim_{n\to\infty} (\sigma_{t/n}^{(1)} \sigma_{t/n}^{(2)})^n (A)$$

for $A \in \mathfrak{A}(B)$ proves subadditivity. (This proof is abstracted from Segal[1].)

THEOREM 4.1.2 *In (4.1.1) let* $A \in \mathfrak{A}(B)$ *and let* $g \equiv 1$ *on* B_t. *Then* $\sigma_t(A)$ *is independent of* g *and* σ *extends to a one parameter automorphism group of* $\mathfrak{A}$. *The propagation speed of* σ *equals one.*

Proof In the notation of Proposition 4.1.1, let $H^{(1)} = H_0$, $H^{(2)} = H_I(g)$. Then $s(\sigma^{(1)}) = 1$, as follows from Lorentz covariance of the free field, for example, or from explicit solutions of the classical wave equation. Moreover $s(\sigma^{(2)}) = 0$. In fact for $A \in \mathfrak{A}(C)$, let $g = g_1 + g_2$ where $g_1 = 0$ on $\sim C$ and $g_2 = 0$ on C. Then

$$\sigma_t^{(2)}(A) = \exp(itH_I(g_1)) A \exp(-itH_I(g_1)) \in \mathfrak{A}(C)$$

since $\exp(itH_I(g_2))$ commutes with $\mathfrak{A}(C)$ by (4.1.3) and Proposition 2.1.3. The same reasoning shows that the Trotter approximants in (4.1.7) are independent of g. By Theorem 3.2.1, $H^{(1)} + H^{(2)} = H(g)$ is essentially self adjoint, and by Proposition 4.1.1, $\sigma^{(3)}$ is an automorphism of $\mathfrak{A}$ with propagation speed at most one. Furthermore $\sigma_t^{(3)}(A)$ is independent of g for $A \in \mathfrak{A}(B_t)$ provided $g \equiv 1$ on B_t. The rest of the theorem follows from $\sigma_t^{(1)} = \sigma_t^{(3)} \times \sigma_{-t}^{(2)}$, showing $1 \leqq s(\sigma^{(3)})$.

Space translations can be introduced in a simple fashion. With $P = \int a^*(k)\, ka(k)\, dk$, we have

$$e^{-iPy} \phi(x, t) e^{+iPy} = \phi(x + y, t) \tag{4.1.8}$$

and for bounded observables $A \in \mathfrak{A}$, we have the space translation automorphism $A \to e^{-iPy} A e^{+iPy}$. If $a = (y, t) \in R^2$, we define

$$\sigma_a(A) = e^{-iPy} \sigma_t(A) e^{+iPy}. \tag{4.1.9}$$

One checks that $a \to \sigma_a$ is a representation of R^2 by automorphisms of $\mathfrak{A}$.

Our next goal is to remove the space cutoff in the field ϕ.

LEMMA 4.1.3 *The fields* ϕ_g *and* π_g *of (4.2-3) are bilinear forms on the domain* $C^\infty(H(g)) \times C^\infty(H(g))$, *and as such, they are continuous in* x *and* t.

Remark (Rosen[2]) The fields are C^2 in x and t. For ϕ_2^4 (and presumably for ϕ_2^{2n}) they are C^∞ in x and t.

Proof Let $b_x = e^{ikx}\mu(k)^{\pm 1/2}$. Then $\phi_g(x, 0) = \phi(x)$ and $\pi_g(x, 0)$ are linear combinations of the bilinear forms $a^\#(b_x) = \int b_x(k)\, a^\#(k)\, dk$. By an N_τ estimate,

$$(H_0 + I)^{-1-\varepsilon}\, a^\#(b_x)\, (H_0 + I)^{-1-\varepsilon}$$

is a bounded operator, bounded by $\|\mu(k)^{-1-\varepsilon} b_x\|_2$, and depending continuously on x. Thus by the higher order estimates, Theorem 3.1.3,

$$(H(g) + I)^{-j}\, e^{itH(g)}\, a^\#(b_x)\, e^{-itH(g)}\, (H(g) + I)^{-j}$$

is a bounded operator, depending continuously on x and t. The lemma follows from this fact.

We now consider the space time averaged fields

$$\phi_g(f) = \int f(x, t)\, \phi_g(x, t)\, dx\, dt, \quad \pi_g(f) = \int f(x, t)\, \pi_g(x, t)\, dx\, dt \tag{4.1.10}$$

and the sharp time fields

$$\phi_g(f(\cdot, t)) = \int f(x, t)\, \phi_g(x, t)\, dx, \quad \pi_g(f(\cdot, t)) = \int f(x, t)\, \pi_g(x, t)\, dx. \tag{4.1.11}$$

LEMMA 4.1.4 *Let f be a real function in $\mathscr{S}(R^2)$. The fields (4.1.10-11) are symmetric operators on the domain $C^\infty(H(g))$.*

Proof The kernel $b_t(k) = \mu(k)^{\pm 1/2} \int f(x, t)\, e^{ikx}\, dx$ for

$$A_t = \int f(x, t)\, \phi_g(x, 0)\, dx \quad \text{or} \quad \int f(x, t)\, \pi_g(x, 0)\, dx$$

is in L_2. Thus by a N_τ estimate and Cor. 3.1.2

$$\|A_t(H(g) + I)^{-1/2}\| \leq \text{const.}\ \|A_t(N + I)^{-1/2}\| \leq \text{const.}\ \|b_t\|_2. \tag{4.1.12}$$

Hence

$$e^{itH(g)}\, A_t\, e^{-itH(g)}\, (H(g) + I)^{-1/2}$$

is a bounded operator, as is its integral over t. This completes the proof.

Remark It is easy to see that the sharp time fields are essentially self adjoint on $C^\infty(H(g))$. By a unitary equivalence, it is sufficient to show this at $t = 0$. By (4.1.12), we may use the domain

$$\mathscr{D}(H(g)) \supset \mathscr{D}(H_0) \cap \mathscr{D}(N^n)$$

and then Proposition 1.2.5 gives essential self adjointness. The space time averaged fields are also self adjoint (see Glimm-Jaffe[4]), but the proof is more difficult.

THEOREM 4.1.5 *Let B be an open region in R^2, let $f \in C_0^\infty(B)$ and let $g \equiv 1$ on $\{y : |y - x| \leqq t, x, t \in B\}$. Then $\phi_g(f)^- = \phi(f)$ and $\pi_g(f)^- = \pi(f)$ are independent of g.*

Proof The spectral projections of the sharp time fields are independent of g, by Theorem 4.1.2, and thus so are the (closed) sharp time fields themselves. Hence $\phi_g(f) \upharpoonright C^\infty(H_0)$ is independent of g. By Theorem 3.2.1, $H(g)$ is essentially self adjoint on $C^\infty(H_0) \subset \mathscr{D}(H(g))$. Thus by (4.1.12), integrated over t,

$$(\phi_g(f) \upharpoonright C^\infty(H_0))^- \supset \phi_g(f) \upharpoonright C^\infty(H(g)) = \phi_g(f)$$

and the proof is complete.

It is a special feature of the $P(\phi)_2$ model that Wick polynomials in the field $\phi(x, t)$ are defined. We start with the momentum cutoff field

$$\phi_\varkappa(x, t) = \int \delta_\varkappa(x - y)\, \phi(y, t)\, dy$$

where $\delta_\varkappa \to \delta$ as $\varkappa \to \infty$. To fix the arbitrary constants in the Wick powers we use the vacuum Ω_g or the vacuum Ω of Sec. 4.3, and we define

$$:\phi_\varkappa^n(x, t): = \phi_\varkappa^n(x, t) + \sum_{j=0}^{n-1} a_{j,n}\, \phi_\varkappa^j(x, t) \tag{4.1.13}$$

where the constants $a_{j,n}$ are determined inductively by the equations

$$[:\phi_\varkappa^n(x, t):, \pi(y, t)] = in\, \delta_\varkappa(x - y) :\phi_\varkappa^{n-1}(x, t): \tag{4.1.14}$$

and

$$\langle \Omega_g, :\phi_\varkappa^n(x, t): \Omega_g \rangle = 0. \tag{4.1.15g}$$

One can show that these equations determine the $a_{j,n}$, that the limit

$$:\phi^n(x, t): = \lim_{\varkappa \to \infty} :\phi_\varkappa^n(x, t): \tag{4.1.16}$$

exists and that

$$e^{itH(g)} :\phi^n(x, 0): e^{-itH(g)} = :\phi^n(x, t):. \tag{4.1.17g}$$

For polynomials P_1 whose degree is bounded by the degree of P, we could presumably apply Cor. 3.1.2 to define Wick ordering in the physical vacuum Ω of Sec. 4.3, replacing (4.1.15g) by

$$\langle \Omega, :\phi_\varkappa^n(x, t): \Omega \rangle = 0, \quad 0 < n \leqq \deg P \tag{4.1.15}$$

and (4.1.17g) by

$$e^{itH} :P_1(\phi(x, 0)): e^{-itH} = :P_1(\phi(x, t)):. \tag{4.1.17}$$

We have $:P(\phi(x, 0)): = \,:Q(\phi(x, 0)):$ for some polynomial Q determined by the $a_{j,n}$ in (4.1.13) and $\deg P = \deg Q$. We note that the Wick dots : : are determined by the Eqs. (4.1.14-5g) with $g = 0$.

The verification of the field equation

$$\phi_{tt}(x, t) - \phi_{xx}(x, t) + m^2\phi(x, t) + j(x, t) = 0 \tag{4.1.18}$$

is straightforward, using higher order estimates, N_τ estimates and the commutation relations, see Glimm and Jaffe[4]. The nonlinear term j is

$$\begin{aligned} j(x, t) &= \vdots Q'(\phi(x, t)) \vdots \\ &= e^{itH(g)} :P'(\phi(x, 0)): e^{-itH(g)}. \end{aligned}$$

If we remove the linear term from Q' and absorb it into m^2, thereby redefining m^2 in (4.1.18), a formal argument due to Källen and Lehmann (see Schweber[1]) indicates that $m_{\text{phys}} \leqq m$. In particular the new m^2 is then nonnegative.

4.2 The physical vacuum is locally Fock

According to formal perturbation theory, the vacuum Ω_g of the cutoff theory tends weakly to zero as $g \to 1$. Thus we consider the limit of states rather than vectors. We define ω_g as the linear functional

$$\omega_g(A) = \langle \Omega_g, A\,\Omega_g \rangle, \quad A \in \mathfrak{A} \tag{4.2.1}$$

and obtain a physical vacuum ω by a limiting process as $g \to 1$. Let g and h be fixed nonnegative C_0^∞ functions, let

$$\left.\begin{aligned} \operatorname{suppt} h \subset [-1, 1], \quad \int h(x)\,dx = 1 \\ g \equiv 1 \quad \text{on} \quad [-3, +3] \end{aligned}\right\} \tag{4.2.2}$$

and define

$$g_n(x) = g(n^{-1}x)$$

$$\omega_n(A) = n^{-1} \int \omega_{g_n(\cdot + y)}(A)\, h(n^{-1}y)\, dy. \tag{4.2.3}$$

We choose ω as a limit point of the sequence ω_n. (We believe that the generalized sequence $\{\omega_g\}$ of (4.2.1) converges as $g \to 1$, but this seems difficult to prove. If it converges, it necessarily converges to ω.) Then ω is a state on the C^*-algebra $\mathfrak{A}$. Using ω to define an inner product, we construct the Hilbert space $\mathfrak{F}_{\text{ren}}$ of physical states in Sec. 4.3.

The existence of the limit point ω is a trivial consequence of general w^* compactness theorems. This section is devoted to regularity properties of ω, as will be needed in Sec. 4.3 to show that the unitary translation group acts continuously on $\mathfrak{F}_{\text{ren}}$ and that the unbounded field operators ϕ act on $\mathfrak{F}_{\text{ren}}$. Our main result (Glimm-Jaffe[5]) is

THEOREM 4.2.1 *Let B be a bounded open region in space. The sequence* $\omega_n \restriction \mathfrak{A}(B)$ *lies in a norm compact subset of the dual Banach space* $\mathfrak{A}(B)^*$. *A limit point* $\omega \restriction \mathfrak{A}(B)$ *is a normal state of* $\mathfrak{A}(B)$. ω *(without restriction to a subalgebra) is a limit of a subsequence of the* ω_n.

The proof of this theorem is lengthy, and we will only indicate the main ideas. Our starting point is the uniform estimate from Cor. 3.1.2,

$$\omega_{g_n}(H_0) \leqq \text{const. } n. \tag{4.2.4}$$

In order to remove the factor n on the right, we use a localization in configuration space. Let ζ be a nonnegative C_0^∞ function and let ζ act as a multiplication operator on L_2.

Then $\zeta(-\Delta + m^2)^{\tau/2}\,\zeta$ is a localization of $(-\Delta + m^2)^{\tau/2}$. In momentum space this operator has the kernel

$$w(k_1, k_2) = \int \zeta^\sim(k_1 - l)\,\mu^\tau(l)\,\zeta^\sim(l - k_2)\,dl$$

and using w, we introduced the localized fractional energy operator

$$N_{\tau,\zeta} = \int a^*(k_1)\,w(k_1, k_2)\,a(k_2)\,dk.$$

Similarly for $\tau < \frac{1}{4}$, we can replace ζ by the characteristic function χ_B of an interval B, to obtain a sharply localized operator $N_{\tau,B}$.

LEMMA 4.2.2 *For* $\tau < \frac{1}{4}$, $\omega_n(N_{\tau,B}) \leqq$ const. *The constant depends on the diameter of B, but is otherwise independent of B, and it is independent of n.*

Sketch of proof Using classical methods of Fourier analysis, one can show that

$$\chi_B(-\Delta + m^2)^\tau \chi_B \leqq \text{const. } \zeta(-\Delta + m^2)^{2\tau+\varepsilon}\,\zeta$$

if $\tau < \frac{1}{4}$ and if $\zeta \equiv 1$ on a neighborhood of B^-. Furthermore if $\zeta_j(x) = \zeta(x + j)$ then

$$\Sigma_j \zeta_j(-\Delta + m^2)^\tau\,\zeta_j \leqq \text{const. } (-\Delta + m^2)^\tau.$$

These inequalities on the single particle space $L_2(R)$ lead immediately to the estimate

$$\Sigma_j N_{\tau,B+j} \leqq \text{const. } H_0$$

where $B + j = \{x + j : x \in B\}$.

Let $U(y)$ be the unitary operator on $\mathscr{F}$ which implements translation by y, as in (4.1.8). We have

$$U(y)\,H(g_n)\,U(y)^* = H(g_n(\cdot + y))$$

and by the uniqueness of the vacuum Ω_g,

$$U(y)\,\Omega_{g_n} = \Omega_{g_n(\cdot+y)}.$$

Thus

$$\omega_n(N_{\tau,B}) = n^{-1} \int \langle \Omega_{g_n(\cdot+y)}, N_{\tau,B}\Omega_{g_n(\cdot+y)}\rangle\, h(n^{-1}y)\, dy$$

$$= n^{-1} \int \langle \Omega_{g_n}, U(y)^* N_{\tau,B}U(y)\, \Omega_{g_n}\rangle\, h(n^{-1}y)\, dy$$

$$\leqq n^{-1}\, \|h\|_\infty \int \omega_{g_n}(N_{\tau,B+y})\, dy$$

$$= n^{-1}\, \|h\|_\infty \int_0^1 \omega_{g_n}(\Sigma_j N_{\tau,B+\theta+j})\, d\theta$$

$$\leqq \text{const. } n^{-1}\, \omega_{g_n}(H_0)$$

$$\leqq \text{const.}$$

We use the configuration space decomposition

$$L_2(R) = L_2(B) \oplus L_2(\sim B)$$

to obtain the factorization

$$\mathscr{F} \approx \mathscr{F}_B \otimes \mathscr{F}_{\sim B} = \mathfrak{S}(L_2(B)) \otimes \mathfrak{S}(L_2(\sim B))$$

as in Sec. 1.3. In this factorization,

$$N_{\tau,B} = (N_{\tau,B} \restriction \mathscr{F}_B) \otimes I$$

and for $0 < \tau < \frac{1}{4}$, one can also show that $N_{\tau,B} \restriction \mathscr{F}_B$ has a compact resolvent.

Now we return to the sequence ω_n and the proof of Theorem 4.2.1. Each state ω_n can be expressed in terms of a density matrix as a convex linear combination of vector states,

$$\omega_n(A) = \Sigma\alpha_j \langle\theta_j, A\theta_j\rangle, \qquad \theta_j \in \mathscr{F}.$$

Using the uniform estimate of Lemma 4.2.2, we can find a "regularized" state ω_n^ε which approximates ω_n on the full algebra $\mathfrak{A}$:

$$\|\omega_n - \omega_n^\varepsilon\| \leqq \varepsilon.$$

The state ω_n^ε is also a linear combination of certain vector states $A \to \langle\theta_j^\varepsilon, A\theta_j^\varepsilon\rangle$, and the regularization of ω_n^ε consists in a bound (uniform in n) on the number of particles which occur in the vectors θ_j. We are not able to bound the total number of particles—such a bound would make the limit state ω a Fock state—but using Lemma 4.2.2 we may require that the total number of particles localized in the space interval $[-a, a]$ grows no faster than, say, $0(a^3)$. Furthermore, because $\tau > 0$ and because $N_{\tau,B} \restriction \mathscr{F}_B$ has a compact inverse, we may require that the wave functions of particles localized in $[-a, a]$ lie in a compact subset of L_2, as $n \to \infty$. In fact there is a bound, uniform in n, on the L_2 norm of the fractional derivatives of order $\tau/2$ for these wave functions.

Since $\omega_n - \omega_n^\varepsilon$ is small in norm, it is sufficient to prove norm compactness with ω_n replaced by ω_n^ε. In fact, if $\{\omega_n^\varepsilon\}$ lies in a norm compact subset of $\mathfrak{A}(B)^*$, there is a norm convergent subsequence $\{\omega_{n_j}^\varepsilon\}$ by a three ε argument,

$$\begin{aligned}\|(\omega_{n_i} - \omega_{n_j}) \restriction \mathfrak{A}(B)\| &\leqq 2\varepsilon + \|(\omega_{n_i}^\varepsilon - \omega_{n_j}^\varepsilon) \restriction \mathfrak{A}(B)\| \\ &\leqq 2\varepsilon + o(1),\end{aligned}$$

and hence $\{\omega_n\}$ lies in a norm compact subset of $\mathfrak{A}(B)^*$.

We now restrict our attention to the ω_n^ε.

The next step in the proof is to regularize the operators A in which the state ω_n^ε is evaluated. The regularization begins with a general argument based on Kaplansky's density theorem. Let $\mathfrak{A}_0(B)$ be a strongly dense *-subalgebra of $\mathfrak{A}(B)$. By Kaplansky's theorem, the unit ball,

$$\{A : A \in \mathfrak{A}_0(B), \|A\| \leqq 1\}$$

of $\mathfrak{A}_0(B)$ is strongly dense in the unit ball of $\mathfrak{A}(B)$, see Dixmier[1]. As a corollary,

$$\|(\omega_n^\varepsilon - \omega_m^\varepsilon) \restriction \mathfrak{A}(B)\| = \|(\omega_n^\varepsilon - \omega_m^\varepsilon) \restriction \mathfrak{A}_0(B)\|.$$

Thus it is sufficient to consider operators $A \in \mathfrak{A}_0(B)$. We choose $\mathfrak{A}_0(B)$ to be the linear span generated by the operators

$$e^{i(\phi(f_1) + \pi(f_2))}$$

$f_i \in C^\infty$, suppt $f_i \subset B$. For $A \in \mathfrak{A}_0(B)$ the Taylor series expansion of the above exponentials converges in the expression $\omega_n^\varepsilon(A)$. Let A^δ be the (unbounded) operator which results from a truncation of the Taylor series. The most difficult part of the proof is to show that the truncation can be done (independently of n) so that

$$\|\omega_n^\varepsilon(A - A^\delta)\| \leqq \delta\|A\|, \tag{4.2.5}$$

and so that for a subsequence n_i (chosen independently of A)

$$\|(\omega_{n_i}^\varepsilon - \omega_{n_j}^\varepsilon)(A^\delta)\| \leqq o(1)\,\|A\|, \tag{4.2.6}$$

as $i, j \to \infty$. Given (4.2.5)–(4.2.6), norm compactness follows.

The idea behind these estimates is very simple. The operator $A \in \mathfrak{A}(B)$ observes mainly vectors localized near B, and ω_n^ε has only vectors with a finite number of particles localized near B. The notions of localization for vectors in Fock space and for operators in $\mathfrak{A}(B)$ are distinct and differ on account of the factors $\mu(k)^{\pm 1/2}$ that occur in the definition of $\phi(f)$ and $\pi(f)$, e.g. for real f,

$$\phi(f) = a(\mu^{-1/2}\tilde{f}) + a(\mu^{-1/2}\tilde{f})^*.$$

The difference between the two notions of localization arises from the tail of the kernel $k_\pm(x, y)$ of the operator $(-\Delta + m^2)^{\pm 1/4}$. As $|x - y| \to \infty$,

$k_\pm = 0(\exp(-m|x-y|))$, so that the probability of the operator A observing vectors localized far from B is exponentially small, but nonzero. We choose a sufficiently large interval B_1 containing B, and let $\mathscr{F} = \mathscr{F}_{B_1} \otimes \mathscr{F}_{\sim B_1}$. Since our two notions of localization differ only by an exponential tail, we introduce a negligible error by retaining in A^δ only the part acting on on $\mathscr{F}_{B_1}$.

A basic tool in the proof of (4.2.6) is the estimate

$$\omega_n^\varepsilon(N_{\tau,B_1}) \leqq \text{const.}, \quad 0 < \tau < \tfrac{1}{4}, \tag{4.2.7}$$

which follows from Lemma 4.2.2 and the definition of ω_n^ε. Since the operator N_{τ,B_1} has a compact resolvent on $\mathscr{F}_{B_1}$, we use (4.2.7) to establish that a subsequence of the states ω_n^ε is norm convergent on the algebra of all bounded operators on $\mathscr{F}_{B_1}$. Using the fact that the bounded functions of A^δ are bounded operators on $\mathscr{F}_{B_1}$, we are able to establish (4.2.6).

The difficulty in proving (4.2.5) is to obtain an estimate that is uniform in the norm of A. There is no convenient estimate for the norm of an operator in terms of its kernel.

We now complete the proof of the theorem. Normal states of a von Neumann algebra $\mathfrak{A}$ are states which can be expressed as an infinite convex linear combination of vector states. Such states can be characterized in terms of the order property:

$$\omega(\sup A_\alpha) = \sup \omega(A_\alpha)$$

for each monotonically increasing family $\{A_\alpha\}$ which is bounded from above, Dixmier[(1)]. Using the order definition, a $3-\varepsilon$ argument proves that the norm limit of normal states is normal. Any limit point of the ω_n can (when restricted to $\mathfrak{A}(B)$) be obtained as a norm limit of a subsequence. By the diagonal process, we can choose one subsequence of the ω_n that converges in norm on each $\mathfrak{A}(B)$. Hence this subsequence converges w^* on $\mathfrak{A}$.

4.3 The Hilbert space of physical states

Let ω be a limit point of the sequence $\{\omega_n\}$ of approximate vacuums. We use ω to define the inner product

$$A, B \to \omega(A^*B) \tag{4.3.1}$$

on $\mathfrak{A} \times \mathfrak{A}$. We introduce the corresponding (pseudo) norm

$$\|A\| = \omega(A^*A)^{1/2}. \tag{4.3.2}$$

The set $\mathfrak{N}$ of null vectors with respect to this norm is a left ideal in $\mathfrak{A}$. We divide by this ideal and complete the quotient $\mathfrak{A}/\mathfrak{N}$ in the norm (4.3.2). We obtain a new Hilbert space, which we call $\mathscr{F}_{\text{ren}}$; (4.3.1) defines the inner produce in $\mathscr{F}_{\text{ren}}$. The map

$$A: B + \mathfrak{N} \to AB + \mathfrak{N} = \pi(A)(B + \mathfrak{N}) \tag{4.3.3}$$

defines a representation $A \to \pi(A)$ of $\mathfrak{A}$ as operators on $\mathscr{F}_{\text{ren}}$. The above is the GNS construction, see Dixmier[2].

$\mathscr{F}_{\text{ren}}$ is the Hilbert space of physical states. $\Omega = I + \mathfrak{N} \in \mathscr{F}_{\text{ren}}$ is the physical vacuum, and

$$\omega(A) = \langle \Omega, \pi(A)\,\Omega \rangle \tag{4.3.4}$$

as one sees from (4.3.1) and the definition of π. In some weak sense, one should regard Ω as a limit point of the vacuums Ω_g. We use the local Fock property of ω, Theorem 4.2.1, to show that π is also locally Fock.

THEOREM 4.3.1 *Let B be a bounded region of space at time zero. There is a unitary operator $U_B : \mathscr{F} \to \mathscr{F}_{\text{ren}}$ such that for $A \in \mathfrak{A}(B)$,*

$$\pi(A) = U_B A U_B^*.$$

Remark As a corollary, we extend the representation π to the unbounded field operators $\phi(f)$, $f \in C_0^\infty$, and preserve local properties (such as the canonical commutation relations and the field equation) satisfied by ϕ. In order to do this, the normality of the representation π is sufficient. Normality of π does not depend upon the theorems of Araki and of Griffin.

Proof The theorem is a simple consequence of Theorem 4.2.1 and theorems of Araki[2] and Griffin[1]. Araki showed that $\mathfrak{A}(B)$ is a factor of type III in the sense of Murray and von Neumann and Griffin showed that isomorphisms between separable factors of type III are unitarily implementable. In order to apply Griffin's theorem we need only the ultraweak continuity (normality) of $\pi \restriction \mathfrak{A}(B)$. In fact on the one hand if $\pi \restriction \mathfrak{A}(B)$ is continuous it must be an isomorphism because there are no weakly closed ideals in a factor, and on the other hand the continuous image of a factor of type III is a factor of type III. (See Dixmier[1].) The continuity of $\pi \restriction \mathfrak{A}(B)$ which we need is equivalent to a bound of the form

$$|\langle \theta, \pi(A)\,\theta \rangle| \leqq \Sigma\, |\langle \psi_j, A\psi_j \rangle|$$

where $\theta \in \mathscr{F}_{\text{ren}}$, $A \in \mathfrak{A}(B)$ and the ψ_j in $\mathscr{F}$ depend on B and the choice of θ. However, this continuity results from the continuity of $\omega \restriction \mathfrak{A}(B_1)$ for B_1 bounded because

$$|\omega(A)| \leqq \Sigma |\langle \psi_j, A\psi_j \rangle|$$

by Theorem 4.2.1 and so

$$\begin{aligned} |\langle \pi(C)\,\Omega, \pi(A)\,\pi(C)\,\Omega \rangle| &= |\omega(C^* A C)| \\ &\leqq \Sigma\, |\langle C\psi_j, AC\psi_j \rangle| \end{aligned}$$

where $C \in \mathfrak{A}(B_1)$, $B \subset B_1$. As B_1 varies, vectors of the form $\theta = \pi(C)\,\Omega$ are dense in $\mathscr{F}_{\text{ren}}$, and the proof is complete.

THEOREM 4.3.2 *$\mathscr{F}_{\text{ren}}$ is separable and π is an isomorphism. There is a continuous unitary representation U of R^2 such that*

$$U(a)\,\pi(A)\,U(a)^* = \pi(\sigma_a(A)), \tag{4.3.5}$$

$$U(a)\,\Omega = \Omega. \tag{4.3.6}$$

The spectrum of the generator H of time translations is contained in the interval $[0, \infty)$.

Proof The separability results from the facts that $\pi \upharpoonright \mathfrak{A}(B)$ is ultraweakly continuous and that each $\mathfrak{A}(B)$ is ultraweakly separable. π is an isomorphism because $\mathfrak{A}$ is simple as a C^*-algebra. In fact for any representation π of $\mathfrak{A}$, $\pi \upharpoonright \mathfrak{A}(B)$ is an isomorphism because $\mathfrak{A}(B)$, as a separable type III factor, is a simple C^*-algebra. Hence $\pi \upharpoonright \mathfrak{A}(B)$ is isometric, and since this is true for every B, π is isometric, and consequently an isomorphism. We define

$$U(a)\,\pi(A)\,\Omega = \pi(\sigma_a(A))\,\Omega.$$

Then $U(\cdot)$ is a unitary representation of R^2, provided $\omega = \omega \circ \sigma$, i.e. provided ω is a fixed point under the space time translation automorphisms, as one can see by a short calculation. To show that ω is a fixed point for space translations, we must utilise the space integration in the definition of ω_n and to show the invariance under time translations we must choose $H(g_n(\cdot + y))$ to implement the time automorphism. The restriction on the supports of g and h were introduced so that this choice of the cutoff Hamiltonian is permitted for any y in the support of h. We omit further details. (See Glimm-Jaffe[5].)

To show that H is positive, we introduce the spectral resolution $H = \int \mu \, dE(\mu)$ and the function

$$\begin{aligned} F(t) &= \omega(A^*\sigma_{0,t}(A)) = \langle \pi(A)\,\Omega, e^{iHt}\,\pi(A)\,\Omega\rangle \\ &= \int e^{i\mu t}\, d\|E(\mu)\,\pi(A)\,\Omega\|^2, \end{aligned}$$

and we must show that only the interval $[0, \infty)$ contributes to the above integral. In other words we must show that F has a Fourier transform with support in $[0, \infty)$. However, this property is easily established for the approximating functions

$$F_n(t) = \omega_n(A^*\sigma_{0,t}(A))$$

(using the positivity of $H(g_n(\cdot + y))$) and it is preserved under limits. This completes the proof. For further results, see Glimm-Jaffe[11].

5 Lorentz covariance and the Haag-Kastler axioms

5.1 Lorentz covariance

We introduce the algebras of local observables $\mathfrak{A}(B)$, where $B \subset R^2$ is a bounded open region of space time. Self adjoint elements A in $\mathfrak{A}(B)$ are interpreted as observables which can be observed by measurements performed in the region B. We start with the self adjoint field operator

$$\phi(f) = \int \phi(x, t)\, f(x, t)\, dx\, dt$$

where f is a real, C_0^∞ function. This operator measures the field strength, averaged over space time points with the weight function f. We define $\mathfrak{A}(B)$ as the von Neumann algebra generated by bounded functions of the fields localized in B,

$$\mathfrak{A}(B) = \{\exp(i\phi(f)) : \operatorname{supp} f \subset B\}''.$$

Let $\mathfrak{A}$ be the C^* algebra generated by $\cup_B \mathfrak{A}(B)$. $\mathfrak{A}$ coincides with the algebra $\mathfrak{A}$ introduced in Sec. 4.1, but B and $\mathfrak{A}(B)$ are distinct from the time zero regions and algebras of Sec. 4.1.

The Lorentz group $\mathscr{P} = \{a, \Lambda\}$ is the semidirect product of R^2 with R^1,

$$\{a_1, \Lambda_1\}\{a_2, \Lambda_2\} = \{a_1 + \Lambda_1 a_2, \Lambda_1\Lambda_2\}. \tag{5.1.1}$$

Here $a \in R^2$ is a space time translation and Λ is the Lorentz rotation:

$$\Lambda_\beta : (x, t) \to (x \cosh\beta + t \sinh\beta, x \sinh\beta + t \cosh\beta).$$

We formulate Lorentz covariance in terms of automorphisms. Our main result is the following: There exists a representation $\sigma_{\{a,\Lambda\}}$ of $\mathscr{P}$ by *-automorphisms of $\mathfrak{A}$, such that

$$\sigma_{\{a,\Lambda\}}(\mathfrak{A}(B)) = \mathfrak{A}(\{a, \Lambda\} B) \tag{5.1.2}$$

for all bounded open sets B and all $\{a, \Lambda\} \in \mathscr{P}$.

The Lorentz group composition law gives

$$\sigma_{\{a,\Lambda\}} = \sigma_{\{a,I\}}\sigma_{\{0,\Lambda\}}. \tag{5.1.3}$$

The space time covariance is an easy consequence of the results of Chapter 4:

$$\sigma_a(\mathfrak{A}(B)) = \sigma_{\{a,I\}}(\mathfrak{A}(B)) = \mathfrak{A}(\{a, I\} B) = \mathfrak{A}(B + a). \tag{5.1.4}$$

To see this we make use of the fact that $\mathfrak{A}(B)$ coincides with the von Neumann algebra generated by sharp time fields,

$$\mathfrak{A}(B) = \left\{\exp\left(i \int \phi(x, t) f(x, t)\, dx\right) : \operatorname{supp} f \subset B\right\}'',$$

see Glimm-Jaffe[4]. In (5.1.4), the time translation is implemented locally by the unitary operators $\exp(it\, H(g))$, and the space translation is implemented by $\exp(-iPx)$. Since

$$\exp(-iPx)\, H(g) \exp(iPx) = H(g_x)$$

$$g_x(y) = g(y - x),$$

the space and time automorphisms commute. Returning to (5.1.3), we conclude that the existence of the automorphism representation $\sigma_{\{a,\Lambda\}}$ follows easily from the construction of the pure Lorentz transformation $\sigma_{\{0,\Lambda\}} = \sigma_\Lambda$. We obtain σ_Λ by constructing locally correct infinitesimal generators.

We now sketch the formal ideas in the proof of Cannon-Jaffe[1] of Lorentz covariance for the $(\phi^4)_2$ model. The extension to the $P(\phi)_2$ model is given by Rosen[3].

Formally, the operator,

$$M = M_0 + M_I(g)$$
$$= \tfrac{1}{2}\int \{:\pi(x)^2: + :(\nabla\phi(x))^2: + m^2 :\phi(x)^2:\}\, x\, dx + H_I(xg) \qquad (5.1.5)$$

is an infinitesimal generator of Lorentz transformations in a region B if the cutoff function g equals one on a sufficiently large interval. Unfortunately the methods of Sec. 2 and 3 can not be applied to the M of (5.1.5), since M is not formally positive. This problem does not affect the study of M_0, since M_0 acts in a simple fashion. It is easy to prove from the explicit form of M_0 acting on n particle states that M_0 is essentially self adjoint on the domain $\mathscr{D}$. It is not known, however, whether M of (5.1.5) is self adjoint.

To avoid this trouble we restrict our attention to regions B_1 contained in the quadrant $x > |t| + 1$. For such regions B_1, we may replace (5.1.5) by

$$M = \int H(x)\, xg(x)\, dx \qquad (5.1.6)$$

where $xg(x)$ is nonnegative.

Here $H(x)$ is the energy density (usually denoted $T_{00}(x)$),

$$H(x) = \tfrac{1}{2}:\{\pi(x)^2 + (\nabla\phi)^2(x) + m^2(\phi^2)(x)\}: + H_I(x)$$
$$\equiv H_0(x) + H_I(x),$$

and is formally positive. Thus (5.1.6) is formally positive. In fact it is technically convenient to use different spatial cutoffs in the free and the interaction part of M. Our final formula for M is

$$M = M(g_0, g) = \alpha H_0 + H_0(xg_0) + H_I(xg), \qquad (5.1.7)$$

where $0 < \alpha$ and $0 \leqq xg_0(x), xg(x)$. In order that (5.1.7) be formally correct, we require that

$$\alpha + xg_0 = x = xg \quad \text{on} \quad [1, R]$$

with R sufficiently large. For technical reasons we require that

$$\alpha + xg_0 = x \quad \text{on} \quad \operatorname{supp} g. \qquad (5.1.8)$$

By our above restrictions on g_0 and g we have

$$\operatorname{supp} g_0, \quad \operatorname{supp} g \subset \{x : \alpha \leqq x\}.$$

We show that the operator M of (5.1.7) is essentially self adjoint on $\mathscr{D}$, and it generates Lorentz rotations in an algebra $\mathfrak{A}(B_1)$,

$$\exp(i\beta M)\,\mathfrak{A}(B_1)\exp(-i\beta M) \subset \mathfrak{A}(\{0, \Lambda_\beta\}\, B_1) \qquad (5.1.9)$$

provided B_1 and $\{0, \Lambda_\beta\}\, B_1$ are contained in the region

$$\{x, t : 1 + |t| < x < R - |t|\} \tag{5.1.10}$$

where M is formally correct. These results permit us to define the Lorentz rotation automorphism σ_Λ on an arbitrary local algebra $\mathfrak{A}(B)$. Using a space time translation σ_a, we can translate B into a region

$$B + a = B_1 \subset \{x, t : |t| + 1 < x\}.$$

For R large enough, B_1 and $\{0, \Lambda_\beta\}\, B_1$ are contained in (5.1.10), and we define σ_{Λ_β} by the formula

$$\sigma_{\Lambda_\beta} \upharpoonright \mathfrak{A}(B) = \sigma_{\{-\Lambda a, I\}}\sigma_{\{0,\Lambda_\beta\}}\sigma_{\{a,I\}} \upharpoonright \mathfrak{A}(B). \tag{5.1.11}$$

THEOREM 5.1.1 *Let $M(g_0, g)$ be given by (5.1.7), with α, g_0 and g restricted as above. Then M is essentially self adjoint on $\mathscr{D}$, and for the ϕ_2^4 theory M is self adjoint.*

By Theorems 3.1.1 and 3.2.1, $\alpha H_0 + H_I(xg)$ is bounded from below and essentially self adjoint on $\mathscr{D}$; for the ϕ_2^4 theory it is self adjoint on $\mathscr{D}(H_0) \cap \mathscr{D}(H_I(xg))$. For the ϕ_2^4 theory, the second order estimate

$$(H_0(xg_0))^2 \leqq a(\alpha H_0 + \gamma H_0(xg_0) + H_I(xg) + bI)^2$$

is valid uniformly in γ, $0 \leqq \gamma \leqq 1$. In particular, this estimate shows that $H_0(xg_0)$ is a densely defined operator. In fact there is a cancellation between the pair creation terms of $:\pi^2(xg_0):$ and $:(\nabla\phi)^2\,(xg_0):$ and similarly between the pair annihilation terms to give the unbounded operator $H_0(xg_0)$ defined on $\mathscr{D}(H_0)$. More important, this estimate exhibits $M(g_0, g)$ as the final term in a finite sequence of Kato perturbations, namely $\alpha H_0 + \gamma_j H_0(xg_0) + H_I(xg)$.

In order to prove (5.1.9), we prove the corresponding identity for the field operators.

THEOREM 5.1.2 *Let B_1 and $\{0, \Lambda_\beta\}\, B_1$ be contained in the set (5.1.10). We have the identity between self adjoint operators:*

$$\exp(i\beta M)\,\phi(f)\exp(-i\beta M) = \phi(f_{\Lambda_\beta})$$

$$= \int \phi(\Lambda_\beta(x, t))\, f(x, t)\, dx\, dt,$$

provided supp $f \subset B_1$.

The field $\phi(x, t)$ is also transformed correctly, in the sense of bilinear forms. Since the transformations

$$f \rightarrow f_{\{a,\Lambda\}}(\cdot) = f(\{a, \Lambda\}^{-1}\cdot)$$

and

$$\phi(f) \rightarrow \phi(f_{\{a,\Lambda\}})$$

obviously satisfy the group law in $\mathscr{P}$, and since $\mathfrak{A}$ is generated by the field operators, Theorem 5.1.2 implies that the automorphisms $\sigma_{\{a, \Lambda\}}$ defined by (5.1.3), (5.1.9) and (5.1.11) also satisfy the group law in $\mathscr{P}$. In particular (5.1.11) is independent of the choice of a.

The proof of Theorem 5.1.2 is reduced to the verification of the equation

$$\left\{x\frac{\partial}{\partial t} + t\frac{\partial}{\partial x}\right\}\phi(x, t) = [iM, \phi(x, t)] \tag{5.1.12}$$

as an equation for bilinear forms on a suitable domain. Since M is self adjoint, we can integrate (5.1.12) to yield the theorem.

The formal proof of (5.1.12) can be understood as follows: We compute for $H = H_0 + H_I(g)$,

$$\begin{aligned}[iM, \phi(x, t)] &= [iM, e^{itH}\,\phi(x, 0)\,e^{-itH}] \\ &= e^{itH}[iM(-t), \phi(x, 0)]\,e^{-itH},\end{aligned} \tag{5.1.13}$$

where

$$M(-t) = e^{-itH}\,M\,e^{itH}.$$

Again formally

$$M(-t) = \sum_{n=0}^{\infty} \frac{(-t)^n}{n!}\,ad^n(iH)\,(M).$$

If M and H were the correct global Lorentz generator and Hamiltonian they would satisfy

$$\begin{aligned}[iH, M] = ad(iH)\,(M) &= P, \\ [iH, [iH, M]] &= 0,\end{aligned} \tag{5.1.14}$$

and

$$M(-t) = M - Pt.$$

Thus by (5.1.13)

$$[iM, \phi(x, 0)] = [iM_0, \phi(x, 0)] = x\,\pi(x, 0)$$

and

$$[iP, \phi(x, 0)] = -(\nabla\phi)\,(x, 0),$$

we have (5.1.12). The difficulty with this formal argument is that H and M do not obey (5.1.14), since they are correct only in B_1. We have instead the equations

$$\begin{aligned}[iH, M] &= P^{\text{loc}} \\ [iH, [iH, M]] &= R^{\text{loc}},\end{aligned} \tag{5.1.15}$$

where P^{loc} acts like the momentum operator in B_1, i.e.

$$[P^{\text{loc}}, \phi(x, t)] = [P, \phi(x, t)], \quad x, t \in B_1.$$

Hence $[iH, P^{\text{loc}}] = R^{\text{loc}}$ is not identically zero, but commutes with $\mathfrak{A}(B_1)$. Formally, further commutators of R^{loc} with H are localized outside B_1, and (5.1.12) follows formally even for our approximate, but locally correct H and M.

In order to convert this formal argument into a mathematical theorem, more work is necessary. We make a Taylor series expansion for

$$F(-t) = \langle \Omega, [iM(-t), \phi(x, 0)]\, \Omega \rangle, \tag{5.1.16}$$

with $\Omega \in C^\infty(H)$, namely

$$F(-t) = F(0) - tF'(0) + \tfrac{1}{2} t^2 F''(s)$$

where $|s| \leqq |t|$. By (5.1.15), we have

$$F''(-s) = \langle e^{isH}\Omega, [iR^{\mathrm{loc}}, \phi(x, s)]\, e^{isH}\Omega \rangle.$$

We note that $(x, t) \in B_1$, so that with $|s| \leqq |t|$, $(x, s) \in B_1$. Hence $[R^{\mathrm{loc}}, \phi(x,s)] = 0$, which is an operator identity after integration over x with a C_0^∞ function. Thus

$$F''(s) = 0, \qquad |s| \leqq |t|$$

and

$$\begin{aligned} F(-t) &= F(0) - tF'(0) \\ &= \langle \Omega, \{[iM, \phi(x, 0)] - t[iP^{\mathrm{loc}}, \phi(x, 0)]\}\, \Omega \rangle \\ &= \langle \Omega, \{x\pi(x, 0) + t(\nabla\phi)(x, 0)\}\, \Omega \rangle. \end{aligned}$$

We conclude that

$$[iM(-t), \phi(x, 0)] = x\pi(x, 0) + t\, \nabla\phi(x, 0)$$

and inserting this relation in (5.1.13) yields (5.1.12). This completes the formal outline of the proof of Lorentz covariance.

5.2 *The Haag-Kastler axioms*

In the physics literature, one encounters axiom schemes for quantum field theory. These axioms formulate the physical ideas of space time localization, Lorentz covariance and locality. Wightman's axioms[1] deal directly with the field operators, acting on the Hilbert space of physical states. The Haag-Kastler[1] axioms are independent of the choice of Hilbert space and the representation of the operators; hence these axioms involve less structure and in principle they are more general. The Haag-Kastler framework avoids certain technical problems by working with bounded functions of the fields, i.e. bounded observables. On the other hand there is no direct way to formulate the field equations in terms of the bounded observables. Rather than comparing these axiom schemes, we are interested in verifying them in our models.

We describe the Haag-Kastler axioms as follows: To each bounded open region B, we associate a von Neumann algebra $\mathfrak{A}(B)$ and we let $\mathfrak{A}$ be the C^* algebra generated by the union of the algebras $\mathfrak{A}(B)$. The axioms are

1) Isotony: If $B_1 \subset B_2$ then $\mathfrak{A}(B_1) \subset \mathfrak{A}(B_2)$.

2) Locality: If B_1 and B_2 are spacelike separated then $\mathfrak{A}(B_1)$ commutes with $\mathfrak{A}(B_2)$.

3) Lorentz covariance: There is a representation $\{a, \Lambda\} \to \sigma_{\{a,\Lambda\}}$ of the Lorentz group by *-automorphism of $\mathfrak{A}$ with the property

$$\sigma_{\{a,\Lambda\}}\, \mathfrak{A}(B) = \mathfrak{A}(\{a, \Lambda\}\, B).$$

4) $\mathfrak{A}$ has a faithful irreducible representation, i.e. $\mathfrak{A}$ is primitive. (The role of this axiom is to exclude classical field theory.)

The Haag-Kastler axioms are valid for the $P(\phi)_2$ quantum field theory, by results of Chapter 4 and Sec. 5.1.

The Wightman axioms are more detailed.

(a) There is a Hilbert space H and a domain $\mathscr{D}$ which is dense in H and invariant under the space time averaged field operators:

$$\phi(f)\, \mathscr{D} \subset \mathscr{D}.$$

(b) There is a strongly continuous unitary representation $\{a, \Lambda\} \to U(a, \Lambda)$ of the Lorentz group on $\mathscr{H}$, and

$$U(a, \Lambda)\, \phi(f)\, U(a, \Lambda)^* = \phi(f_{\{a,\Lambda\}}).$$

(c) For $a = (x, t) \in R^2$, $U(a, I) = \exp(itH - ixP)$ has a unique fixed vector Ω which lies in $\mathscr{D}$ and is cyclic with respect to polynomials in the fields. The joint spectrum of P and H lies in the forward light cone.

(d) The fields $\phi(f)$ and $\phi(g)$ commute if f and g have space like separated supports.

In the $P(\phi)_2$ quantum field theory, parts of the Wightman axioms have been established; presumably they are all valid. In (a) we would expect to take $\mathscr{D}$ as either $C^\infty(H)$ or else the cyclic space generated by polynomials in the fields, applied to the vacuum Ω. Bounds on derivatives of the field hold see Glimm-Jaffe[10-11]. The estimate

$$\pm\phi(f) \leqq |f|\, (H(g) + I),$$

uniform in g, would be sufficient to establish (a). Here $|f|$ is some Schwartz space norm, independent of g. The stronger estimate,

$$N_\tau^{\mathrm{loc}} \leqq \text{const.}\, (H(g) + I)$$

should be valid, uniformly in g. Here N_τ^{loc} denotes one of the local energy operators introduced in Sec. 4.2.

Axiom (b), Lorentz covariance, is known for the translation subgroup, but not for the full Lorentz group. It is an open problem whether the vacuum ω of Sec. 4.2 is invariant under the Lorentz rotation σ_Λ (or whether a Lorentz invariant average of the family $\{\omega \circ \sigma_\Lambda\}$ of states is locally Fock and so usable as a vacuum state). Given a Lorentz invariant locally Fock state on the C^* algebra $\mathfrak{A}$, the general theory assures the existence of the representation U required by (b).

In axiom (c), the spectral condition was proved in Glimm-Jaffe[(10-11)]. The existence of Ω is known, but not its uniqueness. From a mass gap in the spectrum of $H(g)$, uniform as $g \to 1$, we would conclude the uniqueness of ω as a ground state for H and thus the uniqueness of Ω.

Axiom (d) has been established, see Chapter 4 and Glimm-Jaffe[(4,5)].

II Quantum Field Theory Models: Part II. The Yukawa Model

Part II The Yukawa$_2$ Model

The scalar Yukawa$_2$ coupling involves a fermion field ψ and a boson field ϕ. The classical interaction Lagrangian density $-\lambda\bar{\psi}\psi\phi$ leads to the coupled equations

$$(\gamma_0\partial_t + \gamma_1\partial_x + M)\psi - \lambda\psi\phi = 0,$$

$$(\partial_t^2 - \partial_x^2 + m^2)\phi + \lambda\bar{\psi}\psi = 0.$$

The corresponding quantum equations are more singular, since the interaction produces an infinite shift in the mass of a single boson at rest. Hence the renormalized boson field equation has the form

$$(\partial_t^2 - \partial_x^2 + m^2)\phi + j_{\text{ren}} = 0,$$

where

$$j_{\text{ren}} = \lambda\bar{\psi}\psi - \delta m^2\phi$$

and $\delta m^2 = -\infty$. In the equation above, only the difference j_{ren} has a meaning, and it is defined as a limit of a cutoff expression. In the cutoff expression approximating j_{ren}, the corresponding two terms are well defined and the difference is taken in the ordinary sense. Since the locally correct Hamiltonian $H(g)$ contains terms with infinite coefficients, the Yukawa$_2$ theory is more singular than the $P(\phi)_2$ models of Part I. The Hamiltonian $H(g)$ is defined as a limit of self adjoint approximate Hamiltonians $H(g, \varkappa)$.

The renormalization cancellations which occur agree with those predicted by formal perturbation theory. Thus here (and elsewhere in the theory) we find that simple calculations in low order perturbation theory provide excellent predictions. In addition to predicting correctly the renormalization cancellations, perturbation theory has provided conjectures (subsequently verified) about the domains of operators and the validity of uniform estimates expressing the domination of one operator by another.

Work on the doubly cutoff Yukawa theory was done by Y. Kato and Mugibayashi(1) and by Lanford(1) who proved the existence of the theory with $\varkappa, g$ or $\varkappa$, V cutoffs. The first results on the limit of $H(g, \varkappa)$ as $\varkappa \to \infty$ were obtained by Glimm(1-3) and extended by Hepp(1-2). They show that $H(g)$ is bounded from below and is an operator on an explicitly given domain. The theory was developed by Glimm and Jaffe(1,3,8,9), who proved that $H(g) = H(g)^*$ and that the propagation speed is finite. Refinements and extensions are due to Federbush(2), Eckmann(1) Dimock(1) and Schrader(1-2). Osterwalder(1) has renormalized the more singular (but unphysical) ϕ_4^3 interaction.

6 Preliminaries

6.1 The Yukawa$_2$ Hamiltonian

We consider the Fock space $\mathscr{F} = \mathscr{F}_b \otimes \mathscr{F}_f$ of Chapter 1 for a bose particle of mass $m > 0$ and fermions of mass $M > 0$. For convenience, we assume the stability condition $m < 2M$, although this is not necessary for our results. The free fermion field is $\psi_0 = (\psi_0^{(1)}, \psi_0^{(2)})$,

$$\psi_0(x, t) = \begin{Bmatrix} \psi_0^{(1)}(x, t) \\ \psi_0^{(2)}(x, t) \end{Bmatrix} = (4\pi)^{-1/2} \int e^{-ipx}\Big(e^{i\omega t} b'(p)^* \begin{Bmatrix} \nu(-p) \\ \nu(p) \end{Bmatrix}$$

$$+ e^{-i\omega t} b(-p) \begin{Bmatrix} \nu(p) \\ -\nu(-p) \end{Bmatrix}\Big) \omega(p)^{-1/2}\, dp. \tag{6.2.1}$$

Here $\omega(p) = (p^2 + M^2)^{1/2}$, $\nu(p) = (\omega(p) + p)^{1/2}$ and $\psi_0(x, t)$ satisfies the free Dirac equation

$$\left(\gamma_0 \frac{\partial}{\partial t} + \gamma_1 \frac{\partial}{\partial x} + M\right) \psi_0(x, t) = 0,$$

$$\gamma_0 = \begin{Bmatrix} 0 & i \\ i & 0 \end{Bmatrix}, \qquad \gamma_1 = \begin{Bmatrix} 0 & i \\ -i & 0 \end{Bmatrix},$$

with the canonical anticommutation relations

$$\{\psi_0^{(i)}(x, t), \psi_0^{(j)}(y, t)^*\} = \delta(x - y)\, \delta_{ij}.$$

At time zero, we set the interacting field $\psi(x, t)$ equal to the free field, $\psi(x) = \psi_0(x, 0)$. The conjugate field $\bar\psi$ is defined by

$$\bar\psi(x) = (\psi^{(2)}(x)^*, \psi^{(1)}(x)^*) = -i\psi(x)^* \gamma_0,$$

and the scalar current density at time zero is defined by

$$\begin{aligned} j(x) &= :\bar\psi(x)\, \psi(x): \\ &= :\psi^{(2)}(x)^* \psi^{(1)}(x) + \psi^{(1)}(x)^* \psi^{(2)}(x): \end{aligned}$$

The Wick dots : : indicate that in each monomial of creation and annihilation operators the creation operators are permuted to the left of the annihilation operators. In addition, a factor -1 is introduced for each permutation of a pair of adjacent fermion operators.

We use the scalar boson field of Sec. 1.2. The unrenormalized Hamiltonian with a spatial cutoff g is

$$H_{un}(g) = H_0 + \lambda \int j(x)\, \phi(x)\, g(x)\, dx.$$
$$= H_0 + H_I(g).$$

By the N_τ estimates of Proposition 1.2.2, the Hamiltonian $H_{un}(g)$ is a bilinear form on $\mathscr{D} \times \mathscr{D}$, where $\mathscr{D}$ is the domain of vectors with a finite number of particles and wave functions in the Schwartz space $\mathscr{S}(R^m)$. We write

$$H_I(g) = W^C + W + W^A$$

where the pair creation term W^C equals

$$W^C = \int \{a(k)^* + a(-k)\}\, b(p_1)^*\, b'(p_2)^*\, w^C(k, p_1, p_2)\, dk\, dp_1\, dp_2,$$

and the kernel w^C equals

$$w^C(k, p_1, p_2)$$
$$= -\frac{\lambda}{4\pi}\tilde{g}(k + p_1 + p_2)\,(\mu\omega_1\omega_2)^{-1/2}\,(\omega_1\omega_2 - p_1p_2 - M^2)^{1/2}\,\mathrm{sgn}\,(p_1 - p_2)$$
$$= -\bar{w}^C(-k, -p_1, -p_2) = -w^C(k, p_2, p_1).$$

The pair annihilation terms are $W^A = (W^C)^*$.

$$W^A = \int w^C(k, p_1, p_2)\, b'(-p_1)\, b(-p_2)\, \{a(k)^* + a(-k)\}\, dk\, dp_1\, dp_2.$$

Note that in W^A we have adopted the arbitrary convention that the particle annihilator is placed to the right of the antiparticle annihilator. In W^C, we follow the adjoint of this convention. This gives subsequent formulas a more natural form, by suitably fixing an arbitrary sign in the kernels.

The boson emission and absorption terms W equal

$$W = \int w(k, p_1, p_2)\, \{a(k)^* + a(-k)\}\, \{b(p_1)^*\, b(-p_2) + b'(p_1)^*\, b'(-p_2)\}$$
$$\times\, dk\, dp_1\, dp_2,$$

with

$$w(k, p_1, p_2) = -\frac{\lambda}{4\pi}\tilde{g}(k + p_1 + p_2)\,(\mu\omega_1\omega_2)^{-1/2}\,(\omega_1\omega_2 + p_1p_2 + M^2)^{1/2}.$$

It is convenient to express the interaction in terms of diagrams introduced by Friedrichs[1],

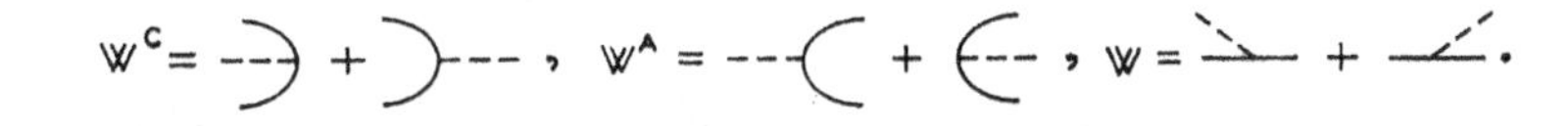

Each diagram represents a Wick monomial W defined in (1.1.9), or possibly a sum of similar monomials. A line pointing to the right stands for an annihilation operator $b(k, \varepsilon)$ while a line pointing to the left stands for a creation form $b(k, \varepsilon)^*$. A solid line denotes a fermion, $\varepsilon = \pm 1$, and a dotted line denotes a boson, $\varepsilon = 0$. A kernel is assigned to each vertex and the diagram is the integral over the momenta of the product of the kernel and the corresponding creation annihilation forms. For clarity, we might designate a variable to be associated with each line in the diagram. For instance, one contribution to W^C is

$$= \int a(k)^* \, b(p_1)^* \, b(p_2)^* \, w^C(k, p_1, p_2) \, dk \, dp_1 \, dp_2,$$

and one contribution to W is

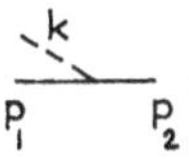

$$= \int a(k)^* \{b(p_1)^* \, b(-p_2) + b'(p_1)^* \, b'(-p_2)\} \, w(k, p_1, p_2) \times dk \, dp_1 \, dp_2 .$$

The use of these Friedrichs diagrams provides a convenient shorthand notation for certain equations. The Friedrichs diagrams differ from Feynman diagrams in the asymmetry between creators and annihilators, depending on the orientation of lines. This asymmetry is useful for the discussion of operators defined on some dense domain. For instance, both W^A and $W^C = (W^A)^*$ are bilinear forms on $\mathscr{D} \times \mathscr{D}$, but only W^A is an operator on this domain.

The product $W_1 W_2$ of two Wick monomials W_1 and W_2 can be expressed as a sum of Wick monomials,

$$W_1 W_2 = \sum_{\substack{\text{finite} \\ \text{sum}}} W_j .$$

Such a representation is obtained by using the commutation relations to permute the creation forms to the left of the annihilation forms. The sum occurs because each use of the commutation relations introduces an extra term $\{b(p_1, \varepsilon), b(p_2, \varepsilon')^*\} = \delta_{\varepsilon\varepsilon'} \delta(p_1 - p_2)$ and $[a(k), a(k')^*] = \delta(k - k')$. These commutation relations lower the number of creation and annihilation operators by two, and yield an integration over the kernels that remain. Such a term is said to be *contracted* and is indicated in the Friedrichs diagram by joining an annihilation leg (pointing right) to a creation leg on its right (the creation leg points left). In fact, Wick's theorem says that $W_1 W_2$ equals the sum over all possible contractions, with the uncontracted legs in Wick order.

Let us take as a particular example, the product of two diagrams in $H_I(g)^2$.

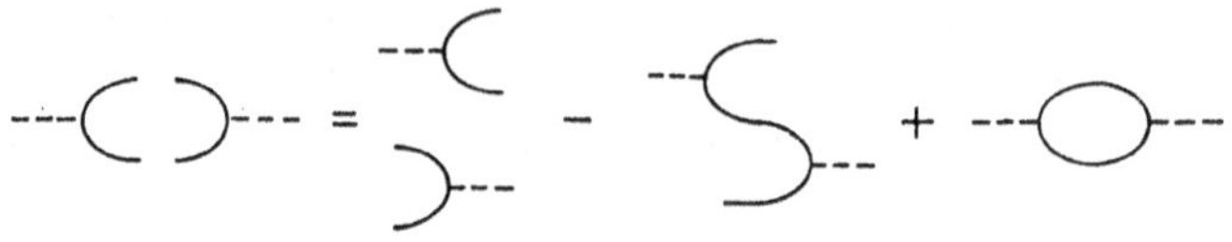

In this example,

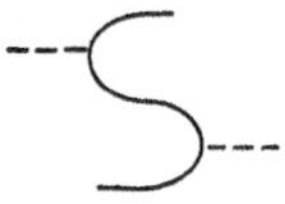

$$= \int dk\, dk'\, dp\, dp' \{b(p)^* b(-p') + b'(p)^* b'(-p')\} a(k)^* a(-k') \times \int dp_1\, w^c(k, p_1, p)\, w^c(k', p', -p_1),$$

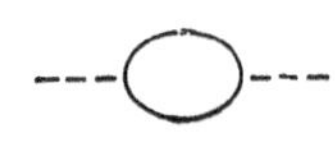

$$= \int dk\, dk'\, a(k)^* a(-k') \times \int dp_1\, dp_2\, w^c(k, p_1, p_2)\, w^c(k', -p_2, -p_1).$$

The kernel of the last term is not defined since for all k, k' the integral

$$\int dp_1\, dp_2\, w^c(k, p_1, p_2)\, w^c(k', -p_2, -p_1),$$

diverges. This divergence is one reason that $H_I(g)$ is a bilinear form on $\mathscr{D} \times \mathscr{D}$ but is not an operator.

The divergences are removed from $H_I(g)$ by the introduction of a cutoff function $\chi_\varkappa(k, p_1, p_2)$. We replace $w^c(k, p_1, p_2)$ and $w(k, p_1, p_2)$ by

$$w_\varkappa^c(k, p_1, p_2) = w^c(k, p_1, p_2)\, \chi_\varkappa(k, p_1, p_2)$$

and

$$w_\varkappa(k, p_1, p_2) = w(k, p_1, p_2)\, \chi_\varkappa(k, p_1, p_2).$$

We choose $\chi_\varkappa$ so that $w_\varkappa^c$ and $w_\varkappa$ are L_2 functions; by Proposition 1.2.3b the corresponding cutoff interaction Hamiltonian $H_I(g, \varkappa)$ is an operator on $\mathscr{D}(N)$. For instance, if $\chi_\varkappa(k, p_1, p_2) = \chi(k/\varkappa)\, \chi(p_1/\varkappa)\, \chi(p_2/\varkappa)$ and $\chi(k)$ has compact support, we say that $\chi_\varkappa$ is a sharp cutoff in momentum space. If the Fourier transform of χ has compact support, we say that $\chi_\varkappa$ is a sharp cutoff in position space. More details will be given in Definition 6.3.2.

We now describe the counterterms $c(g, \varkappa)$ suggested by perturbation theory. The renormalized Hamiltonian

$$H_{\text{ren}}(g, \varkappa) = H_0 + H_I(g, \varkappa) + c(g, \varkappa)$$

is required to be bounded from below, uniformly in $\varkappa$. In addition, a proper choice of counterterms is required to yield an operator in the limit as $\varkappa \to \infty$ rather than a bilinear form.

The no-particle vector is the ground state of H_0 and satisfies $H_0\Omega_0 = 0$. In second order perturbation theory the ground state energy of $H_{un}(g, \varkappa) = H_0 + H_I(g, \varkappa)$ is

$$\begin{aligned} E_2(g, \varkappa) &= -\|H_0^{-1/2} H_I(g, \varkappa)\,\Omega_0\|^2 \\ &= -\int |w_\varkappa^c(k, p_1, p_2)|^2 (\mu + \omega_1 + \omega_2)^{-1}\, dk\, dp_1\, dp_2 . \end{aligned}$$

Since $E_2(g, \varkappa)$ diverges as $\varkappa \to \infty$, and the corresponding first order vectors $\Omega_0 - H_0^{-1} H_I(g, \varkappa)\,\Omega_0$ converge as $\varkappa \to \infty$, the unrenormalized Hamiltonian $H_{un}(g)$ is unbounded from below as a bilinear form on $\mathscr{D} \times \mathscr{D}$. In all orders of perturbation theory greater than two, the additive contribution to the vacuum energy of $H_{un}(g)$ is finite, except for modifications to the mass explained below. Hence our theory is superrenormalizable. In other words the divergences of the renormalization constants become less severe in each higher order of perturbation theory.

In order to obtain a theory with positive energy, we subtract $E_2(g, \varkappa)$ from $H_{un}(g, \varkappa)$. However, there is still an infinite shift in second order perturbation theory for the energy of the one-boson, zero-momentum state δ, namely

$$\begin{aligned} &-\langle \delta, \{H_I(g, \varkappa)\,(H_0 - m)^{-1} H_I(g, \varkappa) - E_2(g, \varkappa)\}\, \delta\rangle \\ &= -\left(\frac{\lambda}{4\pi}\right)^2 \int \tilde{g}|(p_1 + p_2)|^2 \frac{(\omega_1\omega_2 - p_1 p_2 - M^2)}{m\omega_1\omega_2} \\ &\quad \times \left\{\frac{1}{\omega_1 + \omega_2 - m} + \frac{1}{\omega_1 + \omega_2 + m}\right\} \\ &\quad \times |\chi_\varkappa(0, p_1, p_2)|^2\, dp_1\, dp_2 . \end{aligned}$$

Here we neglect the normalization of the continuum state δ. Since

$$\langle \delta, :\phi^2(g^2):\, \delta\rangle = \frac{\|g\|_2^2}{2\pi m},$$

we have for the second order mass shift

$$\begin{aligned} \delta m_2^2(g, \varkappa) = &-\frac{\lambda^2}{4\pi\|g\|_2^2} \int |\tilde{g}(p_1 + p_2)|^2 \left(\frac{\omega_1\omega_2 - p_1 p_2 - M^2}{\omega_1\omega_2}\right) \\ &\times \left(\frac{1}{\omega_1 + \omega_2 + m} + \frac{1}{\omega_1 + \omega_2 - m}\right) |\chi_\varkappa(0, p_1, p_2)|^2\, dp_1\, dp_2 . \end{aligned}$$

We modify this expression by a g dependent constant and define the g independent mass shift

$$\delta m^2(\varkappa) = -\frac{\lambda^2}{2\pi}\int\left|\chi_\varkappa\left(0,\frac{\xi}{2},\frac{-\xi}{2}\right)\right|^2\omega(\xi)^{-1}\,d\xi + \text{const.} + o(1), \tag{6.1.1}$$

where the constant is independent of g, $\varkappa$ and $\chi_\varkappa$, and $o(1)$ vanishes as $\varkappa \to \infty$. (See Lemma 6.3.3.) We define

$$E(g,\varkappa) = E_2(g,\varkappa) + M(g) + o(1), \tag{6.1.2}$$

where the constant $M(g)$ is independent of $\varkappa$ and $\chi_\varkappa$, and $o(1)$ vanishes as $\varkappa \to \infty$.

The renormalized Hamiltonian $H(g,\varkappa)$ is defined by

$$\begin{aligned} H(g,\varkappa) &= H_0 + H_I(g,\varkappa) + c(g,\varkappa) \\ &= H_0 + H_I(g,\varkappa) - \tfrac{1}{2}\,\delta m^2(\varkappa)\int :\phi(x)^2:\,g(x)^2\,dx - E(g,\varkappa). \end{aligned} \tag{6.1.3}$$

In the following for simplicity of notation, we suppress the constants and the term $o(1)$ in (6.1.1) and (6.1.2). Their possible inclusion is elementary, and we use them in Chapter 8.

We also introduce a Hamiltonian $H(g,\varkappa,\varrho)$ with a lower cutoff ϱ. For $\varrho = 0$, $H(g,\varkappa,0) = H(g,\varkappa)$, while $H(g,\varkappa,\infty) = H_0$. To obtain the lower cutoff interaction Hamiltonian $H_I(g,\varkappa,\varrho)$, we replace the kernels $w_\varkappa$ and $w_\varkappa^C$ by

$$w_{\varkappa,\varrho}(k,p_1,p_2) = w_\varkappa(k,p_1,p_2)\,\{1-\theta_\varrho(p_1-p_2)\}$$

and

$$w_{\varkappa,\varrho}^C(k,p_1,p_2) = w_\varkappa^C(k,p_1,p_2)\,\{1-\theta_\varrho(p_1-p_2)\}$$

where $\theta_\varrho(p)$ is the characteristic function of the interval $|p| \leq \varrho$. The counterterms corresponding to the lower cutoff ϱ are

$$c(g,\varkappa,\varrho) = -\tfrac{1}{2}\,\delta m^2(\varkappa,\varrho)\int :\phi(x)^2:\,g(x)^2\,dx - E(g,\varkappa,\varrho), \tag{6.1.4}$$

$$\delta m^2(\varkappa,\varrho) = -\frac{\lambda^2}{2\pi}\int_{|\xi|>\varrho}\left|\chi_\varkappa\left(0,\frac{\xi}{2},\frac{-\xi}{2}\right)\right|^2\omega(\xi)^{-1}\,d\xi, \tag{6.1.5}$$

$$E(g,\varkappa,\varrho) = \int |w_{\varkappa,\varrho}^C(k,p_1,p_2)|^2\,(\mu+\omega_1+\omega_2)^{-1}\,dk\,dp_1\,dp_2. \tag{6.1.6}$$

Then we define

$$H(g,\varkappa,\varrho) = H_0 + H_I(g,\varkappa,\varrho) + c(g,\varkappa,\varrho). \tag{6.1.7}$$

6.2 Limits of self adjoint operators

We often encounter a sequence H_n of self adjoint operators. For H_n, we might choose a sequence of approximate Hamiltonians, or approximate field operators. In this section, we study notions of convergence that yield a self adjoint limit as $n \to \infty$. We choose types of convergence that we are able to verify directly in our models. The results generalize Glimm and Jaffe[3].

One type of convergence is strong convergence of the resolvents $R_n(\zeta) = (H_n - \zeta)^{-1} \to R$ and their adjoints. However, the limit R need not have an inverse (e.g. $H_n = nI$). Whenever R^{-1} is densely defined, R is the resolvent of a self adjoint operator H. In order to assure the invertibility of R, we introduce the notion of *graph convergence* (and later *dense boundedness*) of the sequence of operators H_n. We prove in Theorem 6.2.6 that convergence of the resolvents plus dense boundedness of $\{H_n\}$ yields a self adjoint limit. From functional analysis, one obtains the equivalence of the standard notions of convergence, for example the convergence of the resolvents $R_n(\zeta)$ *to the resolvent of a self adjoint operator* or the strong convergence of the unitary groups $U_n(t) = \exp(itH_n)$. In the following sections we prove dense boundedness and resolvent convergence for approximate Hamiltonians in the Yukawa$_2$ field theory. Analogous results for the $P(\phi)_2$ theory were proved in Part I.

Let H_n be a sequence of densely defined operators on a Hilbert space $\mathscr{H}$. The graph $\mathscr{G}(H_n) \subset \mathscr{H} \oplus \mathscr{H}$ is the set of pairs $\{\psi, H_n\psi : \psi \in \mathscr{D}(H_n)\}$. Let

$$\mathscr{G}_\infty = \{\psi, \chi : \psi = \lim \psi_n, \psi_n \in \mathscr{D}(H_n), \quad \chi = \text{weak lim } H_n\psi_n\},$$

$$\mathscr{D}_\infty = \{\psi : \{\psi, \chi\} \in \mathscr{G}_\infty \quad \text{for some } \chi\}.$$

In general, $\mathscr{G}_\infty$ will not be the graph of an operator. If $\mathscr{G}_\infty$ is the graph of an operator H, then $\mathscr{D}_\infty$ is the domain of H.

DEFINITION 6.2.1 *A sequence H_n of densely defined operators has a weak graph limit H, if $\mathscr{G}_\infty$ is the graph of a densely defined operator H.*

PROPOSITION 6.2.2 *Let $H_n = H_n^*$. The weak graph limit H of $\{H_n\}$ exists if and only if $\mathscr{D}_\infty$ is dense. In that case H is symmetric.*

Proof If the weak graph limit exists, $\mathscr{D}(H) = \mathscr{D}_\infty$ is dense by definition. Conversely, suppose that $\mathscr{D}_\infty$ is dense and $\{0, \chi\} \in \mathscr{G}_\infty$. We prove $\chi = 0$, so that $\mathscr{G}_\infty$ is the graph of an operator H.

Let $w. \lim \phi_n = \phi$, st. $\lim \psi_n = \psi$. We show that $\langle \psi_n, \phi_n \rangle \to \langle \psi, \phi \rangle$. By the principle of uniform boundedness, $\|\phi_n\| \leq$ const. Hence

$$\langle \psi_n, \phi_n \rangle - \langle \psi, \phi \rangle = \langle \psi, \phi_n - \phi \rangle + \langle \psi_n - \psi, \phi_n \rangle \to 0.$$

Let $\psi_n \in \mathscr{D}(H_n)$, $\lim \psi_n = \psi = 0$, $w. \lim H_n\psi_n = \chi$. For $\theta \in \mathscr{D}_\infty$, there is a sequence $\theta_n \in \mathscr{D}(H_n)$ with $\theta = \lim \theta_n$ and $\Lambda = w. \lim H_n\theta_n$. Thus we have $0 = \langle \psi, \Lambda \rangle = \lim \langle \psi_n, H_n\theta_n \rangle = \lim \langle H_n\psi_n, \theta_n \rangle = \langle \chi, \theta \rangle$. Since $\mathscr{D}_\infty$ is dense, $\chi = 0$ and $\mathscr{G}_\infty = \mathscr{G}(H)$.

To prove that H is symmetric, we note for $\theta \in \mathscr{D}_\infty$,

$$\langle H\theta, \theta \rangle = \lim \langle H_n\theta_n, \theta_n \rangle = \lim \langle \theta_n, H_n\theta_n \rangle = \langle \theta, H\theta \rangle.$$

While the graph limit of a sequence of self adjoint operators is symmetric, it is not necessarily self adjoint. We combine graph convergence of the H_n with convergence of the resolvents $R_n(\zeta) = (H_n - \zeta)^{-1}$. Our main results are Theorems 6.2.3 and 6.2.6.

THEOREM 6.2.3 *Let H be the weak graph limit of a sequence $\{H_n\}$ of self adjoint operators and for some ζ, let the resolvents and their adjoints converge strongly,*

$$R_n(\zeta) \to R \quad \text{and} \quad R_n(\zeta)^* = R_n(\bar{\zeta}) \to R^*.$$

Then H is self adjoint and $R = (H - \zeta)^{-1}$.

Proof We show that R has a densely defined inverse, namely Null $R = 0 = (\text{Range } R)^{\perp}$. We use a simplification of Kleinstein[1]. Suppose $\chi \in \text{Null } R$, $\theta \in \mathscr{D}(H)$. Using $\theta = \lim \theta_n$, $H\theta = w.\lim H_n\theta_n$, we have

$$0 = \langle (H - \bar{\zeta})\theta, R\chi\rangle = \lim \langle (H_n - \bar{\zeta})\theta_n, R_n\chi\rangle = \lim \langle \theta_n, \chi\rangle = \langle \theta, \chi\rangle.$$

Since $\mathscr{D}(H)$ is dense, $\chi = 0$ and Null $R = 0$.

We now prove $\mathscr{D}(H) \subset \text{Range } R$, so $(\text{Range } R)^{\perp} = 0$. Let $\chi \in \mathscr{H}$, $\theta \in \mathscr{D}(H)$.

$$\begin{aligned}\langle \chi, R(H - \zeta)\theta\rangle &= \langle R^*\chi, (H - \zeta)\theta\rangle = \lim \langle R_n(\bar{\zeta})\chi, (H_n - \zeta)\theta_n\rangle \\ &= \lim \langle \chi, \theta_n\rangle = \langle \chi, \theta\rangle.\end{aligned}$$

Thus $R(H - \zeta)\theta = \theta$, $\mathscr{D}(H) \subset \text{Range } R$ and R^{-1} is densely defined.

On the other hand, for all χ,

$$R_n(\zeta)\chi \to R\chi \quad \text{and} \quad (H_n - \zeta)R_n(\zeta)\chi = \chi.$$

Thus $R\chi \in \mathscr{D}(H)$ and $(H - \zeta)R\chi = \chi$. Therefore $\mathscr{D}(H) = \text{Range } R$ and $R = (H - \zeta)^{-1}$. Likewise $R^* = (H - \bar{\zeta})^{-1}$. Since

$$\text{Range } (H - \zeta) = \mathscr{D}(R) = \mathscr{H} = \mathscr{D}(R^*) = \text{Range } (H - \bar{\zeta}),$$

H is self adjoint.

LEMMA 6.2.4 *Let $H_n = H_n^*$, and let $\mathscr{D}$ be a dense set of vectors with the property that for each $\psi \in \mathscr{D}$, there exists a sequence ψ_n with*

$$\psi_n \in \mathscr{D}(H_n), \qquad \psi_n \to \psi, \qquad \|H_n\psi_n\| \leq \text{const}.$$

Suppose furthermore that for the sequence $\{\chi_n\}$ approximating a χ in $\mathscr{D}$ as above, the inner products $\langle \chi_n, H_n\psi_n\rangle$ converge as $n \to \infty$. Then $\{H_n\}$ has a weak graph limit H, and $\mathscr{D} \subset \mathscr{D}(H)$.

Proof Given $\varepsilon > 0$ and $\theta \in \mathscr{H}$, choose $\bar{\theta} \in \mathscr{D}$ such that $\|\theta - \bar{\theta}\| < \varepsilon$. Thus for $\bar{\theta}_n \to \bar{\theta}$, $\|H_n\bar{\theta}_n\| \leq$ const.,

$$\langle \theta, H_n\psi_n\rangle = \langle \theta - \bar{\theta}, H_n\psi_n\rangle + \langle \bar{\theta} - \bar{\theta}_n, H_n\psi_n\rangle + \langle \bar{\theta}_n, H_n\psi_n\rangle.$$

Hence

$$|\langle\theta, H_n\psi_n - H_m\psi_m\rangle| \leqq \|\theta - \bar{\theta}\| \, \{\|H_n\psi_n\| + \|H_m\psi_m\|\}$$
$$+ \|\bar{\theta} - \bar{\theta}_n\| \, \|H_n\psi_n\| + \|\bar{\theta} - \bar{\theta}_m\| \, \|H_m\psi_m\|$$
$$+ |\langle\bar{\theta}_n, H_n\psi_n\rangle - \langle\bar{\theta}_m, H_m\psi_m\rangle|$$
$$\leqq o(1)$$

since $\|\theta - \bar{\theta}\| < \varepsilon$, $\|\bar{\theta} - \bar{\theta}_n\| \leqq o(1)$ and the last term is small by assumption. Hence $H_n\psi_n$ converges weakly, and the proposition follows by the density of $\mathscr{D}$ and Proposition 6.2.2.

DEFINITION 6.2.5 *The sequence* $\{H_n\}$ *is densely bounded if there is a dense set* $\mathscr{D} \subset \mathscr{H}$, *such that for each* $\psi \in \mathscr{D}$ *there exists a sequence* $\psi_n \in \mathscr{D}(H_n)$ *with*

$$\lim_{n\to\infty} \psi_n = \psi, \qquad \langle\psi_n, |H_n| \, \psi_n\rangle \leqq \text{const.}$$

Our main result is the following:

THEOREM 6.2.6 *Let* $\mathscr{H}$ *be separable and let* $H_n = H_n^*$. *Assume that the sequence* $\{H_n\}$ *is densely bounded and that for some complex* ζ, *the strong limits* $R = \lim R_n(\zeta)$ *and* $R^* = \lim R_n(\zeta)^*$ *exist. Then* $R = (H - \zeta)^{-1}$ *and*

$$H = H^* = \text{graph} \lim_{n\to\infty} H_n.$$

Proof We assume at first that $0 \leqq H_n$. In this case we may take ζ negative by a simple Neumann series argument (Kato [1, p. 427]) so that

$$I \leqq (H_n - \zeta).$$

We choose in $\mathscr{D}$ a countable dense subset, $\mathscr{D}_1$. For $\psi_n \to \psi \in \mathscr{D}_1$ and $\chi_n \to \chi \in \mathscr{D}_1$ as in Definition 6.2.5, the inner products

$$M_n = \langle\chi_n, (H_n - \zeta)^{1/2} \, \psi_n\rangle$$

are bound in magnitude, uniformly in n, by

$$\|\chi_n\| \, \|(H_n - \zeta)^{1/2} \, \psi_n\| \leqq \text{const.}$$

Thus there exists a convergent subsequence M_{n_j} of the M_n, and by the diagonal process, one subsequence n_j converging for every pair of limits $\psi, \chi \in \mathscr{D}_1$.

Let $B_j = (H_{n_j} - \zeta)^{1/2}$. By Lemma 6.2.4, $B =$ weak graph limit B_j exists. Also $R_n(\zeta)^{1/2} \to R^{1/2}$, so by Theorem 6.2.3 $B = B^* = R(\zeta)^{-1/2}$. The square of a self adjoint operator is self adjoint so $R(\zeta)^{-1} = B^2$ is self adjoint. Since the full sequence $R_n(\zeta)$ converges strongly to $R(\zeta)$, and since Range $R(\zeta) = \mathscr{D}(B^2)$ is dense, the graph limit H of the H_n exists, and $R(\zeta) = (H - \zeta)^{-1}$.

5*

We have used convergence of $R_n(\zeta)^{1/2}$. Since $R_n(\zeta)$ is positive, $R_n(\eta)$ converges strongly for all $\eta < \zeta$. Hence the representation

$$R_n(\zeta)^{1/2} = \pi^{-1} \int_0^\infty \lambda^{-1/2} R_n(\zeta - \lambda)\, d\lambda$$

proves the convergence of the square root.

To prove the theorem in the general case, we use the operators $|H_n| = (H_n^2)^{1/2} = H_n'$, and

$$\begin{aligned} R_n'(\pm i) &= (H_n' \mp i)^{-1} \\ &= |\mathrm{Re}\, R_n(\pm i)| + i \,\mathrm{Im}\, R_n(\pm i). \end{aligned}$$

The resolvents $R_n'(\pm i)$ converge strongly, because by the Neumann series argument refered to above, we may take $\zeta = \pm i$ without loss of generality. Convergence of $R_n(\pm i)$ yields convergence of $|\mathrm{Re}\, R_n(\pm i)|$, etc, and convergence of $R_n'(\pm i)$. By the theorem as proved so far,

$$R'(\pm i) = |\mathrm{Re}\, R(\pm i)| + i \,\mathrm{Im}\, R(\pm i)$$

is the resolvent of a self adjoint operator. Thus

$$0 = \mathrm{Null}\, R'(\pm i) = \mathrm{Null}\, R(\pm i)$$

and so

$$0 = \{\mathrm{Range}\, R(\mp i)\}^\perp.$$

As a consequence of the strong convergence of $R_n(\mp i)$ to an operator $R(\mp i)$ with dense range, it follows that

$$H = \text{graph} \lim_{n\to\infty} H_n$$

exists. By Theorem 6.2.3, $H = H^* = R(i)^{-1} + i$.

We now define the strong graph limit. Let

$$\mathscr{G}^s_\infty = \{\psi, \chi : \psi = \lim \psi_n, \psi_n \in \mathscr{D}(H_n), \chi = \lim H_n\psi_n\}.$$

$$\mathscr{D}^s_\infty = \{\psi : \{\psi, \chi\} \in \mathscr{G}^s_\infty \quad \text{for some } \chi\}.$$

DEFINITION 6.2.7 *The sequence H_n has a strong graph limit H if $\mathscr{G}^s_\infty$ is the graph of a densely defined operator H.*

The existence of a strong graph limit of a sequence H_n implies the existence of the weak graph limit.

LEMMA 6.2.8 *The strong graph limit of a sequence H_n exists if and only if $\mathscr{D}^s_\infty$ is dense. If H = st. graph $\lim H_n = H_n^*$, then H is closed and symmetric.*

Proof We follow the proof of Proposition 6.2.2 except to establish that H is closed. Let $\{\psi_m, \chi_m\} \to \{\psi, \chi\}$ be a Cauchy sequence in $\mathscr{G}(H)$.

Let π_n be the projection in $\mathscr{H} \oplus \mathscr{H}$ onto $\mathscr{G}(H_n)$. To show $\{\psi, \chi\} \in \mathscr{G}(H)$, we prove $\|\{\psi, \chi\} - \pi_n\{\psi, \chi\}\| \to 0$.

$$\begin{aligned}\|\{\psi, \chi\} - \pi_n\{\psi, \chi\}\| &\leqq \|\{\psi, \chi\} - \{\psi_m, \chi_m\}\| + \|\{\psi_m, \chi_m\} - \pi_n\{\psi_m, \chi_m\}\| \\ &\quad + \|\pi_n\{\psi_m, \chi_m\} - \pi_n\{\psi, \chi\}\| \\ &\leqq 2\|\{\psi, \chi\} - \{\psi_m, \chi_m\}\| + \|\{\psi_m, \chi_m\} - \pi_n\{\psi_m, \chi_m\}\|.\end{aligned}$$

The first term is $o(1)$ for m large, by assumption. The second term is the distance from $\{\psi_m, \chi_m\}$ to $\mathscr{G}(H_n)$. Since $\{\psi_m, \chi_m\} \in \mathscr{G}(H)$, there is a sequence $\{\psi_{mn}, \chi_{mn}\}$ in $\mathscr{G}(H_n)$ converging to $\{\psi_m, \chi_m\}$ for m fixed. Thus

$$\begin{aligned}\|\{\psi_m, \chi_m\} - \pi_n\{\psi_m, \chi_m\}\| &\leqq \|\{\psi_m, \chi_m\} - \{\psi_{mn}, \chi_{mn}\}\| \\ &\leqq o(1)\end{aligned}$$

for m fixed. Hence $\mathscr{G}(H)$ is closed.

Question Is the weak graph limit of a sequence H_n closed?

Remark If Theorem 6.2.3 or Theorem 6.2.6 is valid, then the strong graph limit of the sequence H_n exists and

$$w.\text{ graph lim } H_n = \text{st. graph lim } H_n = H = H^*.$$

The assumption that the sequence $\{H_n\}$ is densely bounded may be replaced by the assumption that $\{|H_n|^\varepsilon\}$ is densely bounded, for some $\varepsilon > 0$.

6.3 Properties of the cutoffs

In this section we discuss elementary properties of the allowed momentum cutoff functions $\chi_\varkappa$ and the corresponding cutoff operators $H_I(g, \varkappa, \varrho)$ and $c(g, \varkappa, \varrho)$. We start with a function $\chi(p)$ in $\mathscr{S}(R^1)$, for which $\chi(0) = 1$, $\chi(p) = \chi(-p)$, and we define

$$\chi_\varkappa(k, p_1, p_2) = \chi\left(\frac{k}{\varkappa}\right)\chi\left(\frac{p_1}{\varkappa}\right)\chi\left(\frac{p_2}{\varkappa}\right). \tag{6.3.1}$$

LEMMA 6.3.1 *The cutoff (6.3.1) satisfies*

$$|\chi_\varkappa| \leqq O(1). \tag{a}$$

$$|\chi_\varkappa| \leqq O(\varkappa)(\mu + \omega_1 + \omega_2)^{-1} \tag{b}$$

$$|\chi_\varkappa - 1| \leqq O(\varkappa^{-1})(\mu + \omega_1 + \omega_2). \tag{c$_1$}$$

Let $\eta = p_1 + p_2$, $\xi = p_1 - p_2$. For some $\varepsilon > 0$, (c$_2$)

$$\int_{|\xi| > \varrho} \left|\chi_\varkappa(k, p_1, p_2) - \chi_\varkappa\left(0, \frac{\xi}{2}, \frac{-\xi}{2}\right)\right| \omega(\xi)^{-1}\, d\xi \leqq O(\varrho^{-\varepsilon}\varkappa^{-\varepsilon})(\omega(\eta) + \mu(k))^{2\varepsilon}.$$

$$\chi_\varkappa(k, p_1, p_2) = \chi_\varkappa(k, p_2, p_1) = \chi_\varkappa(-k, -p_1, -p_2). \tag{d}$$

Proof Since $\chi \in \mathscr{S}(R^1)$, properties (a) and (b) follow. Property (c_1) results from

$$\left|\chi\left(\frac{p}{\varkappa}\right) - 1\right| \leqq \int_0^{p\varkappa^{-1}} |\chi'(t)|\, dt \leqq 0(\varkappa^{-1})\,\mu(p).$$

To prove (c_2), we note that the first derivatives of $\chi_\varkappa(k, p_1, p_2)$ are $0(\varkappa^{-1})$. Thus

$$\left|\chi_\varkappa(k, p_1, p_2) - \chi_\varkappa\left(0, \frac{\xi}{2}, \frac{-\xi}{2}\right)\right| \leqq 0(\varkappa^{-1})\,(\mu(k) + \omega(\eta)).$$

Using (a) and (b)

$$\begin{aligned}\left|\chi_\varkappa(k, p_1, p_2) - \chi_\varkappa\left(0, \frac{\xi}{2}, \frac{-\xi}{2}\right)\right| &\leqq 0(\varkappa^{-2\varepsilon})\,(\mu + \omega(\eta))^{2\varepsilon}\, 0(\varkappa^{\varepsilon})\,\omega(\xi)^{-\varepsilon} \\ &= 0(\varkappa^{-\varepsilon})\,\omega(\xi)^{-\varepsilon}\,(\mu + \omega(\eta))^{2\varepsilon},\end{aligned}$$

from which (c_2) follows. Property (d) is obvious.

For each cutoff $\chi_\varkappa$ of the form (6.3.1), we introduce a second cutoff that is sharp in momentum space. Let $\theta(x)$ be the characteristic function of the interval $|x| \leqq 1$, and let

$$\chi'_\varkappa(k, p_1, p_2) = \chi_\varkappa(k, p_1, p_2)\,\theta(k\varkappa^{-1+\alpha})\,\theta(p_1\varkappa^{-1+\alpha})\,\theta(p_2\varkappa^{-1+\alpha}), \tag{6.3.2}$$

for some $0 < \alpha < 1$. Thus the cutoff $\chi'_\varkappa$ is sharp at momenta $0(\varkappa^{1-\alpha})$. The cutoff $\chi'_\varkappa$ of (6.3.2) satisfies the conditions (a)–(d) of Lemma 6.3.1, except that (c_1) must be replaced by

$$(c'_1)\ |\chi'_\varkappa - 1| \leqq 0(\varkappa^{-1})\,(\mu + \omega_1 + \omega_2) \quad \text{on} \quad S_{\varkappa^{1-\alpha}},$$

$$S_r = \{k, p_1, p_2 : |k| \leqq r, |p_1| \leqq r, |p_2| \leqq r\}.$$

We do not prove these elementary estimates that can be found in Glimm-Jaffe [8, Sec. 4.1].

DEFINITION 6.3.2 *A function $\chi_\varkappa(p, k_1, k_2)$ is a general momentum cutoff if $\chi_\varkappa$ satisfies the conditions of Lemma 6.3.1, and is piecewise continuous with at most a finite number of jump discontinuities across cubes S_r.*

Our proofs hold for general cutoffs $\chi_\varkappa$, since the proofs involve only (a)–(d) and the piecewise continuity of $\chi_\varkappa$. Condition (b) ensures that the kernels $w_\varkappa^C$ and $w_\varkappa$ are L_2 functions. Condition (d) makes $H_I(g, \varkappa)$ a symmetric operator. Condition (c_2) is used in estimating the mass renormalization cancellation.

We next show that $\delta m^2(\varkappa)$ of (6.1.1) differs from the second order mass shift $\delta m_2^2(g, \varkappa)$ by a constant plus $o(1)$.

LEMMA 6.3.3 *We have*

$$\delta m^2(\varkappa) = \delta m_2^2(g, \varkappa) + M(g) + o(1)$$

where the constant $M(g)$ *is independent of* $\varkappa$ *and the choice of cutoff* $\chi_\varkappa$, *and where* $o(1)$ *vanishes as* $\varkappa \to \infty$.

Proof We write the integrands defining $\delta m^2(\varkappa)$ and $\delta m_2^2(g, \varkappa)$ as products. Let

$$\begin{aligned}
\alpha_1 &= \frac{1}{2}\left(\frac{\omega_1\omega_2 - p_1p_2 - M^2}{\omega_1\omega_2}\right), & \beta_1 &= 1 \\
\alpha_2 &= \frac{1}{2}\left(\frac{1}{\omega_1 + \omega_2 - m} + \frac{1}{\omega_1 + \omega_2 + m}\right), & \beta_2 &= \omega(\xi)^{-1} \qquad (6.3.3) \\
\alpha_3 &= |\chi_\varkappa(0, p_1, p_2)|^2, & \beta_3 &= \left|\chi_\varkappa\left(0, \frac{\xi}{2}, -\frac{\xi}{2}\right)\right|^2.
\end{aligned}$$

We have the elementary bounds for $0 \leqq \varepsilon \leqq 1$,

$$|\alpha_1 - \beta_1| = \frac{1}{2}\left|\frac{\omega_1\omega_2 + p_1p_2 + M^2}{\omega_1\omega_2}\right| \leqq \text{const.}\ \omega(\eta)^{2\varepsilon}\,\omega(\xi)^{-\varepsilon}$$

$$|\alpha_2 - \beta_2| \leqq \text{const.}\ (\omega_1 + \omega_2)^{-1}\,\omega(\xi)^{-1}\,\omega(\eta). \qquad (6.3.4)$$

Here we use the stability condition $m < 2M$. Using (a) and (c_2) of Lemma 6.3.1, for some $\varepsilon > 0$,

$$\int |\alpha_3 - \beta_3|\,\omega(\xi)^{-1}\,d\xi \leqq 0(\varkappa^{-\varepsilon})\,\omega(\eta)^{2\varepsilon}.$$

Note that

$$\begin{aligned}
\beta_1\beta_2\beta_3 - \alpha_1\alpha_2\alpha_3 &= (\beta_1\beta_2 - \alpha_1\alpha_2) + (\beta_1\beta_2 - \alpha_1\alpha_2)(\beta_3 - 1) \\
&\quad + \alpha_1\alpha_2(\beta_3 - \alpha_3);
\end{aligned}$$

$$\delta m^2(\varkappa) - \delta m_2^2(g, \varkappa) = -\frac{\lambda^2}{2\pi}\|g\|^{-2}\int |\tilde{g}(\eta)|^2\,(\beta_1\beta_2\beta_3 - \alpha_1\alpha_2\alpha_3)\,d\xi\,d\eta.$$

Thus

$$\begin{aligned}
\delta m^2(\varkappa) - \delta m_2^2(g, \varkappa) &= -\frac{\lambda^2}{2\pi}\|g\|^{-2}\int |\tilde{g}(\eta)|^2\,(\beta_1\beta_2 - \alpha_1\alpha_2)\,d\xi\,d\eta \\
&\quad - \frac{\lambda^2}{2\pi}\|g\|^{-2}\int |\tilde{g}(\eta)|^2\,(\beta_1\beta_2 - \alpha_1\alpha_2)(\beta_3 - 1)\,d\xi\,d\eta \\
&\quad + 0(\varkappa^{-\varepsilon}). \qquad (6.3.5)
\end{aligned}$$

By (6.3.4), $|\beta_1\beta_2 - \alpha_1\alpha_2| \leqq \text{const.}\, \omega(\eta)^{2\varepsilon}\omega(\xi)^{-1-\varepsilon}$, the integrals on the right of (6.3.5) converge. Hence by conditions (a) and (c_1),

$$\delta m^2(\varkappa) = \delta m_2^2(g, \varkappa) - \frac{\lambda^2}{2\pi} \|g\|^{-2} \int |\tilde{g}(\eta)|^2 (\beta_1\beta_2 - \alpha_1\alpha_2)\, d\xi\, d\eta + O(\varkappa^{-\varepsilon}),$$

and since $(\beta_1\beta_2 - \alpha_1\alpha_2)$ is independent of $\chi_\varkappa$, the lemma is proved.

LEMMA 6.3.4 *The Hamiltonian $H(g, \varkappa, \varrho)$ is self adjoint on $\mathscr{D}(H_0)$.*

Proof Since $0 < -\delta m^2(\varkappa)$, Proposition 3.2.1 implies that $H_0 + c(g, \varkappa)$ is essentially self adjoint on $\mathscr{D}(H_0)$. This operator is self adjoint on $\mathscr{D}(H_0)$, since Theorem 3.1.3 with deg $P = 2$ yields

$$\|H_0\theta\| \leqq \text{const.}\, \|\{H_0 + c(g, \varkappa)\}\, \theta\|.$$

We now show that $H_I(g, \varkappa, \varrho)$ is infinitesimally small in the sense of Kato [1, p. 287] with respect to H_0, completing the proof of the lemma. We approximate the kernel $w_{\varkappa,\varrho}^C$ of $W_{\varkappa,\varrho}^C$ by a function $r \in \mathscr{S}(R^3)$ so that $\|w_{\varkappa,\varrho}^C - r\|_2 < \varepsilon/2$. Hence by Proposition 1.2.3(b–c), for each $\varkappa$, ϱ we have

$$\begin{aligned} \|W_{\varkappa,\varrho}^C\theta\| &\leqq \|w_{\varkappa,\varrho}^C - r\|\, \|(N + I)\,\theta\| + |r|_s\, \|(N + I)^{1/2}\,\theta\| \\ &\leqq \varepsilon\|N\theta\| + \text{const.}\, \|\theta\|. \end{aligned}$$

The bounds on $W_{\varkappa,\varrho}^A$ and $W_{\varkappa,\varrho}$ are similar.

For use in chapters 7 and 8, we prove related estimates.

LEMMA 6.3.5 *Let $\varepsilon > 0, \tau > 0$. Then*

$$\pm\{H_I(g, \varkappa, \varrho) - H_I(g, \varkappa, \varrho_1)\} \leqq \varepsilon N_\tau + \text{const.},$$

$$\pm\{c(g, \varkappa, \varrho) - c(g, \varkappa, \varrho_1)\} \leqq \text{const.}\, (N_\tau + I) \qquad (6.3.6)$$

with constants uniform in $\varkappa$ and ϱ, $\varrho \leqq \varrho_1$.
For fixed $\varkappa < \infty$, there exists $\delta > 0$ such that

$$\pm H_I(g, \varkappa, \varrho) \pm c(g, \varkappa, \varrho) \leqq O(\varrho^{-\delta})\, (N + I). \qquad (6.3.7)$$

Proof Let

$$H_I(g, \varkappa, \varrho) - H_I(g, \varkappa, \varrho_1) = \delta H_I = \delta W^C + \delta W + \delta W^A.$$

We approximate the kernel $\delta w_{\varkappa,\varrho}$ by $\delta w_{\varkappa,\varrho,V} = \delta w_{\varkappa,\varrho}(k, p_{1V}, p_{2V})$, using the notation of (1.1.3). Since $\gamma = (\omega_1\omega_2 + p_1p_2 + M^2)^{1/2}\,(\omega_1\omega_2)^{-1/2}$ is bounded and has uniformly bounded derivatives, $|\gamma - \gamma_V| \leqq O(V^{-1})$. Also $|\tilde{g}(k + \eta) - \tilde{g}(k + \eta_V)| \leqq O(V)^{-1}\, h(k + \eta)$, where $h(\cdot)$ is rapidly decreasing and $|\chi_\varkappa - \chi_{\varkappa,V}| \leqq O(V^{-1})$. Also, $\theta_\varrho(\xi) - \theta_\varrho(\xi_V)$ is nonzero only in a ξ interval of length $O(V^{-1})$. Thus

$$\begin{aligned} |\delta w_{\varkappa\varrho} - \delta w_{\varkappa\varrho V}| \leqq {} & \{O(V^{-1})\,\theta_{\varrho_1}(\xi) + |\theta_\varrho(\xi) - \theta_\varrho(\xi_V)| + |\theta_{\varrho_1}(\xi) - \theta_{\varrho_1}(\xi_V)|\} \\ & \times \mu^{-1/2}\, h(k + \eta) \end{aligned}$$

and

$$\|\mu^{-\tau/2}(\delta w_{\varkappa\varrho} - \delta w_{\varkappa\varrho V})\|_2 \leqq 0(V^{-1/2}).$$

Hence by Proposition 1.2.3(b),

$$\pm\{\delta W_{\varkappa,\varrho} - \delta W_{\varkappa,\varrho,V}\} \leqq 0(V^{-1/2})(N_\tau + I). \tag{6.3.8}$$

By Proposition 1.2.3(d), we can neglect two fermions in $\delta W_{\varkappa,\varrho,V}$ so

$$|\langle\theta, \delta W_{\varkappa,\varrho,V}\theta\rangle| \leqq \|N_\tau^{1/2}\theta\| \, \|\theta\| \, |\mu^{-\tau/2}\delta w_{\varkappa,\varrho,V}|_{1,2}$$
$$\leqq \varepsilon\langle\theta, N_\tau\theta\rangle + 0(V)\,\|\theta\|^2. \tag{6.3.9}$$

By choosing V large enough to make (6.3.8) small, we obtain by (6.3.9),

$$\pm\delta W \leqq \varepsilon N_\tau + \text{const.},$$

uniformly in $\varkappa$ and $\varrho \leqq \varrho_1$. Identical bounds hold on δW^C and δW^A, so the bound (6.3.6) on δH_I is proved.

For fixed $\varkappa$, $w^C_{\varkappa,\varrho}$ and $w_{\varkappa,\varrho}$ are L_2 functions with L_2 norms equal $O(\varrho^{-\delta})$. Hence the bound (6.3.7) on H_I follows from Corollary 1.2.4.

We now study the counterterms. Since

$$-\delta E = -E(g, \varkappa, \varrho) + E(g, \varkappa, \varrho_1)$$
$$= \|(\mu + \omega_1 + \omega_2)^{-1/2} \, w^C_\varkappa\{\theta_\varrho(\xi) - \theta_{\varrho_1}(\xi)\}\|_2^2$$
$$\leqq \|(\mu + \omega_1 + \omega_2)^{-1/2} \, w^C_\varkappa\theta_{\varrho_1}(\xi)\|_2^2,$$

and since for $0 < \delta \leqq 1$,

$$|w^C_\varkappa| \leqq \text{const.}\, |\tilde{g}(k + \eta)| \, \mu^{-1/2} \, \omega(\xi)^\sigma \, \omega(\eta)^{-\delta/2},$$

we have $|\delta E| \leqq$ const., uniformly in $\varkappa$. Also

$$|\delta m^2(\varkappa, \varrho) - \delta m^2(\varkappa, \varrho_1)| \leqq |\delta m^2(\varkappa, \varrho_1)| \leqq \text{const.},$$

so by Proposition 1.2.3 we have a $\varkappa$ and ϱ independent estimate

$$\pm\delta c \leqq \text{const.}\,(N + I).$$

For fixed $\varkappa$, $|\delta m^2(\varkappa, \varrho)| \leqq 0(\varrho^{-\delta})$, and $|E(g, \varkappa, \varrho)| \leqq 0(\varrho^{-\delta})$. The remaining bounds of the lemma follow.

7 First and Second Order Estimates

We prove estimates for the Yukawa$_2$ Hamiltonian of the form

$$N_\tau \leqq \text{const.}\,(H(g, \varkappa) + \text{const}).\,, \qquad \tau < 1 \tag{7.1}$$

and

$$N_\tau^2 = \text{const.}\,(H(g, \varkappa) + \text{const.})^2, \qquad \tau < 1/2, \tag{7.2}$$

with constants independent of $\varkappa$. We call these inequalities first and second order estimates respectively, since they involve first and second powers of the Hamiltonian. The results of Sec. 6.3 yield $\varkappa$ dependent estimates for $\tau \leqq 1$. However, according to perturbation theory, the estimates (7.1) and (7.2) are the best possible estimates uniform in $\varkappa$.

We now verify in perturbation theory for which values of τ the ground state of $H(g, \varkappa)$ gives expectation values of N_τ or N_τ^2 that are uniformly bounded in $\varkappa$.

To first order perturbation theory, the ground state of $H(g,\varkappa)$ is $\Omega^{(1)} = \Omega_0 - H_0^{-1} H_I(g, \varkappa)\, \Omega_0$. Thus

$$\begin{aligned}\langle \Omega^{(1)}, N_\tau \Omega^{(1)} \rangle &= \| N_\tau^{1/2} H_0^{-1} H_I(g, \varkappa)\, \Omega_0 \|^2 \\ &= \| (\mu^\tau + \omega_1^\tau + \omega_2^\tau)^{1/2} (\mu + \omega_1 + \omega_2)^{-1} w_\varkappa^C \|_2^2 .\end{aligned}$$

The integral diverges logarithmically for $\tau = 1$, since the integral then equals the second order vacuum energy $E^{(2)}(g, \varkappa)$. Thus $\langle \Omega^{(1)}, N_\tau \Omega^{(1)} \rangle$ is bounded uniformly in $\varkappa$ for $\tau < 1$, and we expect a uniform first order estimate to be valid for $\tau < 1$.

Likewise, we compute

$$\begin{aligned}\langle \Omega^{(1)}, N_\tau^2 \Omega^{(1)} \rangle &= \| N_\tau H_0^{-1} H_I(g, \varkappa)\, \Omega_0 \|^2 \\ &= \| (\mu^\tau + \omega_1^\tau + \omega_2^\tau) (\mu + \omega_1 + \omega_2)^{-1} w_\varkappa^C \|_2^2 ,\end{aligned}$$

which is bounded uniformly in $\varkappa$ for $\tau < \frac{1}{2}$. Hence we expect a uniform second order estimate to hold for $\tau < \frac{1}{2}$. In fact, uniform first order estimates do not hold for $\tau = 1$ and uniform second order estimates do not hold for $\tau = \frac{1}{2}$.

7.1 Positivity of the Yukawa$_2$ Hamiltonian

THEOREM 7.1.1 *Glimm*[2]. *Let $\tau < 1$, g be fixed. There are constants a, b independent of $\varkappa$ and ϱ such that*

$$N_\tau \leqq a\big(H(g, \varkappa, \varrho) + bI\big). \tag{7.1.1}$$

Formal motivation Before starting the proof of (7.1.1), we motivate the ideas involved. We use an approximate diagonalization of $H(g, \varkappa, \varrho)$ that exhibits the Hamiltonian as a positive operator plus a small error. Formal perturbation theory suggests that there is a unitary operator T, the wave operator, for which

$$H(g, \varkappa, \varrho) = T H_0 T^* . \tag{7.1.2}$$

Since H_0 is positive, the positivity of H would follow.

The wave operator $T_\pm$ is by definition the limit of operators $T(t) = \exp(-iHt)\exp(iH_0 t)$ if the limits $t \to \infty$ or as $t \to -\infty$ exist. For well behaved problems in potential scattering such limits exist. Under further restrictions (repulsive forces) the limit is unitary. Assuming further that $\frac{d}{dt}T(t) \to 0$, we have

$$\frac{d}{dt}T(t) = -i(HT(t) - T(t)H_0) \to HT - TH_0 = 0.$$

While such a unitary operator T does not exist for relativistic field theory, the formal perturbation expansion for T provides a useful tool.

Since

$$H_0 = \sum_{\varepsilon=0,\pm 1} b(k,\varepsilon)^* \mu(k,\varepsilon)\, b(k,\varepsilon)\, dk,$$

the relation (7.1.2) would yield

$$H(g,\varkappa,\varrho) = \sum_{\varepsilon=0,\pm 1} \int \hat{b}(k,\varepsilon)^* \mu(k,\varepsilon)\, \hat{b}(k,\varepsilon)\, dk \tag{7.1.3}$$

where $\hat{b}(k,\varepsilon) = Tb(k,\varepsilon)T^*$ are the annihilation operators of the incoming or outgoing asymptotic particles.

Divergences occur in perturbation theory for (7.1.3) only in terms involving $\hat{b}$ to first order. Hence we only need to calculate T to first order in order to generate the renormalization counterterms. We estimate the error introduced in (7.1.3) by retaining only first order terms in T.

We write $T = I + X$, and compute X to first order. Since T is assumed unitary, $TT^* = I$, and to first order $X^* = -X$. Thus to first order, we have by (7.1.2), and the fact that $c(g,\varkappa,\varrho)$ is second order,

$$\begin{aligned} H_0 + H_I &= TH_0T^* = (I + X)H_0(1 - X) \\ &= H_0 + [X, H_0] \end{aligned}$$

or $T = I - \Gamma H_I$, where $[H_0, \Gamma H_I] = H_I$. Hence to first order

$$\hat{b}(k,\varepsilon) = b(k,\varepsilon) + [b(k,\varepsilon), \Gamma H_I(g,\varkappa,\varrho)].$$

We now give a proof based on these formal ideas.

Proof of Theorem 7.1.1 We first prove the theorem for ϱ sufficiently large. It is no loss of generality to take $\tau < 1$ close to one, since $N_\sigma \leqq$ const. $\times N_\tau$ for $\sigma \leqq \tau$. We use an approximate form of (7.1.2-3). Only the pair creation terms W^C yield divergent contributions to (7.1.2-3), so we only retain this term in the definition of $\hat{b}$. To simplify computations, we define ΓW^C so that $[H_0, \Gamma W^C] = W^C +$ error. In particular,

$$\begin{aligned} \Gamma W^C_{\varkappa,\varrho} = \int w^C_{\varkappa,\varrho}\{a(k)^*(\mu + \omega_1 + \omega_2)^{-1} + a(-k)(\omega_1 + \omega_2)^{-1}\} \\ \times b(p_1)^*\, b'(p_2)^*\, dk\, dp_1\, dp_2\,. \end{aligned}$$

Diagramatically, we write

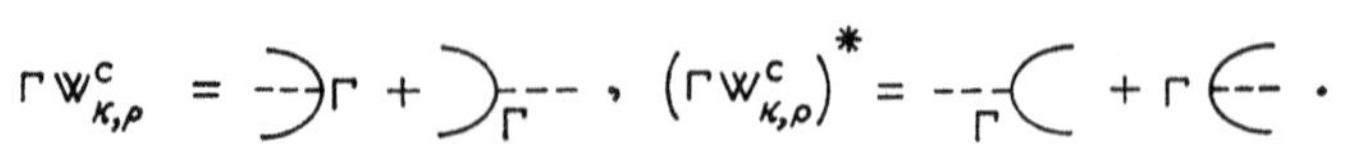

We have

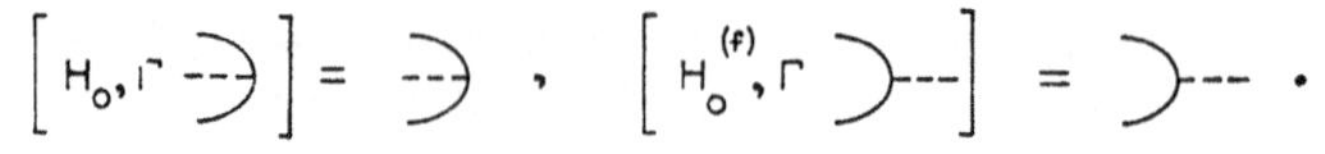

We define

$$\hat{b}(p, \varepsilon) = b(p, \varepsilon) + [b(p, \varepsilon), \Gamma W^C_{\varkappa,\varrho}]. \tag{7.1.4}$$

In particular

$$\hat{b}(p, +1) = b(p, 1) + \int w^C(k, p, p_2)\, b'(p_2)^* \times \left\{ a(k)^* \frac{1}{\mu + \omega + \omega_2} + a(-k) \frac{1}{\omega + \omega_2} \right\} dk\, dp_2,$$

$$\hat{b}(p, 0) = b(p, 0) + \int w^C(p, p_1, p_2)\, (\mu + \omega_1 + \omega_2)^{-1}\, b(p_1)^*\, b'(p_2)^*\, dp, dp_2,$$

$$\hat{b}(p, -1) = b(p, -1) - \int w^C(k, p_1, p)\, b(p_1)^* \times \left\{ a(k)^* \frac{1}{\mu + \omega + \omega_1} + a(-k) \frac{1}{\omega + \omega_1} \right\} dk\, dp_1.$$

Let $1 - \tau$ and c be sufficiently small so that

$$0 \leqq \bar{\mu}(k, \varepsilon) \equiv \mu(k, \varepsilon) - c\mu(k, \varepsilon)^\tau, \qquad \varepsilon = 0, \pm 1.$$

We define a positive approximation to $H(g, \varkappa, \varrho)$ by

$$\hat{H} = \sum_{\varepsilon = 0, \pm 1} \int \hat{b}(k, \varepsilon)^*\, \bar{\mu}(k, \varepsilon)\, \hat{b}(k, \varepsilon)\, dk. \tag{7.1.5}$$

Expanding (7.1.5) with (7.1.4) yields many terms. We write

$$0 \leqq \hat{H} = -cN_\tau + H_1 + H_2. \tag{7.1.6}$$

We choose H_1 to be the terms in (7.1.5) that resemble $H(g, \varkappa, \varrho)$, and we will prove for $\varepsilon > 0$ and $\varrho > \varrho_1$ sufficiently large,

$$\pm\{H_1 - H(g, \varkappa, \varrho)\} \leqq \varepsilon(N_\tau + I). \tag{7.1.7}$$

We also estimate the error term H_2 by

$$H_2 \leqq \varepsilon(N_\tau + I). \tag{7.1.8}$$

Adding (7.1.7-8) and using (7.1.6) we obtain the desired result,

$$(c - 2\varepsilon)\, N_\tau \leqq H_1 + H_2 - 2\varepsilon N_\tau \leqq H(g, \varkappa, \varrho) + I.$$

In terms of diagrams we define

$$H_1 = H_0 + W^C_{\varkappa,\varrho} + W^A_{\varkappa,\varrho}$$

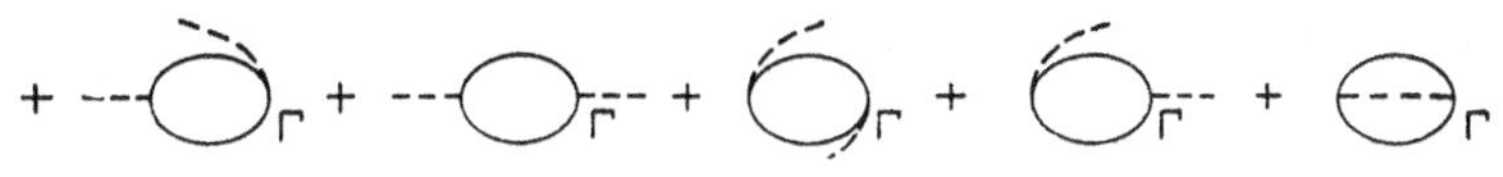

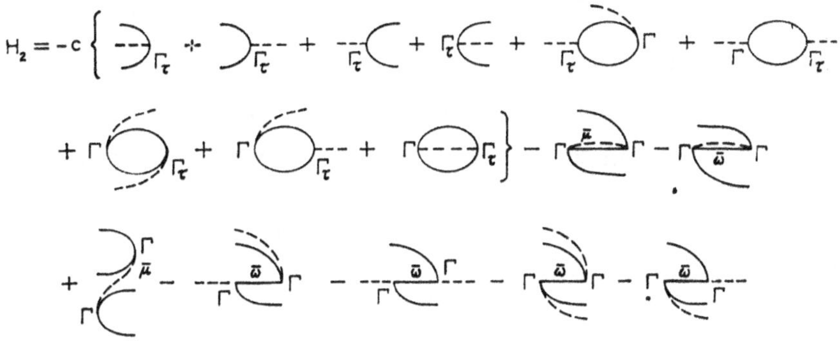

$$= Z_1 + Z_2 + \cdots + Z_{16}.$$

We now explain the notation in the above equations. The vertices that appear in H_1 and H_2 are the vertices in $W^C_{\varkappa,\varrho}$, $\Gamma W^C_{\varkappa,\varrho}$, $\Gamma_\tau W^C_{\varkappa,\varrho}$ and their adjoints-with the exception of Z_7 as explained below. The Γ_τ vertices are defined by the equations

In Z_{10} and Z_{12} the boson line has an extra factor $\bar{\mu}$. In $Z_{11}, Z_{13}, \ldots, Z_{16}$ the contracted fermion (or antifermion) line has an extra factor $\bar{\omega} = \bar{\mu}(\cdot, 1)$. The derivation of these expansions for H_1 and H_2, where $\hat{H} = -cN_\tau + H_1 + H_2$, is elementary. We insert the definition (7.1.4) into (7.1.5) and use relations such as

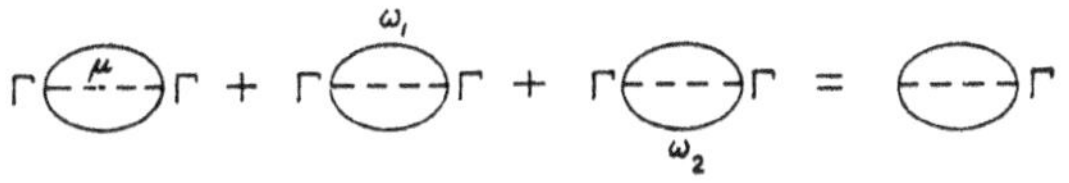

and

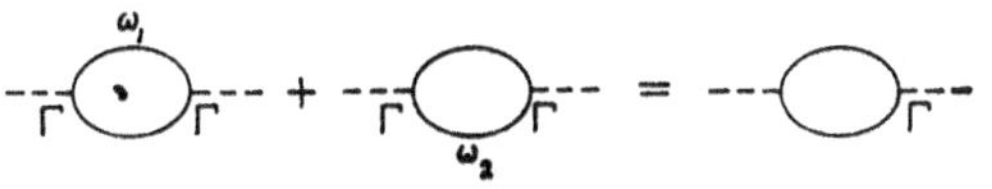

Finally, the kernel of Z_7 is

$$\int \frac{w^C_{\varkappa,\varrho}(k, p_1, p_2)\, w^C_{\varkappa,\varrho}(k', -p_1, -p_2)}{(\mu + \omega_1 + \omega_2)(\mu' + \omega_1 + \omega_2)} \{\omega_1^\tau + \omega_2^\tau + c^{-1}\mu\}\, dp_1\, dp_2.$$

Lemma 7.1.2 *The inequalities (7.1.7-8) are valid.*

Proof To prove (7.1.7) we estimate the difference

$$H_1 - H(g, \varkappa, \varrho) = -W_{\varkappa,\varrho} + \tfrac{1}{2}\,\delta m^2(\varkappa, \varrho) \int :\phi(x)^2: g(x)^2 + E_2(g, \varkappa, \varrho)$$

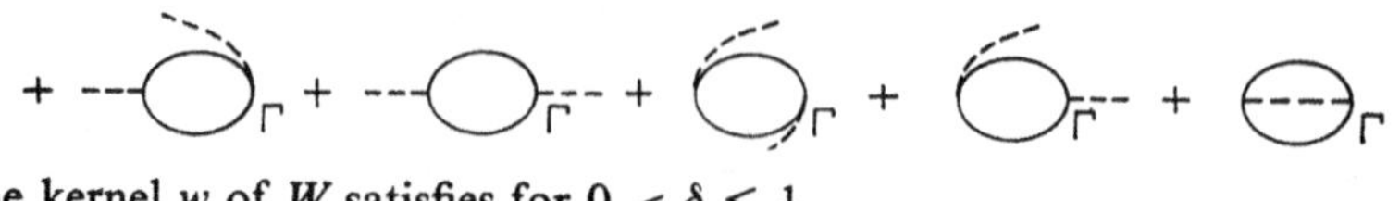

The kernel w of W satisfies for $0 < \delta \leqq \frac{1}{2}$,

$$|w| \leqq \text{const.}\ \mu^{-1/2}\,|\tilde{g}(k + \eta)|\,\omega(\eta)^{2\delta}\,\omega(\xi)^{-\delta}. \tag{7.1.9}$$

For τ sufficiently close to 1, we conclude that $\|(\mu\omega_2)^{-\tau/2}\, w\|_2 < \infty$, and for $0 < \varepsilon$ sufficiently small,

$$\|(\mu\omega_2)^{-\tau/2}\, w_{\varkappa,\varrho}\|_2 \leqq 0(\varrho^{-\varepsilon}).$$

Thus by Proposition 1.2.2,

$$\pm W_{\varkappa,\varrho} \leqq 0(\varrho^{-\varepsilon})\,(N_\tau + I) \tag{7.1.10}$$

where $0(\varrho^{-\varepsilon})$ is independent of $\varkappa$.

We now bound the counterterms in (7.17.). We note that the vacuum energy $E_2(g, \varkappa, \varrho)$ cancels exactly. The four mass counterterms in $c(g, \varkappa, \varrho)$ have the form

$$\int \beta(k, k') : a(k)^{\#}\, a(k')^{\#} : dk\, dk'$$

where $\beta(k, k') = -\frac{1}{2}\,\delta m^2(\varkappa, \varrho)\,(4\pi)^{-1}\,(\mu\mu')^{-1/2}\,(\tilde{g} * \tilde{g})\,(k + k')$. The corresponding term in H_1 only differs by the kernel $\gamma_i(k, k')$ replacing $\beta(k, k')$, $1 \leqq i \leqq 4$. Here

$$\gamma_i(k, k') = \int w^C_{\varkappa,\varrho}(k, p_1, p_2)\, w^C_{\varkappa,\varrho}(k', -p_2, -p_1)\,\delta_i\, dp_1\, dp_2,$$

where

$$\delta_1 = (\omega_1 + \omega_2 + \mu')^{-1}, \qquad \delta_2 = (\omega_1 + \omega_2)^{-1},$$

$$\delta_3 = (\omega_1 + \omega_2 + \mu)^{-1}, \qquad \delta_4 = (\omega_1 + \omega_2 + \mu')^{-1}.$$

We show that $\|\beta - \gamma_i\|_2 \leqq 0(\varrho^{-\varepsilon})$, independent of $\varkappa$. Hence by Proposition 1.2.2,

$$\pm \int (\gamma_i - \beta)\,(k, k') : a(k)^{\#}\, a(k')^{\#} : dk\, dk' \leqq 0(\varrho^{-\varepsilon})\,(N_\tau + I). \tag{7.1.11}$$

The bounds (7.1.10-11) prove (7.1.7).

We now estimate $\|\beta - \gamma_1\|_2$. We write with the notation of (6.3.3),

$$\beta(k, k') = \frac{\lambda^2}{4\pi}\,(\mu\mu')^{-1/2} \int \beta_1\beta_2\beta_3\, \tilde{g}(k + \eta)\, \tilde{g}(k' - \eta)\,\{1 - \theta_\varrho(\xi)\}\, d\xi\, d\eta,$$

and

$$\gamma_1(k, k') = \frac{\lambda^2}{4\pi}\,(\mu\mu')^{-1/2} \int \beta_1\beta_2\bar{\beta}_3\, \tilde{g}(k + \eta)\, \tilde{g}(k' - \eta)\,\{1 - \theta_\varrho(\xi)\}\, d\xi\, d\eta,$$

where $\bar{\beta}_1 = \alpha_1, \bar{\beta}_2 = \delta_1, \bar{\beta}_3 = \chi_\varkappa(k, p_1, p_2)\,\chi_\varkappa(k', -p_1, -p_2)$. We use the bounds (6.3.4) and

$$|\beta_2 - \bar{\beta}_2| \leq \text{const.}\,(\omega_1 + \omega_2 + \mu')^{-1}\,\omega(\xi)^{-1}\,(\omega(\eta) + \mu')$$
$$\leq \text{const.}\,\omega(\xi)^{-1-2\varepsilon}\,(\omega(\eta) + \mu')^{2\varepsilon},$$
$$\int |\beta_3 - \bar{\beta}_3|\,\omega(\xi)^{-1}\,\{1 - \theta_\varrho(\xi)\}\,d\xi \leq 0(\varrho^{-\varepsilon})\,(\omega(\eta) + \mu + \mu')^{3\varepsilon},$$

the latter following by condition (c_2) of Lemma 6.1.1. Using these estimates we obtain from the identity

$$\beta_1\beta_2\beta_3 - \bar{\beta}_1\bar{\beta}_2\bar{\beta}_3 = (\beta_1 - \bar{\beta}_1)\,\bar{\beta}_2\bar{\beta}_3 + \beta_1(\beta_2 - \bar{\beta}_2)\,\bar{\beta}_3 + \beta_1\beta_2(\beta_3 - \bar{\beta}_3)$$

the estimate

$$|(\beta - \gamma_1)\,(k, k')| \leq 0(\varrho^{-\varepsilon})\,(\mu\mu')^{-1/2+3\varepsilon}\,h(k + k'),$$

where $h(\cdot)$ is rapidly decreasing. Taking L_2 norms gives $\|\beta - \gamma_1\|_2 \leq 0(\varrho^{-\varepsilon})$. The proof that $\|\beta - \gamma_i\|_2 \leq 0(\varrho^{-\varepsilon})$ for $i = 2, 3, 4$ is similar, and (7.1.7) is proved.

We next prove (7.1.8). The terms $Z_1, \ldots, Z_6, Z_8, Z_9$ are the same as $-c(H_1 - H_0)$, except that each diagram also has one Γ_τ vertex. For $0 < \delta = 1 - \tau$, we estimate $Z_1, \ldots, Z_4$ by

$$\left\|(\mu + \omega_1 + \omega_2)^{-\tau/2}\left(\frac{\mu^\tau + \omega_1^\tau + \omega_2^\tau}{\mu + \omega_1 + \omega_2}\right)w^C_{\varkappa,\varrho}\right\|_2$$
$$\leq \text{const.}\,\|(\mu + \omega_1 + \omega_2)^{-(1+\delta)/2}\,w^C_{\varkappa,\varrho}\|_2 \leq 0(\varrho^{-\delta/4}),$$

uniformly in $\varkappa$. Hence for ϱ sufficiently large,

$$\pm Z_i \leq \varepsilon(N_\tau + I), \qquad i = 1, 2, 3, 4 \tag{7.1.12}$$

by Corollary 1.2.4.

The extra Γ_τ factor yields L_2 kernels for $Z_5, \ldots, Z_9$. These bounds are uniform in $\varkappa$. Using the lower cutoff ϱ, we obtain norms that are $0(\varrho^{-\varepsilon})$. Hence (7.1.12) is also valid for $Z_5, \ldots, Z_9$. The estimates on Z_{10} and Z_{11} are similar, since the kernel of Z_{10} or Z_{11} is dominated by

$$|z_{10}(p_1, p_2)| + |z_{11}(p_1, p_2)| \leq$$
$$\leq 0(\varrho^{-\varepsilon})\int \mu^{-1}(\mu + \omega + \omega_1)^{-1+\varepsilon}\,|\tilde{g}(k + p_1 + p)\,\tilde{g}(-k - p + p_2)|\,dp\,dk$$
$$\leq 0(\varrho^{-\varepsilon})\,\omega_1^{-1+2\varepsilon}\,h(p_1 + p_2),$$

where $h(\cdot)$ is rapidly decreasing. Thus $\|z_{10}\|_2 + \|z_{11}\|_2 \leq 0(\varrho^{-\varepsilon})$.

We now estimate the singly contracted terms, $Z_{12}, \ldots, Z_{16}$. For vectors $\theta \in \mathscr{D}$,

$$\langle\theta, Z_{12}\theta\rangle \int = \|Z_{17}(k)\,\theta\|^2\,dk$$

where

$$Z_{17}(k) = \bar{\mu}(k)^{1/2} \int b'(-p_1)\, b(-p_2)\, w^C_{\varkappa,\varrho}(k, p_1, p_2)\, (\omega_1 + \omega_2)^{-1}\, dp_1\, dp_2 .$$

By Corollary 1.2.4,

$$\begin{aligned}\|Z_{17}(k)\,(N_\tau + I)^{-1/2}\| &\leq \text{const.}\, \mu^{1/2} \|\, (\omega_1 + \omega_2)^{-1-\tau/2}\, w^C_{\varkappa,\varrho}\|_2 \\ &\leq 0(\varrho^{-\varepsilon})\, \mu^{-(1+\tau-2\varepsilon)/2} \in L_2 .\end{aligned}$$

Hence the bound (7.1.12) is valid for Z_{12}.

Thus the terms Z_i, $i = 1, 2, \ldots, 12$ satisfy (7.1.12). The terms $Z_{13} + \cdots + Z_{16}$ are negative, aside from a small error satisfying (7.1.12). We now prove this. Let

$$\begin{aligned} Z_{18}(p) = \int b(-p_1)\, w^C_{\varkappa,\varrho}(k, -p, p_1) \bar{\omega}(p)^{1/2} \Big\{ a(k)^* \frac{1}{\omega(p) + \omega(p_1)} \\ + a(-k) \frac{1}{\mu(k) + \omega(p) + \omega(p_1)} \Big\}\, dp_1\, dk \end{aligned}$$

and let $Z'_{18}(p)$ have $b'(p_1)$ in place of $b(p_1)$. Then

$$\begin{aligned} Z_{13} + \cdots + Z_{16} = &- \int Z_{18}(p)^*\, Z_{18}(p)\, dp \\ &+ \int b(p_1)^*\, b(-p_2)\, K(p_1, p_2)\, dp_1\, dp_2 \\ &- \int Z'_{18}(p)^*\, Z'_{18}(p)\, dp \\ &+ \int b'(p_1)^*\, b'(-p_2)\, K(p_1, p_2)\, dp_1\, dp_2 . \end{aligned}$$

Here

$$\begin{aligned} &|K(p_1, p_2)| \\ &= \left| \int dk\, dp\, w^C_{\varkappa,\varrho}(-k, p_1, -p)\, w^C_{\varkappa,\varrho}(k, p_2, p) \bar{\omega}(p) (\omega + \omega_1 + \mu)^{-1} (\omega + \omega_2 + \mu)^{-1} \right| \\ &\leq 0(\varrho^{-\delta})\, (\omega_1 \omega_2)^{-1/2 + 3\delta}\, h(p_1 + p_2), \end{aligned}$$

where $h(\cdot)$ is rapidly decreasing. Since $K(p_1, p_2) \in L_2$, it follows that for small δ,

$$Z_{13} + \cdots + Z_{16} \leq 0(\varrho^{-\delta})\, (N_\tau + I).$$

and the proof of (7.1.8) is complete. This proves the lemma and the theorem for ϱ sufficiently large.

To complete the proof of the theorem, we estimate $H(g, \varkappa, \varrho)$ for $\varrho \leq \varrho_1$, with the theorem valid for ϱ_1. Hence

$$\begin{aligned} H(g, \varkappa, \varrho) &= H(g, \varkappa, \varrho_1) + \{H(g, \varkappa, \varrho) - H(g, \varkappa, \varrho_1)\} \\ &\geq \text{const.}\, N_\tau - I + \{H(g, \varkappa, \varrho) - H(g, \varkappa, \varrho_1)\} . \end{aligned}$$

The next lemma completes the proof of the theorem.

LEMMA 7.1.3 *Let $\varepsilon > 0$. Then*

$$0 \leqq \varepsilon N_\tau + \{H(g, \varkappa, \varrho) - H(g, \varkappa, \varrho_1)\} + \text{const.},$$

with a constant uniform in $\varkappa$ and $\varrho \leqq \varrho_1$.

Proof Since

$$0 \leqq -\tfrac{1}{2}\{\delta m^2(\varkappa, \varrho) - \delta m^2(\varkappa, \varrho_1)\} = M,$$

we have by Theorem 3.1.1

$$0 \leqq \varepsilon N + M \int :\phi(x)^2: g(x)^2\, dx + \text{const.}$$

The lemma now follows by Lemma 6.3.5.

7.2. The second order estimate

THEOREM 7.2.1 *Glimm and Jaffe[8], Dimock[1]. Let $\tau < 1$ and let g be fixed. There are constants a and b independent of $\varkappa$ and ϱ such that*

$$(N_{\tau/2})^2 \leqq a(H(g, \varkappa, \varrho) + bI)^2. \tag{7.2.1}$$

We introduce the resolvent $R(\zeta) = R_{\varkappa,\varrho}(\zeta) = (H(g, \varkappa, \varrho) - \zeta)^{-1}$ and we prove that $\|N_{\tau/2} R(\zeta)\| \leqq$ const., for ζ sufficiently negative. Our two basic tools are the first order estimate of Theorem 7.1.1 yielding $\|N_\tau^{1/2} R(\zeta)^{1/2}\| \leqq$ const., and the pull-through formulas,

$$b(k, \varepsilon)\, R(\zeta) = R(\zeta - \mu(k, \varepsilon))\, b(k, \varepsilon) - R(\zeta - \mu(k, \varepsilon))\, X(k, \varepsilon)\, R(\zeta), \tag{7.2.2}$$

where

$$X(k, \varepsilon) = [b(k, \varepsilon), H_I(g, \varkappa, \varrho) + c(g, \varkappa, \varrho)]. \tag{7.2.3}$$

A complete treatment of the domains of these equations is given in Glimm and Jaffe[8].

LEMMA 7.2.2 *Let $\delta > 0$. Then*

$$\|(N_\tau + I)^{-1/2}\, X(p, \pm 1)\, (N_\tau + I)^{-1/2}\| \leqq 0(\omega(p)^{-\tau/2+\delta}).$$

Proof We apply Corollary 1.2.4 to the various terms in $X(p, \pm 1)$. For instance, the pure creation term has a kernel $w^C_{\varkappa,\varrho}(k, p_1, p_2)$ satisfying

$$\begin{aligned}\|\omega_2^{-\tau/2}\, w^C_{\varkappa,\varrho}(\cdot, p, \cdot)\|_2^2 &\leqq \text{const.} \int dp_2\, \omega(p + p_2)^{-1-\delta/2}\, \omega(p_2)^{-\tau+\delta/2} \\ &\leqq \text{const.}\, \omega(p)^{-\tau+\delta},\end{aligned}$$

and this yields a bound of the desired form. The analogous bound does not hold for $\varepsilon = 0$, and the $\varepsilon = 0$ term requires a renormalization cancellation.

We now see how Lemma 7.2.2 is useful in the proof in the theorem. With $N_\tau = \sum_\varepsilon N_\tau^{(\varepsilon)}$, it is sufficient to bound each $(N_\tau^{(\varepsilon)})^2$. Since

$$(N_{\tau/2}^{(\varepsilon)})^2 = N_\tau^{(\varepsilon)} + \int b(k, \varepsilon)^*\, N_{\tau/2}^{(\varepsilon)}\, b(k, \varepsilon)\, \mu(k, \varepsilon)^{\tau/2}\, dk,$$

and since $R(\zeta)\, N_\tau R(\zeta)$ is uniformly bounded by Theorem 7.1.1, it is sufficient to prove that for $\theta \in \mathscr{D}$, and for $\varepsilon = 0, \pm 1$,

$$M_\varepsilon = \int \|(N_{\tau/2}^{(\varepsilon)})^{1/2}\, b(k,\varepsilon)\, R(\zeta)\,\theta\|^2\, \mu(k,\varepsilon)^{\tau/2}\, dk \leqq \text{const. } \|\theta\|^2.$$

Using (7.2.2) we obtain

$$M_\varepsilon \leqq 2\int \|(N_{\tau/2}^{(\varepsilon)})^{1/2}\, R(\zeta - \mu(k,\varepsilon))\, b(k,\varepsilon)\,\theta\|^2\, \mu(k,\varepsilon)^{\tau/2}\, dk$$
$$+ 2\int \|(N_{\tau/2}^{(\varepsilon)})^{1/2}\, R(\zeta - \mu(k,\varepsilon))\, X(k,\varepsilon)\, R(\zeta)\,\theta\|^2\, \mu(k,\varepsilon)^{\tau/2}\, dk.$$

Since

$$\|(N_{\tau/2}^{(\varepsilon)})^{1/2}\, R(\zeta - \mu(k,\varepsilon))\,(N_{\tau/2} + \mu(k,\varepsilon)^{\tau/2})^{1/2}\| \leqq \text{const.},$$

and

$$\int \|(N_{\tau/2} + \mu(k,\varepsilon)^{\tau/2})^{-1/2}\, b(k,\varepsilon)\,\theta\|^2\, \mu^{\tau/2}\, dk$$
$$= \int \|b(k,\varepsilon)\, N_{\tau/2}^{-1/2}\,\theta\|^2\, \mu^{\tau/2}\, dk \leqq \|\theta\|^2,$$

we have

$$M_\varepsilon \leqq \text{const. } \|\theta\|^2 + \int \|(N_{\tau/2}^{(\varepsilon)})^{1/2}\, R(\zeta - \mu(k,\varepsilon))\, X(k,\varepsilon)\, R(\zeta)\,\theta\|^2\, \mu(k,\varepsilon)^{\tau/2}\, dk. \tag{7.2.4}$$

For $\varepsilon = \pm 1$, we use (7.2.2), Lemma 7.2.2 and the relation $\{b(k,\varepsilon), X(k',\varepsilon)\} = 0$ to obtain the bound

$$\int \|(N_{\tau/2}^{(\varepsilon)})^{1/2}\, R(\zeta - \mu(k,\varepsilon))\, X(k,\varepsilon)\, R(\zeta)\,\theta\|^2\, \mu(k,\varepsilon)^{\tau/2}\, dk$$
$$= \int \|b(k',\varepsilon)\, R(\zeta - \mu(k,\varepsilon))\, X(k,\varepsilon)\, R(\zeta)\,\theta\|^2\, \mu(k',\varepsilon)^{\tau/2}\, \mu(k,\varepsilon)^{\tau/2}\, dk\, dk'$$
$$\leqq 2\int \|R(\zeta - \mu - \mu')\, X(k,\varepsilon)\, R(\zeta - \mu')\, b(k',\varepsilon)\,\theta\|^2\, (\mu\mu')^{\tau/2}\, dk\, dk'$$
$$+ 4\int \|R(\zeta - \mu - \mu')\, X(k,\varepsilon)\, R(\zeta - \mu')\, X(k',\varepsilon)\, R(\zeta)\,\theta\|^2\, (\mu\mu')^{\tau/2}\, dk dk'$$
$$\leqq \text{const. } \|\theta\|^2.$$

Hence $M_{\pm 1} \leqq \text{const. } \|\theta\|^2$.

The proof of the theorem is now reduced to the case $\varepsilon = 0$. We remark that it is easy to prove a divergent estimate

$$\|(N_\tau + I)^{-1/2}\, X(k,0)\,(N_\tau + I)^{-1/2}\| \leqq 0(\varkappa^\delta \mu^{-1/2}), \quad \delta > 0, \tag{7.2.5}$$

yielding $M_0 \leqq 0(\varkappa^{2\delta})\,\|\theta\|^2$, and therefore by the above bounds

$$(N_{\tau/2})^2 \leqq 0(\varkappa^{2\delta})\,(H(g,\varkappa,\varrho) + bI)^2. \tag{7.2.6}$$

To prove (7.2.5), we note that

$$X(k,0) = [a(k), H_I(g,\varkappa,\varrho)] - \delta m^2(\varkappa,\varrho)\,(4\pi\mu)^{-1/2} \int e^{-ikx}\, \phi(x)\, g(x)^2\, dx. \tag{7.2.7}$$

By conditions (a), (b) of Lemma 6.1.1, for $1 - \tau < \delta$,

$$\|(\omega_1 + \omega_2)^{-\tau/2} \left(|w^C_{\varkappa,\varrho}(k, \cdot, \cdot)| + |w_{\varkappa,\varrho}(k, \cdot, \cdot)|\right)\|_2 \leqq 0(\varkappa^\delta \mu^{-1/2}).$$

Thus by Corollary 1.2.4,

$$\|(N_\tau + I)^{-1/2} [a(k), H_I(g, \varkappa, \varrho)] (N_\tau + I)^{-1/2}\| \leqq 0(\varkappa^\delta \mu^{-1/2}).$$

Since $\delta m^2(\varkappa, \varrho) \leqq 0(\log \varkappa)$, the bounds (7.2.5-6) hold. A proof of resolvent convergence and self adjointness can be based on this estimate. The proof of (7.2.1) follows by isolating and cancelling the divergence in (7.2.7).

LEMMA 7.2.3 *Dimock*[1]. *For* $\tau < 1$, $0 < \delta$,

$$\|(N^{(0)}_{\tau/2})^{1/2} R(\zeta - \mu) X(k, 0) R(\zeta)\| \leqq 0(\mu^{-1+\delta}). \tag{7.2.8}$$

Inserting (7.2.8) into (7.2.4) yields the theorem. We do not give the lengthy proof of (7.2.8). In order to illustrate the ideas involved, we estimate one particular renormalization cancellation. We write

$$X(k, 0) = X_1(k) + X_2(k) + X_1(-k)^*$$

where

$$X_1(k) = \int w^C_{\varkappa,\varrho}(k, p_1, p_2)\, b'(-p_1)\, b(-p_2)\, dp_1\, dp_2$$
$$- \frac{\delta m^2(\varkappa, \varrho)}{8\pi} \int dk' (\mu\mu')^{-1/2} (\tilde{g} * \tilde{g})(k + k') \{a(k')^* + a(-k')\},$$

$$X_2(k) = \int w_{\varkappa,\varrho}(k, p_1, p_2) \sum_{\varepsilon = \pm 1} b(p_1, \varepsilon)^*\, b(-p_2, \varepsilon)\, dp_1\, dp_2 .$$

The renormalization cancellations occur in X_1 and X_1^*, while X_2 can be estimated by Proposition 1.2.2,

$$\|(N_\tau + I)^{-1/2} X_2(k) (N_\tau + I)^{-1/2}\| \leqq 0(\mu^{-(1+\tau)/2}),$$

which is sufficient for (7.2.8). We now prove the lemma for X_1 replacing X. We simplify $X_1 R$ by using the pull through formula

$$\begin{aligned} b'(-p_1)\, b(-p_2)\, R &= b'(-p_1)\, R(\zeta - \omega_2)\, b(-p_2) \\ &\quad + R(\zeta - \omega_1 - \omega_2)\, X(-p_2, +1)\, R(\zeta - \omega_1)\, b'(-p_1) \\ &\quad - R(\zeta - \omega_1 - \omega_2)\, X(-p_2, +1)\, R(\zeta - \omega_1) \times X(-p_1, -1)\, R(\zeta) \\ &\quad + R(\zeta - \omega_1 - \omega_2)\, X(-p_1, -1)\, R(\zeta - \omega_2) \times X(-p_2, +1)\, R(\zeta) \\ &\quad - R(\zeta - \omega_1 - \omega_2)\, V(-p_1, -p_2)\, R(\zeta), \\ &= B_1(-p_1, -p_2) + \cdots + B_5(-p_1, -p_2), \end{aligned} \tag{7.2.9}$$

where

$$V(p_1, p_2) = \int w^C_{\varkappa,\varrho}(k', p_2, p_1) \{a(k')^* + a(-k')\}\, dk'$$
$$= \{b'(p_1), [b(p_2), H_I(g, \varkappa, \varrho) + c(g, \varkappa, \varrho)]\}.$$

Each B_i, except, B_5, gives a bounded contribution to $R^{1/2}X_1(k)\,R$, namely for $i = 1, 2, 3, 4$,

$$\left\| R(\zeta)^{1/2} \int B_i(-p_1, -p_2)\, w^C_{\varkappa,\varrho}(k, p_1, p_2)\, dp_1\, dp_2 \right\| = 0(\mu^{-1+\delta}).$$

This bound is established by the use of N_τ estimates and Lemma 7.2.2. The bound for $i = 5$ involves a renormalization cancellation. We combine B_5 with the counterterm in X_1. Thus we estimate

$$R^{1/2} \int B_5(p_1, p_2)\, w^C_{\varkappa,\varrho}(k, -p_1, -p_2)\, dp_1\, dp_2\, R - R^{1/2}\left(\frac{1}{8\pi}\right)\delta m^2(\varkappa, \varrho)$$
$$\times \int dk'(\mu\mu')^{-1/2}\, (\tilde{g} * \tilde{g})\, (k + k')\, \{a(k')^* + a(-k')\}\, R$$
$$= R^{1/2} \int \alpha(k, k')\, \{a(k')^* + a(-k')\}\, dk' R,$$

where

$$\int \alpha(k, k')\, \{a(k')^* + a(-k')\}\, dk' = -\int w^C_{\varkappa,\varrho}(k, -p_1, -p_2)\, R(\zeta - \omega_1 - \omega_2)$$
$$\times\ V(p_1, p_2)\, dp_1\, dp_2 - \frac{1}{8\pi}\delta m^2(\varkappa, \varrho) \int dk'(\mu\mu')^{-1/2}\, (\tilde{g} * \tilde{g})\, (k + k')$$
$$\times\ \{a(k')^* + a(-k')\}.$$

By Proposition 1.2.3(a), $\|V(p_1, p_2)\, R^{1/2}\| \leq 0(\omega(p_1 + p_2)^{-1/2})$. Hence we can replace $R(\zeta - \omega_1 - \omega_2)$ by $(\omega_1 + \omega_2)^{-1}$ yielding α' for α above. The error is estimated by

$$\int \left\| R^{1/2} \int w^C_{\varkappa,\varrho}(k, -p_1, -p_2)\, R(\zeta - \omega_1 - \omega_2)\, (\omega_1 + \omega_2)^{-1}\, (H(g, \varkappa, \varrho) - \zeta) \right.$$
$$\left. \times\ V(p_1, p_2)\, R \right\| dp_1\, dp_2$$
$$\leq 0(\mu^{-1/2}) \int |\tilde{g}(k - p_1 - p_2)|\, (\omega_1 + \omega_2)^{-3/2}\, \omega(p_1 + p_2)^{-1/2}\, dp_1\, dp_2$$
$$\leq 0(\mu^{-3/2+\delta}).$$

Thus we need to estimate the renormalization cancellation,

$$R^{1/2} \int \alpha'(k, k')\, \{a(k')^* + a(-k')\}\, dk'\, R,$$

where

$$\alpha'(k, k') = -w^C_{\varkappa,\varrho}(k, p_1, p_2)\, w^C_{\varkappa,\varrho}(k', -p_2, -p_1)\,(\omega_1 + \omega_2)^{-1}\, dp_1\, dp_2$$

$$-\frac{1}{8\pi}\,\delta m^2(\varkappa, \varrho)\,(\mu\mu')^{-1/2}\,(\tilde{g} * \tilde{g})\,(k + k').$$

In Sec. 7.1, in the proof of (7.1.11), we established the inequality $|\alpha'(k, k')| \leqq \text{const.}\,(\mu\mu')^{-1/2+\delta/2}\, h(k + k')$, where h is rapidly decreasing. Hence

$$\left\| R^{1/2} \int \alpha'(k, k')\, \{a(k')^* + a(-k')\}\, dk'\, R \right\| \leqq O(\mu^{-1+\delta}). \tag{7.2.10}$$

Combining (7.2.10) with the estimates on $B_1, \ldots, B_4$ proves

$$\|R^{1/2}\, X_1(k)\, R\| \leqq O(\mu^{-1+\delta}),$$

proving (7.2.8) for X_1 in the place of X. The X_1^* term is more difficult, but the methods are the same.

Using these methods, it is also possible to prove that for $\zeta, \zeta_1 < 0$ and $\delta > 0$,

$$\|R(\zeta + \zeta_1)\, X(k, 0)\, R(\zeta)\| \leqq O((|\zeta_1|\,\mu)^{-1/2+\delta}). \tag{7.2.11}$$

8 Resolvent Convergence and Self Adjointness

8.1 Resolvent convergence for the Hamiltonians

We prove norm convergence of the resolvents $R_{\varkappa,\varrho}(\zeta) = (H(g, \varkappa, \varrho) - \zeta)^{-1}$ of the cutoff Hamiltonians $H(g, \varkappa, \varrho)$. Let ζ be real and sufficiently negative.

THEOREM 8.1.1 *There exist bounded self adjoint operators $R_\varrho(\zeta)$ independent of $\chi_\varkappa$ such that*

$$\|R_{\varkappa,\varrho}(\zeta) - R_\varrho(\zeta)\| \to 0 \quad \text{as} \quad \varkappa \to \infty \tag{8.1.1}$$

uniformly in ϱ and

$$\|R_\varrho(\zeta) - (H_0 - \zeta)^{-1}\| \to 0 \quad \text{as} \quad \varrho \to \infty. \tag{8.1.2}$$

Remarks The proof of resolvent convergence follows Glimm-Jaffe[8]. We inspect the difference $R_{\varkappa_1,\varrho} - R_{\varkappa_2,\varrho} = -R_{\varkappa_1,\varrho}\, \delta H\, R_{\varkappa_2,\varrho}$, where $\delta H = H(g, \varkappa_1, \varrho) - H(g, \varkappa_2, \varrho)$. The first step is to perform explicit cancellations between the divergent counterterms δc and certain parts of the interaction Hamiltonian δH_I. We then estimate the remainder, and show that it converges to zero in norm (uniformly in ϱ). The main tool in the renormalization cancellations is the pull through formula. We use this formula to generate a perturbation series with a finite number of terms. The leading terms exhibit the renormalization cancellations, while the remaining terms can be estimated directly by combining the first and second order energy estimates of Chapter 7 with the N_τ estimates of Chapter 1. We thereby show that $R_{\varkappa,\varrho}$ is a Cauchy sequence (in norm) as $\varkappa \to \infty$, and therefore has a limit R_ϱ.

In order to prove the cutoff independence of the limit R_ϱ, we apply the same methods to show that $\|R_{1,\varkappa,\varrho} - R_{2,\varkappa,\varrho}\| \to 0$, where the subscripts 1 and 2 refer to distinct cutoff functions $\chi_{1,\varkappa}$ and $\chi_{2,\varkappa}$ satisfying the conditions of Lemma 6.3.1. This proves that the limit R_ϱ is independent of the family $\{\chi_\varkappa\}$ of cutoff functions used to define the approximate Hamiltonians $H(g, \varkappa)$.

To establish (8.1.2), we use the fact that the convergence (8.1.1) is uniform in ϱ, and write

$$\|R_\varrho(\zeta) - (H_0 - \zeta)^{-1}\| \leqq \|R_\varrho(\zeta) - R_{\varkappa,\varrho}(\zeta)\| + \|R_{\varkappa,\varrho}(\zeta) - (H_0 - \zeta)^{-1}\|.$$

For $\varkappa$ large, the first term is small uniformly in ϱ. With $\varkappa$ fixed, the second term is seen to converge from Lemma 6.3.5:

$$\|R_{\varkappa,\varrho}(\zeta) - (H_0 - \zeta)^{-1}\| \leqq \| R_{\varkappa,\varrho}(\zeta)\,(H_I(g, \varkappa, \varrho) + c(\varkappa, \varrho))\,(H_0 - \zeta)^{-1}\|$$

$$\leqq \text{const.}\ \|(N + I)^{-1}\,(H_I(g, \varkappa, \varrho) + c(\varkappa, \varrho))\,(N + I)^{-1}\|$$

$$\leqq 0(\varrho^{-\delta}).$$

In order to study the convergence of $R_{\varkappa_1,\varrho}\,\delta H R_{\varkappa_2,\varrho}$, we write

$$\delta H = \delta H_I + \delta c = \delta W^C + \delta W + \delta W^A + \delta c,$$

following the notation of Sec. 6.1. In perturbation theory we find that $H_0 + W^C$, W and $W^A + c$ are individually defined as bilinear forms on the domain of $H(g, \varkappa, \varrho)$. Thus we expect that $R_{\varkappa_1}\,\delta W^C R_{\varkappa_2}$ and $R_{\varkappa_1}\,\delta W R_{\varkappa_2}$ are well behaved as $\varkappa_1, \varkappa_2 \to \infty$ so long as $\varkappa_1 \leqq \varkappa_2$. On the other hand, both $R_{\varkappa_1}\,\delta W^A R_{\varkappa_2}$ and $R_{\varkappa_1}\,\delta c\, R_{\varkappa_2}$ diverge as $\varkappa_1, \varkappa_2 \to \infty$, but their sum converges to zero. We must isolate the renormalization cancellations that occur in this sum, and estimate the remainder after cancellation. We note that writing $R_{\varkappa_2} - R_{\varkappa_1} = -R_{\varkappa_2}\,\delta H R_{\varkappa_1}$, $\varkappa_1 \leqq \varkappa_2$, would reverse the role of δW^C and δW^A in the proof.

We choose the first cutoff $\chi_{1,\varkappa}$ to be sharp in momentum space, and to have the form (6.3.2),

$$\chi'_\varkappa = \chi_{1,\varkappa} = \chi_\varkappa h_\varkappa,$$

where

$$h_\varkappa(k, p_1, p_2) = \theta\left(\frac{k}{\varkappa^{1-\alpha}}\right)\theta\left(\frac{p_1}{\varkappa^{1-\alpha}}\right)\theta\left(\frac{p_2}{\varkappa^{1-\alpha}}\right).$$

We define $H'(g, \varkappa, \varrho)$ to be the Hamiltonian corresponding to the cutoff $\chi'_\varkappa = \chi_\varkappa h_\varkappa$, and we let $R'_{\varkappa,\varrho}(\zeta)$ be the resolvent of $H'(g, \varkappa, \varrho)$. We now study the difference between $H'_1(g, \varkappa_1, \varrho)$ (given by a cutoff $\chi'_{\varkappa_1} = \chi_{1,\varkappa_1} h_{\varkappa_1}$) and $H_2(g, \varkappa_2, \varrho)$ (given by some other cutoff $\chi_{2,\varkappa_2}$). Let $\delta H = H'_1(g, \varkappa_1, \varrho)$

$- H_2(g, \varkappa_2, \varrho)$. To simplify the notation, we set $\varkappa_1 = \varkappa, \chi_1 = \chi$ and $R_{2,\varkappa_2} = R_{\varkappa_2}$. We take ζ real and sufficiently negative. We state our three basic lemmas.

LEMMA 8.1.2 *Let $\varkappa \leqq \varkappa_2$. Then for some $\delta > 0$,*

$$\|R'_{\varkappa,\varrho}\, \delta W\, R_{\varkappa_2,\varrho}\| \leqq 0(\varkappa^{-\delta}\varrho^{-\delta}).$$

LEMMA 8.1.3 *Let $\varkappa \leqq \varkappa_2$. Then for some $\delta > 0$,*

$$\|R'_{\varkappa,\varrho}\, \delta W^C\, R_{\varkappa_2,\varrho}\| \leqq 0(\varkappa^{-\delta}\varrho^{-\delta}).$$

LEMMA 8.1.4 *Let $\varkappa \leqq \varkappa_2$. Then for some $\delta > 0$,*

$$\|R'_{\varkappa,\varrho}\{\delta W^A + \delta c\}\, R_{\varkappa_2,\varrho}\| \leqq 0(\varkappa^{-\delta}\varrho^{-\delta}).$$

Proof of Theorem 8.1.1 We use the three lemmas to prove the theorem. The first step is to prove convergence of $R'_{\varkappa,\varrho}$ as $\varkappa \to \infty$. Hence we pick $\chi_{2,\varkappa} = \chi'_\varkappa$. Thus $H_2(g, \varkappa, \varrho) = H'_1(g, \varkappa, \varrho)$. By Lemmas 8.1.2–8.1.4.

$$\begin{aligned}\|R'_{\varkappa,\varrho} - R'_{\varkappa_2,\varrho}\| &\leqq \|R'_{\varkappa,\varrho}\, \delta W^C\, R'_{\varkappa_2,\varrho}\| \\ &\quad + \|R'_{\varkappa,\varrho}\, \delta W\, R'_{\varkappa_2,\varrho}\| \\ &\quad + \|R'_{\varkappa,\varrho}\{\delta W^A + \delta c\}\, R'_{\varkappa_2,\varrho}\| \\ &\leqq 0(\varkappa^{-\delta}\varrho^{-\delta}).\end{aligned}$$

Thus for a particular cutoff $\chi'_\varkappa$,

$$R_\varrho(\zeta) = \operatorname*{norm\,limit}_{\varkappa\to\infty} R'_{\varkappa,\varrho}$$

exists, and the limit is uniform in ϱ.

Next we let $\chi_{2,\varkappa}$ be any permissable cutoff, as in Definition 6.3.2. Again, we have by the lemmas

$$\begin{aligned}\|R_\varrho(\zeta) - R_{\varkappa,\varrho}(\zeta)\| &\leqq \|R_\varrho - R'_{\varkappa,\varrho}\| + \|R'_{\varkappa,\varrho} - R_{\varkappa,\varrho}\| \\ &\leqq 0(\varkappa^{-\delta}\varrho^{-\delta}).\end{aligned}$$

Hence $R_{\varkappa,\varrho}(\zeta)$ converges to $R_\varrho(\zeta)$, and $R_\varrho(\zeta)$ is independent of the choice of cutoff function $\chi_\varkappa = \chi_{1,\varkappa}$. We therefore have proved (8.1.1). The proof of (8.1.2) was given above.

We now prove Lemmas 8.1.2–8.1.3 and outline the main ideas in the proof of Lemma 8.1.4.

Proof of Lemma 8.1.2 The kernel δw of δW is bounded by

$$\begin{aligned}|\delta w| &\leqq \text{const.}\, \mu^{-1/2}\, |\tilde{g}(k + \eta)|\, \omega(\eta)^{6\delta}\, \omega(\xi)^{-3\delta}|\, \chi'_\varkappa - \chi_{2,\varkappa_2}|\, \{1 - \theta_\varrho(\xi)\} \\ &\leqq 0(\varkappa^{-\delta(1-\alpha)}\varrho^{-\delta})\, (\mu + \omega_1 + \omega_2)^\delta\, \omega(\xi)^\delta\, \mu^{-1/2}\, |\tilde{g}(k + \eta)|\, \omega(\eta)^{6\delta}\, \omega(\xi)^{-3\delta}.\end{aligned}$$

Here we use condition (c_1) and (c_1') of Lemma 6.3.1 to estimate the change of cutoffs in the region $S_{\varkappa^{1-\alpha}}$, and the bound $(\mu + \omega_1 + \omega_2)^{-\delta} \leqq O(\varkappa^{-\delta(1-\alpha)})$ on the complement of $S_{\varkappa^{1-\alpha}}$. Hence for $\tau > 1$, sufficiently close to one, and for a new δ,

$$\|\mu^{-\tau/2}\, \omega_2^{-\tau/2}\, \delta w\|_2 \leqq O(\varkappa^{-\delta}\varrho^{-\delta}).$$

By Theorem 7.1.1 and Proposition 1.2.3(b),

$$\|R'_{\varkappa,\varrho}\, \delta W\, R_{\varkappa_2,\varrho}\| \leqq \text{const.}\ \|(N_\tau + I)^{-1/2}\, \delta W (N_\tau + I)^{-1/2}\| \leqq O(\varkappa^{-\delta}\varrho^{-\delta}).$$

Proof of Lemma 8.1.3 In this lemma we use the special properties of $\chi'_\varkappa$. We first modify $H'(g, \varkappa, \varrho)$ by using a cutoff mass counterterm

$$-\tfrac{1}{2}\, \delta m^2(\varkappa, \varrho) \int :\phi_\varkappa(x)^2 : g(x)^2\, dx,$$

where $\phi_\varkappa$ has a sharp momentum cutoff at $\varkappa^{1-\alpha}$,

$$\phi_\varkappa(x) = (4\pi)^{-\frac{1}{2}} \int e^{-ikx}\{a(k)^* + a(-k)\}\mu^{-\frac{1}{2}}\theta(k\varkappa^{-1+\alpha})dk.$$

We call the corresponding Hamiltonian $\hat{H}(g, \varkappa, \varrho)$ and let $\hat{R}$ be the resolvent of $\hat{H}$. This approximation has the advantage that $\hat{R}_{\varkappa,\varrho}(\zeta)$ commutes with the high momentum part of N_τ,

$$\hat{N}_{\tau,\varkappa} = \sum_{\varepsilon = \pm 1, 0} \int b(k, \varepsilon)^*\, \hat{\mu}(k, \varepsilon)\, b(k, \varepsilon)\, dk,$$

where

$$\hat{\mu}(k, \varepsilon) = \begin{cases} \mu(k, \varepsilon) & \text{for } \varkappa^{1-\alpha} \leqq |k| \\ 0 & \text{otherwise.} \end{cases}$$

Thus for $\tau < 1$,

$$\|\hat{N}_\tau^{1/2}\, \hat{R}_{\varkappa,\varrho} N_\tau^{1/2}\| = \|\hat{R}_{\varkappa,\varrho} \hat{N}_\tau^{1/2} N_\tau^{1/2}\| \leqq \text{const.},$$

and

$$\|\hat{R}_{\varkappa,\varrho}\, \delta W^C R_{\varkappa_2,\varrho}\| \leqq \text{const.}\ \|(N_\tau + I)^{-1/2}\, (\hat{N}_\tau + I)^{-1/2}\, \delta W^C (N_\tau + I)^{1/2}\|. \tag{8.1.3}$$

We estimate (8.1.3) by N_τ estimates. Such estimates for $\hat{N}_\tau$ involve an energy factor $(1 + \hat{\mu})^{-\tau/2}$ in place of $\mu^{-\tau/2}$. We write $\delta W^C = A_1 + A_2 + B$ with kernels α_1, α_2 and β which are nonzero in disjoint regions. We suppose α_1 has support in the region where $|p_1| \geqq \varkappa^{1-\alpha}$, α_2 has support in the region where $|p_2| \geqq \varkappa^{1-\alpha}$, and β has support in $|p_1| \leqq \varkappa^{1-\alpha}$, $|p_2| \leqq \varkappa^{1-\alpha}$. We note that for δ sufficiently small and τ sufficiently close to 1,

$$\begin{aligned} \|(\hat{\omega}_1\omega_2)^{-\tau/2}\, \alpha_1\|^2 + \|(\omega_1\hat{\omega}_2)^{-\tau/2}\, \alpha_2\|^2 &= \|(\omega_1\omega_2)^{-\tau/2}(\alpha_1 + \alpha_2)\|^2 \\ &\leqq O(\varkappa^{-\delta}\varrho^{-\delta}), \end{aligned}$$

since by condition (c_1) and (c_1') on the cutoffs $\chi_\varkappa, \chi'_\varkappa$,

$$\begin{aligned} |\alpha_1| + |\alpha_2| &\leqq \text{const}\, \mu^{-1/2}\, |\tilde{g}(k + \eta)|\, |\chi'_\varkappa - \chi_{2,\varkappa_2}|\, \{1 - \theta_\varrho(\xi)\} \\ &\leqq O(\varkappa^{-\delta(1-\alpha)}\varrho^{-\delta})\, (\mu + \omega_1 + \omega_2)^{2\delta}\, \mu^{-1/2}\, |\tilde{g}(k + \eta)|. \end{aligned} \tag{8.1.4}$$

Thus by Proposition 1.2.3(a), with one $\hat{\omega}$ variable, and a new δ,

$$\|(N_\tau + I)^{-1/2}\, (\hat{N}_\tau + I)^{-1/2}\, (A_1 + A_2)\, (N_\tau + I)^{-1/2}\| \leqq O(\varkappa^{-\delta}\varrho^{-\delta}). \tag{8.1.5}$$

We next bound the contribution B to δW^C. On the support of β, both $|p_1|$ and $|p_2|$ are less than $\varkappa^{1-\alpha}$. We write $B = B_1 + B_2$ with kernels β_1 and β_2, where $\beta_1 = \beta\theta\left(\frac{k}{\varkappa^{1-\alpha}}\right)$ and $\beta_2 = \beta\left(1 - \theta\left(\frac{k}{\varkappa^{1-\alpha}}\right)\right)$. Since $\hat{\mu}\beta_2 = \mu\beta_2$, we have

$$\begin{aligned}\|(N_\tau + I)^{-1/2}(\hat{N}_\tau + I)^{-1/2} B_2(N_\tau + I)^{-1/2}\| &\leqq \text{const.}\ \|\mu^{-\tau/2}(\omega_1 + \omega_2)^{-\tau/2}\beta_2\|_2 \\ &\leqq 0(\varkappa^{-\delta}\varrho^{-\delta})\end{aligned}$$

as in (8.1.4)–(8.1.5). Hence the $\hat{N}_\tau$ factor is useful for estimating the high momentum parts of δW^C, namely A_1, A_2 and B_2.

For the low momentum contribution B_1 we use the standard N_τ estimate

$$\|(N_\tau + I)^{-1/2} B_1(N_\tau + I)^{-1/2}\| \leqq \text{const.}\ \|(\mu + \omega_1 + \omega_2)^{-\tau/2}\beta_1\|_2. \tag{8.1.6}$$

On $S_{\varkappa^{1-\alpha}}$ we have

$$\begin{aligned}|\delta\chi| = |\chi'_\varkappa - \chi_{2,\varkappa_2}| = |\chi_\varkappa - \chi_{2,\varkappa_2}| &\leqq 0((\mu + \omega_1 + \omega_2)\varkappa^{-1}). \\ &\leqq 0(\varkappa^{-\alpha}).\end{aligned}$$

Hence by (8.1.4),

$$\begin{aligned}\|(\mu + \omega_1 + \omega_2)^{-\tau/2}\beta_1\|_2^2 &\leqq 0(\varkappa^{-2\alpha}) \int_{S_{\varkappa^{1-\alpha}}} (\mu + \omega_1 + \omega_2)^{-\tau} |\beta_1|^2\, dk\, dp_1\, dp_2 \\ &\leqq 0(\varrho^{-\delta}\varkappa^{-2\alpha})\,\varkappa^{2\delta+(1-\tau)} \\ &\leqq 0(\varkappa^{-\delta}\varrho^{-\delta}),\end{aligned}$$

for δ and $1 - \tau$ sufficiently small. Combining these estimates,

$$\|\hat{R}_{\varkappa,\varrho}\,\delta W^C R_{\varkappa_2,\varrho}\| \leqq 0(\varkappa^{-\delta}\varrho^{-\delta}).$$

To complete the proof of the lemma, we must estimate the error introduced by replacing R' with $\hat{R}$.

$$\begin{aligned}\|R'_{\varkappa,\varrho}\,\delta W^C R_{\varkappa_2,\varrho}\| &\leqq \|\hat{R}_{\varkappa,\varrho}\,\delta W^C R_{\varkappa_2,\varrho}\| + \|\hat{R}_{\varkappa,\varrho}\,\delta\hat{H} R'_{\varkappa,\varrho}\,\delta W^C R_{\varkappa_2\varrho}\|, \\ &\leqq 0(\varkappa^{-\delta}\varrho^{-\delta}) + \|\hat{R}_{\varkappa,\varrho}\,\delta\hat{H} R'_{\varkappa,\varrho}\,\delta W^C R_{\varkappa_2,\varrho}\|.\end{aligned} \tag{8.1.8}$$

Here $\delta\hat{H} = \hat{H} - H' = -\frac{1}{2}\,\delta m^2(\varkappa, \varrho) \int :\phi_\varkappa^2 - \phi^2: g(x)^2\, dx$.

We now use standard estimates to prove that (8.1.8) is small.

$$-\frac{1}{2}\delta m^2(\varkappa, \varrho) = \int_{|\xi|>\varrho} \left|\chi'_\varkappa\left(0, \frac{\xi}{2}, \frac{-\xi}{2}\right)\right|^2 \omega(\xi)^{-1}\, d\xi \leqq 0(\varkappa^{\delta}\varrho^{-\delta}),$$

$$\|(N_\tau + I)^{-1/2}\, W^C(N_\tau + I)^{-1/2}\| \leqq 0(\varkappa^{\delta}),$$

$$\left\|(N + I)^{-1/2} \int :\phi_\varkappa^2 - \phi^2: g(x)^2\, dx(N_\tau + I)^{-1/2}\right\| \leqq 0(\varkappa^{-(1-\alpha)/2}).$$

Hence for δ sufficiently small

$$\|\hat{R}_{\varkappa,\varrho}\,\delta H R'_{\varkappa,\varrho}\,W^C R_{\varkappa_2,\varrho}\| \leqq 0(\varkappa^{-(1-\alpha)/2+2\delta}\varrho^{-\delta})$$
$$\leqq 0(\varkappa^{-\delta}\varrho^{-\delta}),$$

and the lemma follows by (8.1.8).

We do not give the lengthy proof of Lemma 8.2.4, see Glimm-Jaffe [8, Sec. 4], but we illustrate the type of cancellations that occur, and how some of the convergent terms are estimated. There are two parts in W^A,

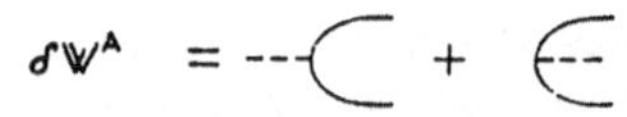

The first of these diagrams cancels two of the mass counterterms. The second diagram cancels the remaining two mass counterterms and the vacuum energy $-\delta E$. We use the expansion (7.2.9) for $\delta W^A R_{\varkappa_2,\varrho}$, and we consider only the term corresponding to B_5 in (7.2.9),

$$Z = -\int \{a(-k) + a(k)^*\}\, R_{\varkappa_2,\varrho}(\zeta - \omega_1 - \omega_2)\, V(-p_1, -p_2)\, R_{\varkappa_2,\varrho}(\zeta)$$
$$\times\ \delta w^C(k, p_1, p_2)\, dk\, dp_1\, dp_2$$

because all the cancellations occur between the counterterms and $R'Z$. We study the $a(-k)$ contribution to $R'Z$. It contains the vacuum energy cancellation and part of the mass renormalization cancellation. Let

$$C = -R'_{\varkappa,\varrho}\int a(-k)\, R_{\varkappa_2,\varrho}(\zeta - \omega_1 - \omega_2)\, a(k')^*\, R_{\varkappa_2,\varrho}(\zeta)$$
$$\times\ w^C_{\varkappa_2,\varrho}(k', -p_1, -p_2)\, \delta w^C(k, p_1, p_2)\, dk\, dk'\, dp_1\, dp_2.$$

Using the pull through formula, we have

$$C = -R'_{\varkappa,\varrho}\int R_{\varkappa_2,\varrho}(\zeta - \omega_1 - \omega_2 - \mu)\, |w^C_\varrho(k, p_1, p_2)|^2\, \delta\chi(k, p_1, p_2)$$
$$\times\ \chi_{\varkappa_2}(-k, -p_1, -p_2)\, dk\, dp_1\, dp_2\, R_{\varkappa_2,\varrho}(\zeta),$$
$$-R'_{\varkappa,\varrho}\int a(k')^*\, a(-k)\, R_{\varkappa_2,\varrho}(\zeta - \omega_1 - \omega_2 - \mu')\, R_{\varkappa_2,\varrho}(\zeta)$$
$$\times\ w^C_{\varkappa_2,\varrho}(k', -p_1, -p_2)\, \delta w^C(k, p_1, p_2)\, dk\, dk'\, dp_1\, dp_2,$$
$$+R'_{\varkappa,\varrho}\int a(-k)\, R_{\varkappa_2,\varrho}(\zeta - \omega_1 - \omega_2)\, X(k', 0)$$
$$\times\ R_{\varkappa_2,\varrho}(\zeta - \omega_1 - \omega_2 - \mu')\, R_{\varkappa_2,\varrho}(\zeta)$$
$$w^C_{\varkappa_2,\varrho}(k', -p_1, -p_2)\, \delta w^C(k, p_1, p_2)\, dk\, dk'\, dp_1\, dp_2$$
$$= C_1 + C_2 + C_3.$$

In this expression, $\delta\chi(k, p_1, p_2) = \chi'_\varkappa(k, p_1, p_2) - \chi_{2,\varkappa_2}(k, p_1, p_2)$. The term C_1 cancels the vacuum energy counterterm. The term C_2 cancels the two mass terms

$$R'_{\varkappa,\varrho}\{-\tfrac{1}{2}\,\delta m_1^2(\varkappa) + \tfrac{1}{2}\,\delta m_2^2(\varkappa_2)\}$$
$$\times \int a(k')^* \, a(-k)\,(4\pi)^{-1}(\mu\mu')^{-1/2}\,(\tilde{g} * \tilde{g})\,(k + k')\,dk\,dk'\,R_{\varkappa_2,\varrho},$$

while C_3 converges to zero in norm.

$$\left\|R'_{\varkappa,\varrho}\int a(-k)\,\delta w^C_{\varkappa,\varrho}(k, p_1, p_2)\,dk\right\| \leqq \text{const.}\ \|\mu^{-\tau/2}\,\delta w^C(\cdot, p_1, p_2)\|_2$$
$$\leqq O(\varkappa^{-\delta}\varrho^{-\delta})\,\omega(\eta)^{-1+3\delta}(\omega_1\omega_2)^\delta.$$

Hence by (7.2.11) we have

$$\|C_3\| \leqq O(\varkappa^{-\delta}\varrho^{-\delta})\int \omega(\eta)^{-1+3\delta}(\omega_1\omega_2)^\delta\,(\omega_1 + \omega_2)^{-(3/2)+\delta}$$
$$\times\,(\mu')^{-(1/2)+\delta}\,|\tilde{g}(k' + \eta)|\,dk'\,dp_1\,dp_2$$
$$\leqq O(\varkappa^{-\delta}\varrho^{-\delta}).$$

We now show how C_1 cancels. We prove

$$\|C_1 - R'_{\varkappa,\varrho}\,\delta E\,R_{\varkappa_2,\varrho}\| \leqq O(\varkappa^{-\delta}\varrho^{-\delta}).$$

We note that $-\delta E = -E_1'(g, \varkappa, \varrho) + E_2(g, \varkappa_2, \varrho)$,

$$-\delta E = \int |w^C_\varrho(k, p_1, p_2)|^2\,(\mu + \omega_1 + \omega_2)^{-1}\,(|\chi'_\varkappa|^2 - |\chi_{2,\varkappa_2}|^2)\,dk\,dp_1\,dp_2,$$

$$|\chi_{2,\varkappa_2}|^2 - |\chi'_\varkappa|^2 = (\delta\chi)\,\chi_{2,\varkappa_2}(-k, -p_1, -p_2) + \delta\chi(-k, -p_1, -p_2)\,\chi'_\varkappa.$$

The term $(\delta\chi)\,\chi_{2,\varkappa_2}(-k, -p_1 - p_2)$ will cancel with C_1, while the low momentum part $(\delta\chi)\,(-k, -p_1, -p_2)\,\chi'_\varkappa$ will converge to zero.

$$C_1 - R'_{\varkappa,\varrho}\,\delta E\,R_{\varkappa_2,\varrho} = R'_{\varkappa,\varrho}\int |w^C_\varrho|^2\,\delta\chi(-k, -p_1, -p_2)$$
$$\times\,\chi'_\varkappa(\mu + \omega_1 + \omega_2)^{-1}\,dk\,dp_1\,dp_2\,R_{\varkappa_2,\varrho}$$
$$+\,R'_{\varkappa,\varrho}\int |w^C_\varrho|^2\,R_{\varkappa_2,\varrho}(\zeta - \mu - \omega_1 - \omega_2)$$
$$\times\,(\omega_1 + \omega_2 + \mu)^{-1}\,(\delta\chi)\,\chi_{2,\varkappa_2}(-k, -p_1, -p_2)$$
$$\times\,dk\,dp_1\,dp_2. \tag{8.1.9}$$

The first term in (8.1.9) vanishes because $|\chi'_\varkappa| \leqq O((\mu + \omega_1 + \omega_2)^{-\delta}\varkappa^{(1-\alpha)\delta})$, and $\chi'_\varkappa$ has support on $S_{\varkappa^{1-\alpha}}$. On that set $|\delta\chi| \leqq O((\mu + \omega_1 + \omega_2)\,\varkappa^{-1})$, so

$$\int_{S_{\varkappa^{1-\alpha}}} |w^C_\varrho|^2\,|\delta\chi(-k, -p_1, -p_2)|\,(\mu + \omega_1 + \omega_2)^{-1-\delta}\,dk\,dp_1\,dp_2 \leqq O(\varkappa^{-\delta+\varepsilon}\varrho^{-\varepsilon/2}),$$

where we can choose $\varepsilon < \alpha\,\delta/2$. Hence the first term in (8.1.10) is $0(\varkappa^{-\alpha\delta/2}\varrho^{-\varepsilon/2})$. The second term in (8.1.9) is also $0(\varkappa^{-\delta}\varrho^{-\delta})$, but in this case the kernel has degree of convergence 1. The cancellation of C_2 with the mass counterterm and the other contributions to $R'\{Z + \delta c\}\,R$ are dealt with in a similar fashion.

8.2 Self adjointness and graph convergence

THEOREM 8.2.1 *The limit*

$$R(\zeta) = \operatorname{norm} \lim_{\varkappa \to \infty} (H(g, \varkappa) - \zeta)^{-1}$$

is the resolvent of a self adjoint operator $H(g)$, independent of the cutoff function $\chi_\varkappa$, and

$$H(g) = \text{strong graph} \lim_{\varkappa \to \infty} H(g, \varkappa).$$

Proof In Theorem 8.1.1 we established norm convergence and $\chi_\varkappa$ independence for the resolvents. Self adjointness and graph convergence follow from Theorem 6.2.6. In order to verify the hypothesis of Theorem 6.2.6, we need only to show that the family $H(g, \varkappa)$ is densely bounded as $\varkappa \to \infty$. For some ζ, we let

$$\mathscr{D} \bigcup_{\varrho < \infty} \text{Range } R_\varrho(\zeta).$$

Since Range $(H_0 - \zeta)^{-1} = \mathscr{D}(H_0)$ is dense, and since by Theorem 8.1.1

$$\lim_{\varrho \to \infty} R_\varrho(\zeta)\,\theta = (H_0 - \zeta)^{-1}\,\theta,$$

it follows that $\mathscr{D}$ is dense. For $\chi = R_\varrho(\zeta)\,\theta \in \mathscr{D}$, we define $\chi_\varkappa = R_{\varkappa,\varrho}(\zeta)\,\theta$. Hence $\chi_\varkappa \in \mathscr{D}(H(g, \varkappa, \varrho)) = \mathscr{D}(H_0) = \mathscr{D}(H(g, \varkappa))$ and $\lim \chi_\varkappa = \chi$. To prove dense boundedness we estimate

$$\langle \chi_\varkappa, H(g, \varkappa)\,\chi_\varkappa \rangle = \langle \chi_\varkappa, \theta \rangle + \zeta \|\chi_\varkappa\|^2 + \langle \chi_\varkappa, \{H(g, \varkappa) - H(g, \varkappa, \varrho)\}\,\chi_\varkappa \rangle. \tag{8.2.1}$$

Let

$$\delta H = H(g, \varkappa) - H(g, \varkappa, \varrho) = \delta H_I + \delta c.$$

We have for fixed ϱ, θ,

$$\begin{aligned} |\langle \chi_\varkappa, \delta H\,\chi_\varkappa \rangle| &= |\langle \theta, R_{\varkappa,\varrho}\,\delta H\,R_{\varkappa,\varrho}\,\theta \rangle| \\ &\leqq \text{const.}\ \|(N_\tau + I)^{-1/2}\,(\delta H_I + \delta c)\,(N_\tau + I)^{-1/2}\| \leqq \text{const.}, \end{aligned}$$

uniformly in $\varkappa$ by Lemma 6.3.5. We conclude from (8.2.1) that $H(g, \varkappa)$ is densely bounded, and the proof is complete.

The norm convergence of the resolvents $H(g, \varkappa)$ has immediate consequences in terms of the spectrum of $H(g)$. We now introduce a new definition of $E(g, \varkappa)$. Let

$$E(g, \varkappa) = \inf \text{spectrum} \left\{ H_0 + H_I(g, \varkappa) - \tfrac{1}{2}\,\delta m^2\,(\varkappa) \int :\phi(x)^2:\,g(x)^2\,dx \right\}. \tag{8.2.1}$$

Thus we renormalize $H(g, \varkappa)$ by adding a constant to the vacuum energy, chosen so that

$$\inf \text{spectrum } H(g, \varkappa) = 0.$$

We let $E_2(g, \varkappa)$ be the second order vacuum energy counterterm used so far, namely

$$E_2(g, \varkappa) = -\|(\mu + \omega_1 + \omega_2)^{-1/2} w_\varkappa^C\|^2.$$

COROLLARY 8.2.2 *The limit*

$$\lim_{\varkappa \to \infty} \{E(g, \varkappa) - E_2(g, \varkappa)\}$$

exists and is finite. Hence with our new renormalization,

$$0 = \inf \text{spectrum } H(g).$$

This definition of $E(g, \varkappa)$ is consistent with the definition of Sec. 6.1,

$$E(g, \varkappa) = E_2(g, \varkappa) + \text{const.} + o(1).$$

COROLLARY 8.2.3 *There is a vacuum vector Ω_g for $H(g)$, namely $H(g)\,\Omega_g = 0$. In the spectral interval* $[0, \min\{m, M\})$, *the operator $H(g)$ is compact.*

The proofs of these corollaries follow Theorem 3.3.2. In particular, the space of vacuum vectors is finite dimensional. We note that the convergence of $R_\varkappa(\zeta)$ as $\varkappa \to \infty$ is uniform in λ for bounded intervals of λ. Since each $R_\varkappa(\zeta)$ is continuous in λ, for λ real, the limit $R(\zeta)$ is continuous in λ. Hence for λ sufficiently close to zero, the eigenvalue zero of $H(g)$ is simple, see Kato [1, p. 213]. It is an interesting problem whether the vacuum is unique for all values of λ, or for some bounded, but g independent, interval in λ at least after a suitable finite mass renormalization. Do the vacuum state(s) lie in the charge zero sector?

A dense domain for the operator $H(g)$ has been constructed explicitly in terms of a truncated power series expansions for the wave operator, Glimm(1,4), Hepp(1,2). We finish this section by outlining the construction in closed form of a dense domain for $H(g)^{1/2}$. Let

$$T_\varkappa = \exp(-\Gamma W_{\varkappa,\varrho}^C). \tag{8.2.2}$$

Thus $T_\varkappa$ is the truncation of the perturbation expansion for the wave operator for the Hamiltonian $H(g, \varkappa, \varrho)$.

For ϱ sufficiently large, we prove that $T_\varkappa$ maps $\mathscr{D}$ into vectors $\chi_\varkappa = T_\varkappa\theta$ in $\mathscr{D}(H(g, \varkappa))$, such that

$$\chi = \lim_{\varkappa \to \infty} \chi_\varkappa = \exp(-\Gamma W_\varrho^C)\,\theta \in \mathscr{D}(H(g))^{1/2}.$$

For the vectors $\chi_\varkappa = T_\varkappa\theta$, we prove

$$\langle \chi_\varkappa, H(g, \varkappa)\,\chi_\varkappa \rangle \to \langle \chi, H(g)\,\chi \rangle.$$

We first establish that $T_\varkappa$ is defined on $\mathscr{D}$ for all $\varkappa \leqq \infty$. For $\tau > \frac{1}{2}$ the kernel and $(\omega_1 + \omega_2)^{-\tau}\, w^C_{\varkappa,\varrho}$ are in L_2, and for $\varkappa < \varkappa_1$,

$$\|(\omega_1 + \omega_2)^{-\tau}\, w^C_{\varkappa,\varrho}\|_2 \leqq 0(\varrho^{-\delta}),$$

$$\|(\omega_1 + \omega_2)^{-\tau}\,(w^C_{\varkappa,\varrho} - w^C_{\varkappa_1,\varrho})\|_2 \leqq 0(\varkappa^{-\delta}\varrho^{-\delta}). \tag{8.2.3}$$

Hence by Proposition 1.2.3(b),

$$\|\Gamma W^C_{\varkappa,\varrho}(N+I)^{-1}\| \leqq M\varrho^{-\delta}, \quad \|\Gamma(W^C_{\varkappa,\varrho} - W^C_{\varkappa_1,\varrho})(N+I)^{-1}\| \leqq M\varkappa^{-\delta}\varrho^{-\delta}$$

for a constant M independent of $\varkappa$. For θ a vector with fewer than n particles, $\Gamma W^C_{\varkappa,\varrho}$ has fewer than $(n+3)$ particles. Thus

$$\|(\Gamma W^C_{\varkappa,\varrho})^j\,\theta\| = \|(\Gamma W^C_{\varkappa,\varrho}(N+I)^{-1}\,(N+I))^j\,\theta\|$$
$$\leqq (M\varrho^{-\delta})^j \max\{(2n)^j, (6j)_j\},$$

and for ϱ sufficiently large the exponential series (8.2.2) converges on $\mathscr{D}$, for $\varkappa \leqq \infty$. Furthermore, by (8.2.3), for $\theta \in \mathscr{D}$, and for ϱ sufficiently large,

$$\lim_{\varkappa\to\infty} T_\varkappa\theta = T\theta, \quad \lim_{\varkappa\to\infty} N^j T_\varkappa\theta = N^j T\theta, \tag{8.2.4}$$

where $T\theta = T_\infty\theta$, and j is any integer. Also, for $\tau < \frac{1}{2}$,

$$N_\tau T_\varkappa\theta \to N_\tau T\theta.$$

In order to study $(H_0 + W^C)$, we need to compute $[H_0, T_\varkappa]$, and this involves

$$B_{\varkappa,\varrho} = [W^C, \Gamma W^C] =$$
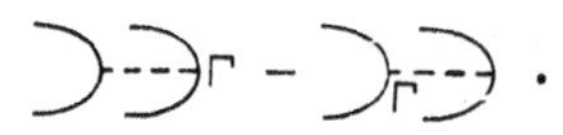

The kernel $b_{\varkappa,\varrho}$ of $B_{\varkappa,\varrho}$ is

$$\int dk\; w^C_{\varkappa,\varrho}(k, p_1, p_2)\, w^C_{\varkappa,\varrho}(-k, p_1', p_2')\,\{(\mu + \omega_1' + \omega_2')^{-1} - (\omega_1 + \omega_2)^{-1}\}. \tag{8.2.5}$$

The norm of $b_{\varkappa,\varrho}$ can be estimated using the relation

$$|(\mu + \omega_1' + \omega_2')^{-1} - (\omega_1 + \omega_2)^{-1}|$$
$$= (\mu + \omega_1' + \omega_2')^{-1}\,(\omega_1 + \omega_2)^{-1}\,|(\mu + \omega_1' + \omega_2' - \omega_1 - \omega_2)|.$$

We conclude that for $\varkappa \leqq \varkappa_1$, and $0 < \tau < \frac{1}{2}$,

$$\|(\omega_1 + \omega_2 + \omega_1' + \omega_2')^{-\tau/2}\,(b_{\varkappa,\varrho} - b_{\varkappa_1,\varrho})\|_2 < 0(\varkappa^{-\delta}).$$

Thus $(N_\tau + I)^{-1} B_{\varkappa,\varrho}(N+I)^{-1}$ is bounded, and norm convergent as $\varkappa \to \infty$.

On the domain $\mathscr{D}$,

$$H_0 T_\varkappa = T_\varkappa H_0 - W^C_{\varkappa,\varrho}\, T_\varkappa - \tfrac{1}{2} B_{\varkappa,\varrho} T_\varkappa.$$

Thus for $\theta \in \mathscr{D}$,

$$(H_0 + W^C_\varkappa)\,\theta = T_\varkappa H_0 \theta + (W^C_\varkappa - W^C_{\varkappa,\varrho})\,T_\varkappa \theta - \tfrac{1}{2} B_{\varkappa,\varrho} T_\varkappa \theta,$$

and the first two terms on the right converge as $\varkappa \to \infty$, since

$$\|(W^C_\varkappa - W^C_{\varkappa,\varrho})\,\theta\| \leqq \text{const.}\ \|w^C_\varkappa \theta_\varrho(\xi)\|_2\ \|(N+I)\,\theta\| \leqq 0(\varkappa^{-\delta}).$$

Also

$$\langle T_\varkappa \theta, B_{\varkappa,\varrho} T_\varkappa \theta\rangle \to \langle T\theta, B_\varrho T\theta\rangle.$$

We prove that $W^A_\varkappa + c(g,\varkappa)$ has a strong graph limit on $T_\varkappa \mathscr{D}$. In fact

$$W^A_\varkappa T_\varkappa = T_\varkappa W^A_\varkappa + D_{\varkappa,\varrho} T_\varkappa,$$

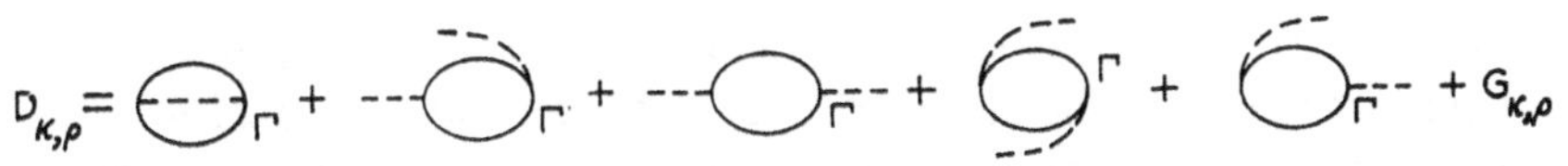

$$= F_{\varkappa,\varrho} + G_{\varkappa,\varrho}$$

where

$$\{F_{\varkappa,\varrho} + c(g,\varkappa)\}\,(N+I)^{-1}$$

is bounded and norm convergent as $\varkappa \to \infty$. The error $G_{\varkappa,\varrho}$ is a finite sum of diagrams $G_{\varkappa,\varrho,j}$, where $G_{\varkappa,\varrho,j}(N+I)^{-i}$ is bounded, for $i = i(j)$ sufficiently large, and norm convergent as $\varkappa \to \infty$. Hence for $\theta \in \mathscr{D}$,

$$W^A_\varkappa T_\varkappa \theta = T_\varkappa W^A_\varkappa (H_0 + I)^{-2}\,(H_0 + I)^2\,\theta + \{D_{\varkappa,\varrho} + c(g,\varkappa)\}\,(N+I)^{-i}\,(N+I)^i\,T_\varkappa \theta$$

and the two terms on the right converge as $\varkappa \to \infty$.

By similar methods, one can prove that for $\theta \in \mathscr{D}$,

$$\langle T_\varkappa \theta, W_\varkappa T_\varkappa \theta\rangle \to \langle T\theta, WT\theta\rangle.$$

Hence $\bigcup_\varrho T\mathscr{D}$ is a dense subset of $\mathscr{D}(H(g)^{1/2})$.

It is possible to use a better dressing transformation and obtain strong graph convergence of $H(g,\varkappa)$ on an explicit domain $T_\varkappa \mathscr{D}$. The transformation

$$T_\varkappa = \exp(-\Gamma V_{\varkappa,\varrho}), \tag{8.2.6}$$

where

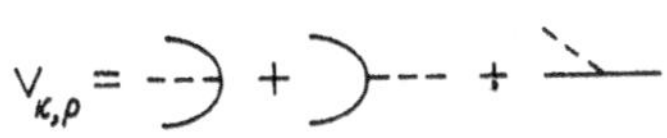

and

$$= \sum_{\varepsilon = \pm 1} \int b(p_1,\varepsilon)^*\, b(-p_2,\varepsilon)\, a(k)^*\, w(k,p_1,p_2)\, \frac{1}{\mu(k) + \omega_1}\, dk\, dp_1\, dp_2$$

seems to be correct, but the details have not been checked. A modification of (8.2.6) is used in Hepp[(2)].

9 The Heisenberg Picture

9.1 Finite propagation speed

We introduce the algebras $\mathfrak{A}(B)$ of time zero observables for the Yukawa$_2$ theory. We study the field algebras $\mathfrak{A}_f(B)$ generated by $\psi(f)$ and $\exp\{i(\phi(f) + \pi(g))\}$, $\operatorname{supp} f, g \subset B$, and also the algebras of observables $\mathfrak{A}(B)$, satisfying $[\mathfrak{A}(B_1), \mathfrak{A}(B_2)] = 0$ if $B_1 \cap B_2 = \emptyset$. Let $\mathfrak{A}$ be the C^* algebra obtained as the norm closure of $\bigcup_B \mathfrak{A}(B)$ as in Sec. 4.1.

We prove that the Hamiltonians $H(g)$ induce a g independent automorphism group σ of A, given locally by the equation

$$\sigma_t(A) = e^{itH(g)} A\, e^{-itH(g)} = A(t). \tag{9.1.1}$$

For the $P(\phi)_2$ theory of Part I, the properties of σ_t followed as a consequence of the essential self adjointness of $H(g)$ and the Trotter product formula. For the Yukawa$_2$ theory, this proof does not apply since the Hamiltonian $H(g)$ is defined only as a limit of the Hamiltonians $H(g, \varkappa)$ with a momentum cutoff $\varkappa$, and it has no mathematical meaning when written as the sum

$$H(g) = H_0 + H_I(g) + c(g). \tag{9.1.2}$$

On the other hand, the approximate Hamiltonians $H(g, \varkappa)$ have infinite propogation speed, so we are faced with proving that a limit of approximate nonlocal theories is local.

THEOREM 9.1.1 *Let A belong to $\mathfrak{A}(B)$ or $\mathfrak{A}_f(B)$, and let g equal one on B_t. Then (9.1.1) defines a g independent automorphism on $\mathfrak{A}$ or $\mathfrak{A}_f$, and the propagation speed of σ is at most one:*

$$\sigma_t : \mathfrak{A}(B) \to \mathfrak{A}(B_t),$$

$$\sigma_t : \mathfrak{A}_f(B) \to \mathfrak{A}_f(B_t).$$

We outline the main steps of the proof. See Glimm and Jaffe[9] for more details. We introduce a new approximate theory, which has the required finite propagation speed. We define a g independent transformation $A \to A_\Lambda(t)$, where $A \in \mathfrak{A}(B)$, $A_\Lambda(t) \in \mathfrak{A}(B_t)$. However, $A_\Lambda(t)$ is not given by the dynamics of $H(g)$, but rather by an approximate dynamics. We then prove that weak $\lim_{\Lambda\to\infty} A_\Lambda(t) = A(t)$, where $A(t)$ is defined in (9.1.1), provided g equals one on B_t. Since the algebras $\mathfrak{A}(B)$ are weakly closed, $A(t) \in \mathfrak{A}(B_t)$ and (9.1.1) does define a g independent automorphism of $\mathfrak{A}$.

In carrying out this procedure, we choose an approximation $A_\Lambda(t)$ which obviously possesses the desired property of finite propagation speed. Thus the difficult feature of the proof to establish the convergence of $A_\Lambda(t)$ to $A(t)$.

In order to minimize the loss of locality in $H(g, \varkappa)$, we choose a sharp cutoff in configuration space.

$$H_I(g, \varkappa) = \lambda \int \phi_\varkappa(x) :\psi_\varkappa(x)\, \psi_\varkappa(x):\, g(x)\, dx,$$

where $\phi_\varkappa(x) = \delta_\varkappa^* \phi$, $\psi_\varkappa(x) = \delta_\varkappa^* \phi$, and

$$\delta_\varkappa(x) = \varkappa\tilde{\chi}(\varkappa x) \in C_0^\infty, \quad \int \delta_\varkappa(x)\, dx = 1 .$$

With this cutoff, $\phi_\varkappa(x)$ is localized (in the sense of fields) in an interval of diameter $0(\varkappa^{-1})$ centered at x. In other words, $\exp(i\phi_\varkappa(x)) \in \mathfrak{A}(B)$, where $B = (x - 0(\varkappa^{-1}), x + 0(\varkappa^{-1}))$. For definiteness, let us suppose that diam. supp. $\delta_\varkappa \leqq (2\varkappa)^{-1}$.

If we attempt to apply Proposition 4.1.1 to the Hamiltonian $H(g, \varkappa) = H_0 + H_I(g, \varkappa) + c(g, \varkappa)$, we wish to write $H(g, \varkappa) = H_1 + H_2$ where H_1 has propagation speed $s(\sigma^{(1)}) \leqq 1$ and H_2 has propagation speed $s(\sigma^{(2)}) = 0$. Since the ultraviolet cutoff enters the counterterms only in the coefficients $\delta m^2(\varkappa)$ and $E(g, \varkappa)$, Proposition 4.1.1 shows that $H_1 = H_0 + c(g, \varkappa)$ generates an automorphism with propagation speed $s(\sigma^{(1)}) \leqq 1$. However, the interaction term $H_2 = H_I(g, \varkappa)$ has an infinite propagation speed. If for each s it were possible to write

$$H_I(g, \varkappa) = A_1 + A_2$$

where

$$\exp(iA_1 t) \in \mathfrak{A}(B_s) \quad \text{and} \quad \exp(iA_2 t) \in \mathfrak{A}(\sim B_s),$$

then H_2 would generate an automorphism with propagation speed zero. The difficulty with writing such a decomposition for H_I arises from the contribution

$$\delta H_I(s) = \lambda \int_{I_s} \phi_\varkappa(x) :\psi_\varkappa(x)\, \psi_\varkappa(x):\, g(x)\, dx, \tag{9.1.2}$$

where I_s is the set of points within the distance $2\varkappa^{-1}$ from the boundary of B_s. Following these ideas, we replace $H_I(g, \varkappa)$ by the (time dependent) interaction Hamiltonian

$$H_I(s) = H_I(g, \varkappa) - \delta H_I(s)$$

and we replace $H(g, \varkappa)$ by the time dependent Hamiltonian $H(s) = H(g, \varkappa) - \delta H_I(s)$ as a step in defining the approximate but local dynamics $A \to A_A(t)$, see Figure 2.

We construct a unitary propagator $W_\varkappa(t)$ generated by a sequence of Hamiltonians $H(t_j)$. Each Hamiltonian $H(t_j)$ is used to propagate for a time interval of length $0(\varkappa^{-1})$ and the $H(t_j)$ dynamics carries $\mathfrak{A}(B_{t_j})$ into $\mathfrak{A}(B_{t_j+\tau})$ in an elapsed time τ, $|\tau| \leqq 0(\varkappa^{-1})$. The propagator $W_\varkappa(t)$ implements a g independent map of $A \in \mathfrak{A}(B)$ into $\hat{A}(t) \in \mathfrak{A}(B_t)$,

$$\hat{A}(t) = W_\varkappa^*(t)\, A\, W_\varkappa(t). \tag{9.1.3}$$

We let n equal the integer part of $|\varkappa t|$, that is $n = [|\varkappa t|]$, and we define

$$W_\varkappa(t) = e^{-i\frac{t}{n}H(0)}\, e^{-i\frac{t}{n}H\left(\frac{t}{n}\right)} e^{-i\frac{t}{n}H\left(\frac{2t}{n}\right)} \dots e^{-i\frac{t}{n}H\left(\frac{(n-1)}{n}t\right)}. \quad (9.1.4)$$

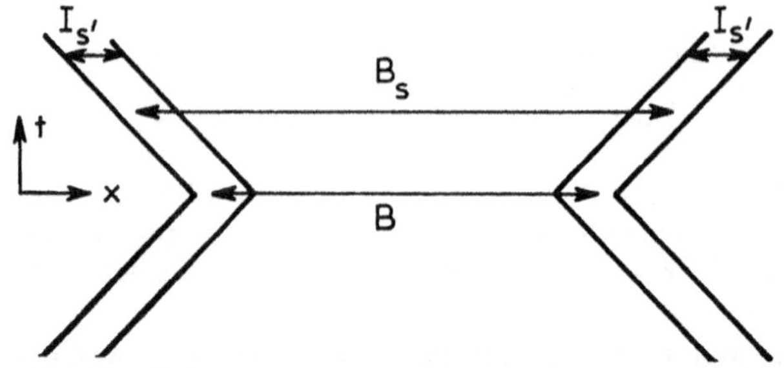

Figure 2. Localization of the time dependent Hamiltonian $H(s)$. $H(s)$ is localized outside I_s and $\delta H(s)$ is localized inside I_s.

There are $n = 0(\varkappa)$ factors in this product. We note that the propagator $W_\varkappa(t)$ is defined separately for each bounded open set B. The most general open subset of the real line is a countable disjoint union of open intervals. To simplify the proof, we take B to be the interval $(-1, 1)$. The same proof applies to any finite union of intervals, and the general case is covered by a limiting argument. In fact if B is a countable disjoint union of of open intervals and if supp $f \subset B$ then supp f is contained in a finite union of open intervals contained in B. Thus the proof we give shows that σ_t maps a set of generators for $\mathfrak{A}(B)$ into $\mathfrak{A}(B_t)$ and consequently $\sigma_t\mathfrak{A}(B) \subset \mathfrak{A}(B_t)$ for a general open set B.

LEMMA 9.1.2 *Let $A \in \mathfrak{A}(B)$ and let $\hat{A}(t)$ be defined by (9.1.3-4). Then $\hat{A}(t) \in \mathfrak{A}(B_t)$ and $\hat{A}(t)$ is independent of g.*

Proof As in Proposition 4.1.1, the Hamiltonian $H(s)$ induces a g independent mapping

$$A \to \exp(i\tau H(s))\, A \exp(-i\tau H(s))$$

of $\mathfrak{A}(B_s)$ into $\mathfrak{A}(B_{|s|+|\tau|})$, for $|\tau| \leqq tn^{-1}$. Composing n such maps, we obtain the lemma.

For technical reasons, to aid the proof of convergence as $\varkappa \to \infty$, we smooth out the approximation $\hat{A}(t)$ of (9.1.3-4). Let $0 \leqq x \leqq 1$. We scale the length of time propagation by x and define

$$W_\varkappa(t; x) = e^{-i\frac{tx}{n}H(0)}\, e^{-i\frac{tx}{n}H\left(\frac{t}{n}\right)}\, e^{-i\frac{tx}{n}H\left(\frac{2t}{n}\right)} \dots e^{-i\frac{tx}{n}H\left(\frac{n-1}{n}t\right)}. \quad (9.1.5)$$

For $A \in \mathfrak{A}(B)$, the operator

$$W_\varkappa^*(t; x)\, A\, W_\varkappa(t; x)$$

also belongs to $\mathfrak{A}(B_t)$ and is g independent. This is easily understood, since we use each Hamiltonian $H(j\,t/n)$, $j = 0, 1, \dots, n-1$, for a time interval of length $|x\, t/n| \leqq |t/n|$. (Note that $x > 1$ can not be used.) We can average over $0 \leqq x \leqq 1$ to obtain our final approximation

$$A_\Lambda(t) = \int W_\varkappa^*(t; x)\, A W_\varkappa(t; x) f_\Lambda(x)\, dx, \tag{9.1.6}$$

where f_Λ is a positive continuous function such that

$$\int f_\Lambda(x)\, dx = 1, \quad \text{suppt}\, f_\Lambda(x) \subset [1 - \Lambda^{-1}, 1].$$

In (9.1.6) we will let $\varkappa = \varkappa(\Lambda)$ depend on Λ. Theorem 9.1.1 follows from

THEOREM 9.1.3 *For a suitable choice of the function* $\varkappa(\Lambda)$,

$$\text{weak}\lim_{\Lambda\to\infty} A_\Lambda(t) = A(t).$$

The proof of Theorem 9.1.3 is divided into several main steps:

Step 1 We show that $W_\varkappa(t; x)$ is the boundary value of an analytic function $W_\varkappa(t; z)$, analytic in the half plane $\text{Im}\, tz < 0$ and bounded uniformly in $\varkappa$ and z.

Step 2 We prove that for $\text{Im}\, tz < 0$,

$$\text{weak}\lim_{\varkappa\to\infty} W_\varkappa(t; z) = W(t; z)$$

exists and equals $e^{-itzH(g)}$.

Step 3 We use step 2 to prove weak convergence of $W_\varkappa(t; x)$ as measures (with respect to the variable x) to $e^{-itxH(g)} = U(tx)$.

Step 4 We prove weak convergence of

$$A_\Lambda(t) - A(t)$$

to zero.

The first two steps depend crucially on properties of $\delta H(s)$. In particular, we use first and second order estimates on $\delta H(s)$ to show that $\delta H(s)$ is small as $\varkappa \to \infty$. The first order estimate is

$$\pm\delta H(s) \leqq 0(\varkappa^{-\tau/2+\delta})(N_\tau + I). \tag{9.1.7}$$

The second order estimate is divergent. For any positive ε, there is a constant M such that for all large $\varkappa$

$$(\delta H(s))^2 \leqq \varepsilon H^2 + M \log \varkappa. \tag{9.1.8}$$

We also use the convergence of $R_\varkappa(\zeta) = (H(g, \varkappa) - \zeta)^{-1}$ to $(H(g) - \zeta)^{-1}$ proved in Sec. 8. After proving steps one and two, steps three and four follow by general methods.

We prove the first order estimate (9.1.7) by using an N_τ estimate. We have

$$H_I(s) = H_I(g, \varkappa) - H_I(\delta g, \varkappa) = H_I(g, \varkappa) - \delta H_I(s),$$

where

$$\delta g = g(x)\,\theta_{2/\varkappa}(|x| - |s| - 1).$$

Hence

$$|(\delta g)^{\sim}| \leqq \|\delta g\|_1 \leqq 2\|\theta_{2/\varkappa}\|_1 = O(\varkappa^{-1}).$$

Since $\delta H_I(s)$ is localized in a region of space of size $O(\varkappa^{-1})$, we lose momentum conservation as $\varkappa \to \infty$, but we have gained a factor $\varkappa^{-1}$ from the size $O(\varkappa^{-1})$ of the region I_s. By N_τ estimates of Proposition 1.2.3,

$$\|(N_\tau + I)^{-1/2}\,\delta H_I(s)\,(N_\tau + I)^{-1/2}\| \leqq \text{const.}\ \|\omega_1^{-\tau/2}(|\delta w| + |\delta w^C|)\|_2$$

$$\leqq \text{const.}\left(\int \omega_1^{-\tau}\mu^{-1}\,O(\varkappa^{-2})\,|\chi_\varkappa|^2\,dk\,dp_1\,dp_2\right)^{1/2}$$

$$= O(\varkappa^{-\tau/2+\delta}),$$

which proves (9.1.7).

To prove (9.1.8), we expand $\delta(H_I(s))^2$ as a sum of Wick monomials,

$(\delta H_I(s))^2 =$ 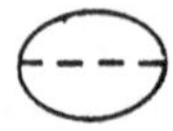 $+$ other terms.

The first term above is a divergent constant,

$0 \leqq$ 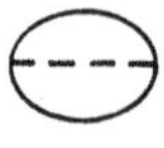 $\leqq \int \mu^{-1}\varkappa^{-2}|\chi_\varkappa|^2\,dk\,dp_1\,dp_2 \leqq M\log\varkappa,$

while the other terms are convergent and can be dominated by εH^2 + const., using the first and second order estimates of Sec. 7. (See Glimm-Jaffe[(9)] for details.) We remark that (9.1.8) can be avoided in the proof of self adjointness of $H(s)$, but it will be needed in Lemma 9.1.5 below.

LEMMA 9.1.4. *The Hamiltonian $H(s)$ is self adjoint, and bounded from below uniformly in $\varkappa$ and s.*

Proof By (9.1.8), for $\varepsilon < 1$, the perturbation $\delta H(s)$ is small in the sense of Kato with respect to $H(g, \varkappa)$. The self adjointness of $H(s)$ follows. The uniform lower bound on

$$H(s) = H(g, \varkappa) - \varepsilon N_\tau + \varepsilon N_\tau - \delta H(s)$$

follows from Theorem 7.1.1 and the first order estimate (9.1.7) on $\delta H(s)$.

Since $H(s)$ is bounded uniformly from below, we add a constant so that $I \leqq H(s)$. Similarly we let $I \leqq H(g, \varkappa)$. For convenience of notation, we let $R = H(g, \varkappa)^{-1}$, $H = H(g, \varkappa)$. With this normalization $\|W_\varkappa(t, z)\| \leqq 1$ for $\operatorname{Im} tz < 0$ and step 1 of the proof is completed.

As preparation for step 2 of Theorem 9.1.3, i.e. weak convergence of the propagators for $\operatorname{Im} tz < 0$, we prove three lemmas.

LEMMA 9.1.5 *Let $t \geqq 0$. Then*

$$\|H^{1/2}e^{-tH(s)}R^{1/2}\| \leqq (1 + Mt\log\varkappa).$$

Proof We prove that

$$F(t) = \|H^{1/2} e^{-tH(s)} R^{1/2}\theta\|^2$$

satisfies the inequality $\frac{dF}{dt} \leqq M \log \varkappa \|\theta\|^2$, which yields the lemma on integration. This bound on dF/dt follows from

$$\frac{dF}{dt} = -\langle e^{-tH(s)}R^{1/2}\theta, \{HH(s) + H(s)H\} e^{-tH(s)} R^{1/2}\theta\rangle$$

and

$$\begin{aligned} -(HH(s) + H(s)H) &= -2H^2 + H\,\delta H_I(s) + \delta H_I(s)\,H \\ &\leqq -H^2 + (\delta H_I(s))^2 \leqq M \log \varkappa. \end{aligned}$$

The final inequality uses the second order estimate (9.1.8).

LEMMA 9.1.6 *Let* $0 \leqq t \leqq T$ *and let* $\tau < 1$. *Then*

$$\|R^{1/2}\{e^{-tH} - e^{-tH(s)}\} R^{1/2}\| \leqq O(t\varkappa^{-\tau/2}).$$

Proof Let $0 < \varepsilon, \tau + \varepsilon < 1$. Then

$$\begin{aligned} \|R^{1/2}\{e^{-tH} - e^{-tH(s)}\} R^{1/2}\| &\leqq \int_0^t \|e^{-(t-r)H} R^{1/2}\,\delta H_I(s)\, R^{1/2} H^{1/2} e^{-rH(s)} R^{1/2}\|\, dr \\ &\leqq \int_0^t O(\varkappa^{-(\tau+\varepsilon)/2})\,(1 + Mr\log\varkappa)\, dr \\ &\leqq O(\varkappa^{-\tau/2} t), \end{aligned}$$

by (9.1.7) and Lemma 9.1.5.

Let $U_\varkappa(t) = \exp(-itH(g, \varkappa))$, $U(t) = \exp(-itH(g))$.

LEMMA 9.1.7 *For* ty *positive but sufficiently small, there is a positive* δ *such that*

$$\|R^{1/2}\{W_\varkappa(t; -iy) - U_\varkappa(-iyt)\} R^{1/2}\| \leqq O(\varkappa^{-\delta}).$$

Proof The difference $R^{1/2}\{W_\varkappa(t; -iy) - U_\varkappa(-iyt)\} R^{1/2}$ is a sum of $n = O(\varkappa)$ terms of the form

$$\prod_{j=0}^{r-1} \left(R^{1/2} e^{-y\left(\frac{t}{n}\right)H(jt/n)} H^{1/2}\right) R^{1/2}\{e^{-ytH/n} - e^{-ytH(rt/n)/n}\} R^{1/2} U(-iyt/n)^{n-r-1}.$$

Using Lemmas 9.1.5 and 9.1.6, we obtain the bound

$$\begin{aligned} \|R^{1/2}\{W_\varkappa(t; -iy) - U_\varkappa(-iyt)\} R^{1/2}\| &\leqq O(n)\left(1 + M\frac{ty}{n}\log\varkappa\right)^n O(\varkappa^{-\tau/2})\left(\frac{ty}{n}\right) \\ &\leqq O(\varkappa^{-\tau/2 + Mty}), \end{aligned}$$

which is $O(\varkappa^{-\delta})$ for y sufficiently small.

THEOREM 9.1.8 *Let* Im $tz < 0$. *Then*

$$\lim_{\varkappa\to\infty} \langle\theta, \{W_\varkappa(t; z) - U_\varkappa(tz)\}\,\theta\rangle = 0,$$

and convergence is uniform on compact sets of z.

Proof Since $W_\varkappa(t; z)$ and $U_\varkappa(tz)$ are bounded uniformly in $\varkappa$ convergence on a dense set is sufficient. For $\theta \in D(H(g))$, there is a sequence of vectors $\theta_\varkappa \in D(H(g, \varkappa))$ such that $\theta_\varkappa \to \theta$ and $H(g, \varkappa)\,\theta(\varkappa) \to H(g)\,\theta$. Let

$$F_\varkappa(z) = \langle\theta, \{W_\varkappa(t; z) - U_\varkappa(tz)\}\,\theta\rangle.$$

For $z = -iy$ and $0 < ty$, we have

$$\begin{aligned} |F_\varkappa(z)| &\leqq \text{const.}\ \|\theta - \theta_\varkappa\| + |\langle H_\varkappa\theta_\varkappa, R_\varkappa\{W_\varkappa(t; z) - U_\varkappa(tz)\}\, R_\varkappa H_\varkappa\theta_\varkappa\rangle| \\ &\leqq \text{const.}\ \|\theta - \theta_\varkappa\| + O(\varkappa^{-\tau/2+Mty}) \\ &\to 0. \end{aligned}$$

For fixed θ, the functions $F_\varkappa(z)$ are bounded uniformly in the half plane Im $tz < 0$, and are analytic in z. Hence convergence for small $y = iz$ implies convergence for all z in the half plane, by Vitali's convergence theorem for normal families of analytic functions. This completes the second step of the proof of Theorem 9.1.3.

COROLLARY 9.1.9 *For some* $\varkappa(\Lambda)$ *such that* $\varkappa(\Lambda) \to \infty$,

$$\operatorname*{weak\,lim}_{\Lambda\to\infty} \int \{W_{\varkappa(\Lambda)}(t; x) - U_{\varkappa(\Lambda)}(tx)\} f_\Lambda(x)\, dx = 0.$$

Proof The corollary is a consequence of the convergence of the boundary values of $F_\varkappa(z)$ as measures, as $\varkappa \to \infty$, namely for fixed $f \in C([0, 1])$ and fixed θ,

$$\int F_\varkappa(x) f(x)\, dx \to 0 \quad \text{as} \quad \varkappa \to \infty. \tag{9.1.10}$$

It follows that for some choice $\varkappa = \varkappa(\Lambda)$, $\varkappa(\Lambda) \to \infty$ as $\Lambda \to \infty$ and

$$\int F_{\varkappa(\Lambda)}(x) f_\Lambda(x)\, dx \to 0. \tag{9.1.11}$$

By the diagonal process, we can pick one function $\varkappa(\Lambda)$ yielding convergence for a countable dense set of vectors θ_i. Since $|F_\varkappa(x)| \leqq 2\|\theta\|^2$, this proves that (9.1.11) is valid for all θ, completing the proof of the corollary.

To establish (9.1.10), we approximate f by a function h analytic in a complex neighborhood of [0, 1], and such that

$$\int_0^1 |f(x) - h(x)|\, dx \equiv \|f - h\|_1 \leqq \varepsilon.$$

Thus by the Cauchy integral theorem,

$$\left| \int F_\varkappa(x) f(x)\, dx \right| \leqq \|f - h\|_1 + \left| \int_C F_\varkappa(z)\, h(z)\, dz \right|,$$

where C is a path from 0 to 1, lying in the half space Im $tz > 0$ and in the domain of analyticity of h. By Theorem 9.1.8, $F_\varkappa(z)$ converges to zero, pointwise on C. Thus (9.1.10) is valid and the proof is complete.

We now take up the fourth step, completing the proof of Theorem 9.1.3.

Proof of Theorem 9.1.3 It is sufficient to prove that $A_\Lambda(t) - U^*_{\varkappa(\Lambda)}(t) \times AU_{\varkappa(\Lambda)}(t)$ converges weakly to zero, and since $\{U_{\varkappa(\Lambda)}(tx) - U_{\varkappa(\Lambda)}(t)\}$ converges strongly to zero for $x \in \operatorname{supp} f_\Lambda$, it is sufficient to prove

$$0 = \underset{\Lambda \to \infty}{\text{weak lim}} \int dx\, f_\Lambda(x) \{W^*_{\varkappa(\Lambda)}(t; x)\, A W_{\varkappa(\Lambda)}(t; x) - U^*_{\varkappa(\Lambda)}(tx)\, A U_{\varkappa(\Lambda)}(tx)\}$$

$$= \underset{\Lambda \to \infty}{\text{weak lim}} \{A_\Lambda(t) - \bar{A}_\Lambda(t)\}$$

for any $A \in \mathfrak{A}(B)$. The equation above defines $\bar{A}_\Lambda(t)$. By the Schwarz inequality,

$$\langle \Omega, \{A_\Lambda(t) - \bar{A}_\Lambda(t)\}\, \Omega\rangle| \leqq 2\|A\|\, \|\Omega\| \int dx\, f_\Lambda(x)\, \|\{W_\varkappa(t; x) - U_\varkappa(tx)\}\Omega\|$$

$$\leqq 2\|A\|\, \|\Omega\| \left(\int dx\, f_\Lambda(x)\, \|\{W_\varkappa(t; x) - U_\varkappa(tx)\}\, \Omega\|^2\right)^{1/2}.$$

Since

$$\|(W - U)\,\Omega\|^2 = 2 \operatorname{Re} \langle (U - W)\,\Omega, U\Omega\rangle,$$

the theorem follows from Corollary 9.1.9.

9.2 The field operators

We use the preceeding results to construct local quantum fields $\phi(x, t)$ and $\psi(x, t)$ that are independent of the spatial cutoff function g. These fields obey equations of motion defined as limits (as $\varkappa \to \infty$) of cutoff equations. When averaged with C_0^∞ functions, the field bilinear forms uniquely determine operators $\phi(f)$ and $\psi(f)$.

We define the sharp time fields

$$\phi(f, t) = e^{itH(g)}\, \phi(f, 0)\, e^{-itH(g)},$$

$$\psi(f, t) = e^{itH(g)}\, \psi(f, 0)\, e^{-itH(g)}.$$

THEOREM 9.2.1 (*Glimm and Jaffe*[(9)]). *The sharp time fields $\phi(f, t)$, $\psi(f, t)$ are g independent if g equals 1 on a sufficiently large set. The domain of $H(g)$ is a core for $\phi(f, t)$. The fields obey canonical equal time commutation and anticommutation relations.*

THEOREM 9.2.2 *The bilinear form $\phi(f, t)$ is a C^∞ function of t on $C^\infty(H(g)) \times C^\infty(H(g))$. The weak limits of*

$$\left(\frac{d}{dt}\right)^n R_\varkappa(\zeta)^{n+1} e^{itH(g,\varkappa)} \phi(f, 0) e^{-itH(g,\varkappa)} R_\varkappa(\zeta)^{n+1}$$

exist as $\varkappa \to \infty$. The corresponding strong limits exist for the fermion fields.

The field equations can be defined as the limits, as $\varkappa \to \infty$, of the space averaged, cutoff equations. In fact the cutoff equations

$$\left(\frac{\partial^2}{\partial t^2} - \frac{\partial^2}{\partial x^2} + m^2\right)\phi_{g,\varkappa}(x, t) + j_{g,\varkappa}(x, t) = 0,$$

$$\left(\gamma_0 \frac{\partial}{\partial t} + \gamma_1 \frac{\partial}{\partial x} + M\right)\psi_{g,\varkappa}(x, t) + J_{g,\varkappa}(x, t) = 0$$

after averaging with a C_0^∞ function have weak limits as $\varkappa \to \infty$ on $C^\infty(H(g)) \times C^\infty(H(g))$. See Glimm-Jaffe[9]. See Dimock[1] for further properties of the fields and currents.

We also define the space time averaged fields $\phi(f)$, where $f = \bar{f} \in C_0^\infty(R^2)$.

$$\phi(f) = \int \phi(x, t) f(x, t)\, dx\, dt.$$

Following the proof for the $P(\phi)_2$ theory (Glimm and Jaffe [4, Sec. 3.2]) we conclude that $\phi(f)$ is essentially self adjoint on $\mathscr{D}(H(g))$ provided g equals 1 on a sufficiently large set.

THEOREM 9.2.3 *Schrader*[1]. *Let f be a real C_0^∞ function. The self adjoint operator $\phi(f)$ is independent of g.*

Proof The idea of the proof is to approximate $\phi(f)$ by bounded self adjoint operators C_n that are independent of g. For this approximation, and for $\theta \in \mathscr{D}(H(g))$,

$$\lim_{n \to \infty} C_n \theta = \phi(f)\, \theta.$$

Since $\mathscr{D}(H(g))$ is a core for $\phi(f)$, the resolvents $R_n(\zeta) = (C_n - \zeta)^{-1}$ converge strongly to $(\phi(f) - \zeta)^{-1}$. Since $R_n(\zeta)$ is independent of g, so is the limit.

To choose the approximation C_n, let

$$A(t) = \left(\int \phi(x, 0) f(x, t)\, dx\right)^- = \int \lambda\, dE(\lambda, t).$$

By Theorem 9.2.1,

$$e^{itH(g)} A(t)\, e^{-itH(g)}$$

is independent of g, and hence so is the spectral measure

$$E^t(\lambda, t) = e^{itH(g)} E(\lambda, t)\, e^{-itH(g)}.$$

We define the bounded g-independent operator C_n by the equation

$$C_n = \int dt \int_{|\lambda| \leq n} \lambda \, dE^t(\lambda, t),$$

and for $\theta \in \mathscr{D}(H(g))$

$$\{C_n - \phi(f)\}\, \theta = -\int dt \int_{|\lambda| \geq n} \lambda \, dE^t(\lambda, t)\, \theta .$$

This yields the estimate

$$\begin{aligned} \|\{C_n - \phi(f)\}\, \theta\| &\leqq \int dt \Big\| \int_{|\lambda| \geq n} \lambda \, dE(\lambda, t) \, (A(t)^2 + I)^{-1}, \\ &\qquad \times \big(A(t)^2 + I\big)\big(H(g) + I\big)^{-1} \big(H(g) + I\big) e^{-itH(g)} \theta \Big\| \\ &\leqq \text{const.} \, \|(H(g) + I)\, \theta\| \int dt \Big\| \int_{|\lambda| \geq n} \lambda(\lambda^2 + 1)^{-1} \, dE(\lambda, t) \Big\| \end{aligned}$$

where we use the second order estimate, Theorem 7.2.1, and the N_τ estimate,

$$\|\big(A(t)^2 + I\big)\, (N + I)^{-1}\| \leqq \text{const.}$$

Hence $\|\{C_n - \phi(f)\}\, \theta\| \to 0$, and the theorem is proved.

Bibliography

H. Araki, [1] Von Neumann algebras of local observables for the free scalar field, *J. Math. Phys.*, **5**, 1–13 (1964).

[2] The type of von Neumann algebra associated with the free field, *Prog. Theor. Phys.*, **32**, 956–965 (1964).

[3] On the algebra of all local observables, *Prog. Theor. Phys.* **32**, 844–854 (1964).

V. Bargmann, [1] On a Hilbert space of analytic functions and an associated integral transform. Part I, *Comm. Pure Appl. Math.*, **14**, 187–214 (1961).

J. Bjorken and S. Drell, [1] *Relativistic quantum fields*, McGraw Hill, New York (1965).

N. Bogoliubov and D. Shirkov, [1] *Introduction to the theory of quantized fields*, Interscience, New York (1959).

J. Cannon, [1] Quantum field theoretic properties of a model of Nelson: Domain and eigenvector stability for perturbed linear operators, *J. Funct. Analy.*, to appear.

[2] Convergence criteria for a sequence of semigroups, to appear.

J. Cannon and A. Jaffe, [1] Lorentz covariance of the $(\phi^4)_2$ quantum field theory, *Commun. Math. Phys.*, **17**, 261–321 (1970).

J. Dixmier, [1] *Les algèbres d'opérateurs dans l'espace Hilbertien (algèbres de von Neumann)*, Gauthier-Villars, Paris (1957).

[2] *Les C*-algèbras et leurs représentations*, Gauthier-Villars, Paris (1964).

J. Dimock, [1] Estimates, renormalized currents and field equations for the Yukawa$_2$ field theory, Harvard University thesis (1971).

N. Dunford and J. T. Schwartz, [1] *Linear operators, Part I: general theory*, Interscience, New York (1958).

J.-P. Eckmann, [1] Estimates on graphs (unpublished). J.-P. Eckmann and K. Osterwalder, [1] On the uniqueness of the Hamiltonian and of the representation of the CCR for the quartic boson interaction in three dimensions, *Helo. Phys. Acta*, to appear.

H. Epstein, V. Glaser, and A. Jaffe, [1] Nonpositivity of the energy density in quantized field theories, *Nuovo Cimento*, **36**, 1016–1022 (1965).

J. Fabrey, [1] Exponentical representations of the canonical commutation relations *Commun. Math. Phys.*, **19**, 1–30 (1970).

P. Federbush, [1] A partially alternate derivation of a result of Nelson, *J. Math. Phys.*, **10**, 50–52 (1969).

[2] Renormalization of some one-space dimensional quantum field theories by unitary transformations, preprint.

[3] Unitary renormalization of $(\phi^4)_{2+1}$, to appear.

[4] A convergent expansion for the resolvent of $:\phi^4:_{1+1}$, to appear.

P. Federbush and B. Gidas, [1] Renormalization of the one-space dimensional Yukawa model by unitary transformations, preprint.

J. Fell, [1] The dual spaces of C^*-algebras, *Trans. Amer. Math. Soc.*, **94**, 365–403 (1960).

K. Friedrichs, [1] *Perturbation of spectra in Hilbert space*, American Mathematical Society, Providence (1965).

J. Glimm, [1] Yukawa coupling of quantum fields in two dimensions, I. *Commun. Math. Phys.*, **5**, 343–386 (1967).

[2] Yukawa coupling of quantum fields in two dimensions, II. *Commun. Math. Phys.*, **6**, 120–127 (1967).

[3] Boson fields with nonlinear self interaction in two dimensions, *Commun. Math. Phys.*, **8**, 12–25 (1968).

[4] Boson fields with the $:\phi^4:$ interaction in three dimensions, *Commun. Math. Phys.*, **10**, 1–47 (1968).

[5] The foundations of quantum field theory, *Advances in Math.*, **3**, 101–125 (1969).

[6] Models for quantum field theory, in *Rendiconti di Fisica Internazionale de Fisica "E. Fermi" XLV Corso*, Academic Press, New York (1969).

J. Glimm and A. Jaffe, [1] A Yukawa interaction in infinite volume, *Commun. Math. Phys.*, **11**, 9–18 (1968).

[2] A $\lambda\phi^4$ quantum field theory without cutoffs. I., *Phys. Rev.*, **176**, 1945–1951 (1968).

[3] Singular perturbations of self adjoint operators, *Comm. Pure Appl. Math.*, **22**, 401–414 (1969).

[4] The $\lambda(\phi^4)_2$ quantum field theory without cutoffs. II. The field operators and the approximate vacuum, *Ann. Math.*, **91**, 362–401 (1970).

[5] The $\lambda(\phi^4)_2$ quantum field theory without cutoffs. III. The physical vacuum, *Acta. Math.*, **125**, 203–261 (1970).

[6] Infinite renormalization of the Hamiltonian is necessary, *Jour. Math. Phys.*, **10**, 2213–2214 (1969).

[7] Rigorous quantum field theory models, *Bull. A.M.S.*, **76**, 407–410 (1970).

[8] Self-adjointness of the Yukawa_2 Hamiltonian, *Ann. of Phys.*, **60**, 321–383 (1970).

[9] The Yukawa_2 quantum field theory without cutoffs, *Jour. Funct. Analysis*, **7**, 323–357 (1971).

[10] The energy-momentum spectrum and vacuum expectation values in quantum field theory, *Jour. Math. Phys.*, **11**, 3335–3338 (1970).

[11] The energy momentum spectrum and vacuum expectation values in quantum field theory. II, *Commun. Math. Phys.* **22**, 1–22 (1971).

[12] Positivity and self adjointness of the $P(\phi)_2$ Hamiltonian, *Commun. Math. Phys.*, to appear.

E. Griffin, [1] Some contributions to the theory of rings of operators II, *Trans. Amer. Math. Soc.*, **79**, 389–400 (1955).

L. Gross, [1] Existence and uniqueness of physical ground states, to appear.

M. Guenin, [1] On the interaction picture, *Comm. Math. Phys.*, **3**, 120–132 (1966).

R. Haag and D Kastler, [1] An algebraic approach to quantum field theory, *Jour. Math. Phys.*, **5**, 848–861 (1964).

D. Hall and A. Wightman, [1] A theorem on invariant functions with applications to relativistic quantum field theory, *Mat.-Fys. Medd. Kong. Danske Videns. Skelskab*, **31**, No. 5 (1957).

K. Hepp, [1] Renormalized Hamiltonians for a class of quantum fields with infinite mass and charge renormalizations. In *Anniversary volume for N. N. Bogoliubov*, Nauka, Moscow (1969).

[2] *Théorie de la renormalisation*, Springer-Verlag, Heidelberg (1969).

R. Høegh-Krohn, [1] A general class of quantum fields without cut-offs in two space-time dimensions, *Commun. Math. Phys.*, **21**, 244–255 (1971).

[2] On the spectrum of the space cutoff $:P(\phi):$ Hamiltonian in two space-time dimensions, *Commun. Math. Phys.*, **21**, 256–260 (1971).

R. Høegh-Krohn and B. Simon, [1] Hypercontractive semi-groups and two-dimension self-coupled bose fields, *I. Funct. Analysis*, to appear.

A. Jaffe, [1] Divergence of perturbation theory for bosons, *Commun. Math. Phys.*, **1**, 127–149 (1965).

[2] Existence theorems for a cut-off $\lambda\phi^4$ field theory, in *Mathematical theory of elementary particles*, ed. R. Goodman and I. Segal, M.I.T. Press, Cambridge (1966).

[3] *The dynamics of a cut-off $\lambda\phi^4$ field theory*, Princeton University thesis (1965).

[4] *Constructive quantum field theory*, E.T.H. Lecture Notes, Zürich (1968).

[5] Constructing the $\lambda(\phi^4)_2$ theory, in *Rendiconti di Fisica Internazionale de Fisica "E. Fermi"—XLV Corso*, Academic Press, New York, 120–151 (1969).

[6] Whither axiomatic field theory?, *Reviews of Modern Physics*, **41**, 576–580 (1969).

A. Jaffe, O. Lanford, and A. Wightman, [1] A general class of cut-off model field theories, *Commun. Math. Phys.*, **15**, 47–68 (1969).

A. Jaffe and R. Powers, [1] Infinite volume limit of a $\lambda\phi^4$ field theory, *Commun. Math. Phys.*, **7**, 218–222 (1968).

R. Jost, [1] *General theory of quantum fields*, American Mathematical Society, Providence (1965).

M. Kac, [1] *Probability and related topics in physical sciences*, Interscience Publications, New York (1959).

T. Kato, [1] *Perturbation theory for linear operators*, Springer-Verlag, New York (1966).

Y. Kato and N. Mugibayasi, [1] Regular perturbation and asymptotic limits of operators in quantum field theory, *Prog. Theor. Phys.*, **30**, 103–133 (1963).

[2] Asymptotic fields in the $\lambda(\phi^4)_2$ quantum field theory, *Prog. Theor. Phys.* **45**, 628-639 (1971).

A. Kleinstein, [1] The Hamiltonian in a class of models for a quantum field theory, Columbia University Thesis (1970).

J. Konrady, [1] Almost positive perturbations of positive self adjoint operators, to appear,

R. Kunze and I. Segal, [1] *Integrals and operators*, McGraw-Hill, New York (1968).

O. Lanford, [1] Construction of quantum fields interacting by a cut-off Yukawa coupling, Princeton University Thesis (1966).

D. Masson, [1] Essential self adjointness of semi bounded operators: An extension of the Kato-Rellich theorem, to appear.

D. Masson and W. Mc Clary, [1] On the essential self adjointness of the $(g(x)\phi^4)_2$ Hamiltonian, *Commun. Math. Phys.*, **21**, 71-74 (1971).

M. Naimark, [1] *Normed rings*, P. Noordhoff, Groningen (1964).

E. Nelson, [1] Analytic vectors, *Ann. of Math.*, **70**, 572–615 (1959).

[2] Interaction of nonrelativistic particles with a quantized scalar field, *J. Math. Phys.*, **5**, 1190–1197 (1964).

[3] A quartic interaction in two dimensions, in *Mathematical theory of elementary particles*, ed. R. Goodman and I. Segal, M.I.T. Press, Cambridge (1966).

[4] *Topics in dynamics: I flows*, Princeton University Press, Princeton (1970).

K. Osterwalder, [1] Cubic boson theories in two, three and four dimensions, Fortschr. d. Phys., to appear.

L. Rosen, [1] A $\lambda\phi^{2n}$ field theory without cutoffs, *Commun. Math. Phys.*, **16**, 157–183 (1970).
[2] The $(\phi^{2n})_2$ quantum field theory: higher order estimates, Comm. Pure appl. Math., to appear.
[3] The $(\phi^{2n})_2$ quantum field theory: Lorentz covariance, preprint.

L. Rosen and B. Simon, [1] The $(\phi^{2n})_2$ field Hamiltonian for complex coupling constant, to appear.

L. Schwartz, [1] Théorie des distributions, Hermann, Paris (1966).

S. Schweber, [1] *An introduction to relativistic quantum field theory*, Row, Peterson and Co., New York (1961).

R. Schrader, [1] A remark on Yukawa plus boson selfinteraction in two space dimensions, *Commun. Math. Phys.* **21**, 164–170 (1971).
[2] Yukawa quantum field theory in two space time dimensions without cutoffs, to appear.

I. Segal, [1] Notes toward the construction of nonlinear relativistic quantum fields I; The Hamiltonian in two space-time dimensions as the generator of a C^*-automorphism group, *P.N.A.S.*, **57**, 1178–1183 (1967).
[2] Notes toward the construction of nonlinear relativistic quantum fields III: Properties of the C^*-dynamics for a certain class of interactions, *Bull. A.M.S.*, **75**, 1390–1395 (1969).
[3] Construction of nonlinear local quantum processes: I, *Ann. Math.* **92**, 462–481 (1970).

B. Simon, [1] Borel summability of the ground state energy in spatially cutoff $(\phi^4)_2$, *Phys. Rev. Letters*, **25**, 1583–1586 (1970).

E. Stein, [1] Interpolation of linear operators, *Trans. A.M.S.*, **83**, 482–492 (1956).
[2] *Paley-Littlewood theory and related topics, Ann. Math. Studies.* (1970).

R. Streater and A. Wightman, *PCT, spin* and *statistics, and all that*, W. A. Benjamin, New York (1964).

S. Weinberg, [1] High energy behavior in quantum field theory, *Phys. Rev.*, **118**, 838–849 (1960).

G.-C. Wick, [1] The evaluation of the collision matrix, *Phys. Rev.*, **80**, 268–272 (1950).

A. S. Wightman, [1] Quantum field theory in terms of vacuum expectation values, *Phys. Rev.*, **101**, 860–866 (1956).
[2] Introduction to some aspects of the relativistic dynamics of quantized fields, in *1964 Cargèse Summer School Lectures*, Ed. by M. Lévy, Gordon and Breach, New York (1967), pp. 171–291.

A. S. Wightman and L. Gårding, [1] Fields as operator valued distributions in relativistic quantum theory, *Arch. för Physik*, **28**, 129–184 (1964).

III Boson Quantum Field Models: Part I. General Results

Boson Quantum Field Models

J. GLIMM*

Courant Institute of Mathematical Sciences, New York University, New York, USA

A. JAFFE†

Lyman Laboratory of Physics, Harvard University, Cambridge, Massachussetts, USA

PART I. GENERAL RESULTS

1. INTRODUCTION

Quantum fields, from a mathematical point of view, are highly singular. These fields are believed to describe the interactions of elementary particles. For the interaction of electrons with light (photons), the quantum field description is exact within the limits of experimental accuracy (5 significant figures). For these reasons, i.e. the mathematical difficulties and the importance to physics, the problem of formulating the mathematical foundations of quantum field theory has attracted the attention of both mathematicians and physicists over a period of several decades. On the side of the physicists, the most striking achievements were the calculation in the late 1940's and early 1950's of the Lamb shift and the anomalous magnetic moment of the electron together with the development of the renormalization method on which these calculations were based. Of the mathematicians, J. von Neumann was the first to realize that new mathematical theories would be required to formulate quantum field theory correctly and this realization was one of the motives for developing the theory of operator algebras.

* Supported in part by the National Science Foundation, NSF-GP-24003.

† Supported in part by the Air Force Office of Scientific Research, Contract F44620-70-C-0030, and by the National Science Foundation, Grant GP-31239X.

The calculations referred to above are based on a perturbation method, and they depend essentially on the fact that a numerical coefficient, multiplying the interaction term in the field equation, is small ($\sim 137^{-\frac{1}{2}}$). For strong interactions (protons, neutrons, mesons, etc.), the numerical coefficient is about 15, and perturbation calculations have had only limited success. To obtain a deeper understanding of these problems, Wightman, Haag and Kastler and others have formulated precisely the basic principles (axioms) of quantum field theory, and have sought to derive rigorously consequences of these axioms. High points of this program include the PCT theorem, the spin and statistics theorem and the Haag–Ruelle and the Lehmann, Symanzik and Zimmermann scattering theories.

However the axiomatic program leaves open the question of whether quantum fields with conventional interactions satisfy the axioms, or indeed whether nontrivial fields can be shown to exist in any reasonable sense. In the past five years there has been considerable progress on the existence problem. We have worked in two- and three-dimensional space–time. This is a simplifying assumption, which reduces but does not eliminate the basic singularities of quantum field theory. The results for two dimensions are summarized as follows.

THEOREM *In two space time dimensions, quantum fields exist. These fields are known to satisfy all or most of the Haag–Kastler axioms and most of the Wightman axioms.*

The fields in this theorem have conventional Yukawa or polynomial Boson interactions. In three space time dimensions, the situation looks promising, although the results to date are of a preliminary and technical nature, and apply only to the ϕ^4 boson interaction. The methods, in their present form, do not seem to apply to four dimensions.

We will consider here only the polynomial boson interactions, because they are simpler than the Yukawa interaction. As physics background, we want to assume only a familiarity with ordinary quantum mechanics. In ordinary quantum mechanics, the time evolution

$$\psi(q,t) = (e^{-itH}\psi)(q,0) \tag{1.1}$$

is governed by the Schrödinger equation

$$i\frac{\partial}{\partial t}\psi = H\psi, \tag{1.2}$$

where H is the Hamiltonian or energy operator. For example

$$H = -\frac{1}{2}\sum_{l=1}^{n} m_l^{-1}\frac{\partial^2}{\partial q_l^2} + V(q) \tag{1.3}$$

acts on $L_2(\mathbf{R}^n)$, and $q \in \mathbf{R}^n$. The quantum field differs from the above in having an infinite number of degrees of freedom, i.e. $n = \infty$. The singularities of quantum field theory can all be traced to the fact that $n = \infty$.

For $n = \infty$, we have no satisfactory analogue of Lebesgue measure. It turns out, however, that Gaussian measures, which do generalize to the infinite dimensional case, are more suitable to our purposes. We use a Gaussian measure $d_B q$ defined by a quadratic form B on the infinite dimensional space $Q = \mathscr{S}'(\mathbf{R})$ of configurations of the classical field. Formally $d_B q$ has the density $\exp(-B(q, q))\,dq$. Then $L_2(Q, d_B q)$ replaces $L_2(\mathbf{R}^n)$ above. $\mathscr{S}'$ denotes the space of tempered distributions.

The Hamiltonian H is the second most important operator in quantum field theory. To study H, we will use both the theory of a single self-adjoint operator and the theory of operator algebras. H is self-adjoint, and the use of self-adjoint operator techniques is hardly surprising. In the main, we are concerned with resolvents $R(\zeta) = (H - \zeta)^{-1}$ and semigroups e^{-tH}. We use criteria for the self adjointness of a limit $H = \lim_\kappa H(\kappa)$ of self-adjoint operators, expressed in terms of the convergence of the associated resolvents and semigroups.

The use of operator algebra techniques to study a self-adjoint operator is less standard, and results from the fact that $n = \infty$. Gaussian measures $d_B q$ on an infinite dimensional space are very sensitive to changes in the quadratic form B. If the Jacobian $\det B_1^{\frac{1}{2}} B_2^{-\frac{1}{2}}$ does not exist (e.g. if $\det B_1^{\frac{1}{2}} B_2^{-\frac{1}{2}}$ equals zero or infinity) then the measures $d_{B_1} q$ and $d_{B_2} q$ are mutually singular. Moreover the field operators act in a natural fashion on $L_2(Q, d_{B_i} q)$, $i = 1, 2$, and these two representations of the field operators are then unitarily inequivalent. Thus in many natural approximation or limiting procedures we find operators acting on one Hilbert space, in one representation, converging to operators acting on another Hilbert space, in a unitarily inequivalent representation. Plainly we will have to think of operators independently of the Hilbert spaces on which they act, and this is exactly what operator algebras accomplish. Non-Gaussian measures seem to be equally sensitive to changes in the parameters of the problem. Typically unitary inequivalence of representations is characterized by the divergence of some integral or infinite series.

In addition to the general theories of Gaussian measures, self-adjoint operators and operator algebras, we need some information which is special to the interactions and problems we are considering. For polynomial Boson interactions in one space dimension, we have approximate Hamiltonians

$$H(g) = H_0 + H_I(g) \tag{1.4}$$

where

$$0 \leqslant g \in L_1(\mathbf{R}) \cap L_2(\mathbf{R}), \tag{1.5}$$

and the approximation is removed by the limit $g \to 1$. We show that $H(g)$ is essentially self-adjoint on

$$\mathscr{D}(H_0) \cap \mathscr{D}(H_I(g)). \tag{1.6}$$

$H_I(g)$ is a multiplication operator on $L_2(Q, d_B q)$, and as such, $H_I(g) \in L_p(Q, d_B q)$ for all $p < \infty$. H_0 acts on $L_2(Q, d_B q)$ as a Hermite operator, and it is conveniently studied in terms of a Hermite function expansion for $L_2(Q, d_B q)$. This Hermite expansion is called Fock space.

We require that the polynomial P in the interaction be positive. Let $p = \deg P$. Then $H_I(g)$ is "nearly lower bounded" in the sense that for any positive number κ, we can write

$$H_I(g) = H_I(g, \kappa) + H_I(g, \kappa)' \tag{1.7}$$

where

$$-O(\ln \kappa)^{p/2} \leqslant H_I(g, \kappa) \tag{1.8}$$

and $H_I(g, \kappa)'$ is a "small" unbounded operator, with magnitude

$$|H_I(g, \kappa)'| \sim O(\kappa^{-\frac{1}{2}+\varepsilon}). \tag{1.9}$$

In general, estimates involving positivity, such as (1.8) are performed in the Schrödinger representation, i.e. on $L_2(Q, d_B q)$, while estimates on small but indefinite remainders, such as (1.9), are performed on Fock space.

2. HERMITE OPERATORS

Before passing to the limit $n = \infty$, we study the Hermite operator with a finite number of degrees of freedom. We begin with one degree of freedom, and then the Hermite operator

$$H_0(\mu) = \frac{1}{2}\left[-\left(\frac{d}{dq}\right)^2 + \mu^2 q^2 - \mu\right], \tag{2.1}$$

acting on $L_2(R)$, is the Hamiltonian for the quantized harmonic oscillator. Here μ is an arbitrary positive normalization factor.

We introduce the annihilation and creation operators

$$b = 2^{-\frac{1}{2}}(\mu^{\frac{1}{2}}q + i\mu^{-\frac{1}{2}}p)$$
$$b^* = 2^{-\frac{1}{2}}(\mu^{\frac{1}{2}}q - i\mu^{-\frac{1}{2}}p). \tag{2.2}$$

where $p = -id/dq$. The importance of these operators derives from the representation

$$H_0(\mu) = \mu b^* b. \tag{2.3}$$

As a convenient domain for b and b^*, let

$$\mathscr{D} = \{P(q)\, e^{-\mu q^2/2} : P \text{ is a polynomial}\}. \tag{2.4}$$

Then

$$b\mathscr{D} \subset \mathscr{D}, \qquad b^*\mathscr{D} \subset \mathscr{D} \tag{2.5}$$

and so $\mathscr{D}$ is an invariant domain for b, b^*, p and q. On $\mathscr{D}$, the commutation relations

$$[b, b^*] = I \tag{2.6a}$$

$$[q, p] = iI \tag{2.6b}$$

are valid. We set $e_0(q) = (\mu/\pi)^{1/4} \exp(-\mu q^2/2)$. Then $\|e_0\|_2 = 1$ and

$$be_0 = 0. \tag{2.7}$$

Let

$$e_j = \|b^{*j}e_0\|^{-1} b^{*j}e_0.$$

To justify this definition, we compute

$$\begin{aligned} \langle b^{*j}e_0,\ b^{*l}e_0\rangle &= \langle bb^{*j}e_0,\ b^{*(l-1)}e_0\rangle \\ &= \langle [b, b^{*j}]e_0,\ b^{*(l-1)}e_0\rangle \\ &= j < b^{*(j-1)}e_0,\ b^{*(l-1)}e_0\rangle \\ &= j!\,\delta_{jl} \end{aligned} \tag{2.8}$$

and see that $\|b^{*j}e_0\| \neq 0$. This calculation also shows that the e_j's form an orthogonal family. e_j is the jth Hermite function. From (2.8) we have

$$\left.\begin{aligned} b^*e_j &= (j+1)^{\frac{1}{2}}\, e_{j+1} \\ b\, e_j &= j^{\frac{1}{2}}\, e_{j-1}, \qquad j > 0. \end{aligned}\right\} \tag{2.9}$$

It follows from (2.3) and (2.9) that

$$H_0(\mu)\, e_j = j\mu e_j.$$

Thus $H_0(\mu)$ has eigenvalues $0, \mu, 2\mu, \ldots$ and eigenfunctions $e_0, e_1, \ldots$. Since $e_j \in \mathcal{D}$,

$$e_j(q) = P_j(q)\, e_0(q)$$

for some polynomial P_j. P_j is the jth Hermite polynomial. From (2.2) and (2.9) we see that P_j has degree j and that the coefficient of q^j in P_j is positive.

To complete our analysis of $H_0(\mu)$, we must show that the eigenfunctions e_j are complete in L_2. The linear span of the eigenfunctions is exactly $\mathcal{D}$. $\mathcal{D}^-$ contains the functions $\exp(i\lambda q)\exp(-\mu q^2/2)$, since the Taylor series in powers of λ for such functions converge in L_2. To establish this statement, we invert the relation between b, b^* and q, p, obtaining

$$\begin{aligned} q &= (2\mu)^{-\frac{1}{2}}(b^* + b) \\ p &= i2^{-\frac{1}{2}}\mu^{\frac{1}{2}}(b^* - b). \end{aligned} \tag{2.10}$$

Then

$$\begin{aligned} (\mu/\pi)^{\frac{1}{2}} \int q^j e^{-\mu q^2}\, dq &= (2\mu)^{-j/2}\langle (b^* + b)^j e_0, e_0\rangle \\ &\leqslant (\text{const.})^j\, O(j!)^{\frac{1}{2}}, \end{aligned}$$

and the required L_2 convergence follows. With $\exp(i\lambda q)\exp(-\mu q^2/2) \in \mathcal{D}^-$ for all λ, we have

$$(2\pi)^{-\frac{1}{2}} \int f(\lambda)^{\sim}\, e^{i\lambda q}\, e^{-\mu q^2/2}\, d\lambda = f(q)\, e^{-\mu q^2/2}$$

in $\mathcal{D}^-$ for all $f \in \mathcal{S}$, and so $\mathcal{D}^- = L_2$.

The operators b and b^* are uniquely characterized by the relations (2.5), (2.6a) and (2.7). This uniqueness theorem is valid without any restriction on the number of degrees of freedom. There is another uniqueness theorem, the von Neumann uniqueness theorem, which assumes the commutation relations (2.6) in integrated form

$$e^{isq}\, e^{itp} = e^{itp}\, e^{isq}\, e^{-ist}, \tag{2.11}$$

see the lectures of B. Simon. For a finite number of degrees of freedom, (2.11) determines the action of p and q on a Hilbert space $\mathcal{H}$ uniquely, up to unitary equivalence and multiplicity. However for an infinite number of degrees of freedom, (2.11) does not lead to uniqueness. The failure of the von Neumann uniqueness theorem for an infinite number of degrees of freedom indicates again that operator algebras will play a larger role in field theory than in ordinary quantum mechanics.

Definition 2.1. Let $\mathscr{E}$ be a real prehilbert space (= incomplete Hilbert space). A representation of the canonical commutation relations over $\mathscr{E}$ is a pair of linear maps

$$f \to b(f), \qquad g \to b^*(g)$$

from $\mathscr{E}$ to operators $b(f)$ and $b^*(g)$ defined on a dense domain $\mathscr{D}$ in a (complex) Hilbert space $\mathscr{H}$ such that

$$b(f)\mathscr{D} \subset \mathscr{D}, \qquad b^*(g)\mathscr{D} \subset \mathscr{D}$$

$$[b(f), b^*(g)]\theta = \langle f, g\rangle\theta$$

$$[b(f), b(g)]\theta = 0 = [b^*(f), b^*(g)]\theta$$

and

$$\langle \theta_1, b(f)\theta_2\rangle = \langle b^*(f)\theta_1, \theta_2\rangle$$

for all θ, θ_1 and θ_2 in $\mathscr{D}$ and all f and g in $\mathscr{E}$.

Definition 2.2. A representation of the canonical commutation relations is a Fock representation if there is a unit vector $\Omega \in \mathscr{D}$ such that

$$b(f)\Omega = 0$$

for all $f \in \mathscr{E}$ and such that $\mathscr{D}$ is spanned algebraically by vectors of the form

$$\{b^*(g_1) \ldots b^*(g_m)\Omega : g_l \in \mathscr{E};\ m = 0, 1, \ldots\}.$$

Here Ω is called the Fock vacuum vector.

Example. Let $\mathscr{H} = L_2(\mathbf{R})$, $\mathscr{E} = \mathbf{R}$, $\mathscr{D}$ as above, $\Omega = e_0$, $b(\lambda) = \lambda b$ and $b^*(\lambda) = \lambda b^*$. This defines a Fock representation of the canonical commutation relations.

THEOREM 2.3. *The Fock representation of the canonical commutation relations over $\mathscr{E}$ is unique up to unitary equivalence. If $\{b_i, b_i^*\}$, $i = 1, 2$, are two Fock representations over $\mathscr{E}$ with vacuums Ω_i, then the unitary equivalence operator U is uniquely determined, if we require*

$$U\Omega_1 = \Omega_2.$$

Proof. Let

$$\theta_i = b_i^*(f_1) \ldots b_i^*(f_n)\Omega_i$$

$$\psi_i = b_i^*(g_1) \ldots b_i^*(g_m)\Omega_i.$$

We compute $\langle\psi_i, \theta_i\rangle$, and obtain the same value for $i = 1$ and $i = 2$. Then U, defined by the equations $U\theta_1 = \theta_2$, extends by linearity to a unitary operator from $\mathscr{H}_1$ to $\mathscr{H}_2$ giving the required equivalence. Furthermore if U' is any unitary operator from $\mathscr{H}_1$ onto $\mathscr{H}_2$ and if

$$U'\Omega_1 = \Omega_2$$
$$U'b_1{}^*(f)U'^* = b_2{}^*(f)$$

for all $f \in \mathscr{E}$, then $U'\theta_1 = \theta_2$ and so U is uniquely determined.

From the commutation relations and the fact that $b_i\Omega_i = 0$, we have

$$\langle\psi_i, \theta_i\rangle = \sum_{l=1}^{m} \langle f_1, g_l\rangle$$
$$\times \langle b_i{}^*(g_1) \dots b_i{}^*(g_{l-1})b_i{}^*(g_{l+1}) \dots b_i{}^*(g_m)\Omega_i, b_i{}^*(f_2) \dots b_i{}^*(f_n)\Omega_i\rangle.$$

Continuing by induction, we have

$$\langle\psi_i, \theta_i\rangle = \delta_{m,n} \sum_{\sigma} \prod_{l=1}^{n} \langle f_l, g_{\sigma(l)}\rangle$$

where the sum ranges over all permutations σ on n elements. This completes the proof. ∎

For n degrees of freedom, the oscillator Hamiltonian is a sum

$$H_0 = \sum_{l=1}^{n} H_0(\mu_l) \tag{2.12}$$

of n one-dimensional Hermite operators, each acting on one of the n variables q_l of $Q = \mathbf{R}^n$. The eigenfunctions of H_0 are tensor products

$$e(q) = \prod_{l=1}^{n} e_{j(l)}(q_l)$$

of the one-dimensional Hermite functions. H_0 acts on the Hilbert space $\mathscr{H} = L_2(Q)$ and $\mathscr{H}$ is a tensor product,

$$\mathscr{H} = L_2(Q) = L_2(\mathbf{R}^n) = \bigotimes_{l=1}^{n} L_2(Q_l), \tag{2.13}$$

where Q_l is a copy of the real numbers, $\mathbf{R}$.

In Section 3, we take the limit $n \to \infty$. In order to construct a measure on the resulting infinite tensor product space Q, it is essential to have a measure on each factor whose total mass is one. Furthermore, we want H_0 to be a positive self-adjoint operator on $\mathscr{H}$ in the limit $n \to \infty$, and so it is natural to require that the ground state e_0 of H_0 lie in $\mathscr{H}$. To ensure that $e_0 \in \mathscr{H}$ in the limit $n \to \infty$, we replace p_l, q_l and H_0 by unitarily equivalent operators, before taking the limit. We write $L_2(Q, dq)$ to indicate the use of Lebesgue measure. The multiplication operator

$$Uf = e_0 f$$

is a unitary operator from $L_2(Q, e_0{}^2 dq)$ onto $L_2(Q, dq)$, and $U^* g = e_0{}^{-1} g$. Then

$$U^* q_l U = q_l \tag{2.14}$$

$$U^* p_l U = -i \frac{\partial}{\partial q_l} + i \mu_l q_l \tag{2.15}$$

and $U^* H_0 U$ act on $L_2(Q, e_0{}^2 dq)$. The eigenfunctions of $U^* H_0 U$ are tensor products

$$\prod_{l=1}^{n} P_{j(l)}(q_l)$$

of the one-dimensional Hermite polynomials and the ground state of $U^* H_0 U$ is

$$\prod_{l=1}^{n} P_0(q_l) \equiv 1.$$

Certainly $1 \in L_2(Q, dv(q))$ for any measure dv of total mass one. For comparison with later chapters, we write

$$e_0{}^2\, dq = \text{const}\, e^{-B(q,q)}\, dq$$

where

$$B(q, q) = \sum_{l=1}^{n} \mu_l q_l{}^2.$$

The finite constant μ, subtracted from $H_0(\mu)$ in (2.1), is a renormalization. With this renormalization, the bottom of the spectrum of $H_0(\mu)$, H_0 and $U^* H_0 U$ are all zero. This renormalization is optional for finite n, but if

$$\sum_{l=1}^{\infty} \mu_l = \infty,$$

as is generally the case, this renormalization is crucial in order to take the limit $n \to \infty$.

THEOREM 2.4. *The kernel of* $\exp(-tU^*H_0U)$ *is positive.*

Proof. Because H_0 is a sum of terms, each acting on a distinct factor in (2.13), the above kernel factors also, and so it is sufficient to consider the case of one degree of freedom. We show that in this case the kernel is

$$k(q,q') = 2^{-\frac{1}{2}}(1 - e^{-2\mu t})^{-\frac{1}{2}} \exp\left[-\frac{\mu(q' - e^{-\mu t}q)^2}{2(1 - e^{-\mu t})} + \mu q'^2 \right] \tag{2.16}$$

relative to the measure $e_0{}^2\, dq$ on Q. The theorem follows from (2.16). ∎

We prove equality of Fourier transforms. Let

$$Kf(q) = \int k(q,q') f(q')\, e_0(q')^2\, dq'.$$

We show that $Ke^{i\lambda q} = U^* e^{-tH_0} U e^{i\lambda q}$, and since the vectors $e^{i\lambda q}$ are dense in $L_2(Q, e_0{}^2 dq)$, this completes the proof. An elementary calculation gives

$$Ke^{i\lambda q} = \exp\left[-\lambda^2(1 - e^{-2t\mu})/2\mu\right] \exp\left(i\lambda e^{-t\mu} q\right).$$

However since

$$\begin{aligned} e^{i\lambda q} e_0 &= e^{-\lambda^2/2\mu} \exp\left[i\lambda(2\mu)^{-\frac{1}{2}} b^*\right] e_0 \\ &= e^{-\lambda^2/2\mu} \sum_{n=0}^{\infty} \left(\frac{i\lambda}{\sqrt{2\mu}}\right)^n (n!)^{-\frac{1}{2}}\, e_n, \end{aligned}$$

we have

$$\begin{aligned} e^{-tH_0} e^{i\lambda q} e_0 &= e^{-\lambda^2/2\mu} \sum_{n=0}^{\infty} \left(\frac{i\lambda}{\sqrt{2\mu}} e^{-t\mu}\right)^n (n!)^{-\frac{1}{2}}\, e_0 \\ &= \exp\left[-\lambda^2/2\mu\right] \exp\left[i\lambda(2\mu)^{-\frac{1}{2}} e^{-t\mu} b^*\right] e_0 \\ &= \exp\left[-\lambda^2(1 - e^{-2t\mu})/2\mu\right] \exp\left[i\lambda e^{-t\mu} q\right] e_0. \end{aligned}$$

Hence

$$U^* e^{-tH_0} U e^{i\lambda q} = \exp\left[-\lambda^2(1 - e^{-2t\mu})/2\mu\right] \exp\left[i\lambda e^{-t\mu} q\right]$$

and the proof is complete. ∎

3. GAUSSIAN MEASURES AND THE SCHRÖDINGER REPRESENTATION

In this section we introduce the quantum field operators ϕ and π. They are analogous to the operators q_l and p_l of Section 2, and we go somewhat out of our way to make the analogy evident, so that the definitions should seem natural to mathematicians familiar with ordinary quantum mechanics. For simplicity we confine our attention to the quantization of the Klein–Gordon equation

$$\phi_{tt} - \phi_{xx} + m^2\phi = 0, \tag{3.1}$$

although the methods of this chapter apply to more general linear equations. Let $\pi = \partial\phi/\partial t$. The energy of a solution ϕ of (3.1) is

$$H_0(\phi, \pi) = \tfrac{1}{2}\int (\pi(x,t)^2 + \nabla\phi(x,t)^2 + m^2\phi(x,t)^2)\,dx \tag{3.2}$$

and using (3.1), it is easy to see that H is independent of the time, t. We note that (3.1) is a Hamiltonian system:

$$\dot{\phi}(x,t) = \delta H_0/\delta\pi(x,t)$$

$$\dot{\pi}(x,t) = -\,\delta H_0/\delta\phi(x,t)$$

and that $\pi(x,t)$ is the momentum variable conjugate to $\phi(x,t)$. Configuration space Q is a linear space of functions $\phi(x) = \phi(x,0)$ and phase-space is a linear space of pairs of functions $\{\phi(x), \pi(x)\}$. It is convenient to take Q to be the real† valued Schwartz space, $Q = \mathscr{S}_{\mathbf{R}}'(\mathbf{R})$. Then H_0 is real valued and finite on a dense subspace of phase space.

$H_0(\phi, \pi)$ is a quadratic form. For comparison with Section 2, we want to diagonalize H_0, so we take Fourier transforms. With

$$\pi(x) = (2\pi)^{-\frac{1}{2}}\int e^{-ikx}\pi^{\sim}(k)\,dk, \tag{3.3}$$

one sees that

$$H_0(\phi,\pi) = \tfrac{1}{2}\int [(\mathrm{Re}\,\pi^{\sim}(k))^2 + (\mathrm{Im}\,\pi^{\sim}(k))^2]\,dk$$

$$+ \tfrac{1}{2}\int \mu(k)^2 [(\mathrm{Re}\,\phi^{\sim}(k))^2 + (\mathrm{Im}\,\phi^{\sim}(k))^2]\,dk, \tag{3.4}$$

where

$$\mu(k) = (k^2 + m^2)^{\frac{1}{2}}.$$

Let

$$\mu_x = (-(d/dx)^2 + m^2)^{\frac{1}{2}}. \tag{3.4}$$

We want to quantize the free field Hamiltonian H_0 of (3.2), (3.4) according to the prescription of Section 2. Thus we seek a Gaussian

† See Note 1 on page 143.

measure $d_B q$ on Q which formally has a density proportional to $\exp(-B(q, q))$, where

$$B(f, g) = \int \mu(k)[\operatorname{Re} f(k)^\sim \operatorname{Re} g(k)^\sim + \operatorname{Im} f(k)^\sim \operatorname{Im} g(k)^\sim]\, dk = \langle f, \mu_x g\rangle. \tag{3.5}$$

For each finite dimensional set F of linear coordinate functions on Q, we define a finite dimensional Gaussian measure. These measures satisfy a consistency condition when $F_1 \subset F_2$ and they determine the measure $d_B q$. This measure is countably additive† on Q. The countable additivity is not obvious and it depends on the choice $Q = \mathcal{S}_{\mathbf{R}}'(\mathbf{R})$. Smaller spaces than $\mathcal{S}_{\mathbf{R}}'(\mathbf{R})$ could be chosen, but it seems that Q cannot be a space of functions.

The coordinate functions on Q are just the elements of $Q' = \mathcal{S}_{\mathbf{R}}(\mathbf{R})$. For $f \in Q'$, the coordinate function is

$$Q \ni q \to q(f) = \int q(x) f(x)\, dx. \tag{3.6}$$

A polynomial on Q is just a polynomial in the coordinate functions on Q. If $f_1, \ldots, f_n \in Q$ and if P is a polynomial in n variables, then

$$\Phi(q) = P(q(f_1), \ldots, q(f_n)) \tag{3.7}$$

is a polynomial on Q. If $f_1, \ldots, f_n$ lie in a closed subspace F of Q', then we say that Φ in (3.7) is based on F. If Φ is based on F then the function $\Phi(\mu_x^{-\frac{1}{2}} q)$ is based on $\mu_x^{-\frac{1}{2}} F$. We use this change of variables, $\Phi(q) \to \Phi(\mu_x^{-\frac{1}{2}} q)$ in the definition of $d_B q$. Formally, this change of variables maps a density proportional to $\exp(-B(q, q))$ onto a density proportional to $\exp(-\|q\|_2^2)$, since formally

$$\int \Phi(q) e^{-B(q,q)}\, dq = \text{const} \int \Phi(\mu_x^{-\frac{1}{2}} q) \exp(-\|q\|_2^2)\, dq.$$

Let F be a finite dimensional subspace of Q' and let $\{f_1, \ldots f_n\}$ be a basis for F chosen so that $\{\mu_x^{-\frac{1}{2}} f_1, \ldots, \mu_x^{-\frac{1}{2}} f_n\}$ is an orthonormal basis for $\mu_x^{-\frac{1}{2}} F$. If Φ and P are related by (3.7), we define

$$\int \Phi(q)\, d_B q = \pi^{-n/2} \int P(\lambda_1, \ldots, \lambda_n) \exp[-\Sigma \lambda_i^2]\, d\lambda. \tag{3.8}$$

LEMMA 3.1. *The integral* (3.8) *is independent of the finite dimensional space* F *on which* Φ *is based.*

Proof. If Φ is based on F and G, it is also based on $F + G$, and so we may suppose that $F \subset G$. We choose a basis $\{g_1, \ldots, g_n\}$ for G so that

† See Note 2 on page 143.

$\{g_1, \ldots, g_m\}$ is a basis for F and so that the set $\{\mu_x^{-\frac{1}{2}} g_l\}$ is orthonormal. There are polynomials P_F and P_G such that

$$\Phi(q) = P_F(q(g_1), \ldots, q(g_m)) = P_G(q(g_1), \ldots, q(g_n)).$$

Thus P_G is independent of the final $n - m$ variables and equals P_F as a function of the first m variables. Hence integrating over the final $n - m$ variables, we have

$$\pi^{-n/2} \int P_G(\lambda_1, \ldots, \lambda_n) \exp\left(- \sum_{l=1}^{n} \lambda_l^2 \right) d\lambda$$

$$= \pi^{-m/2} \int P_F(\lambda_1, \ldots, \lambda_m) \exp\left(- \sum_{l=1}^{m} \lambda_l^2 \right) d\lambda$$

and the lemma is proved. ∎

With F held fixed, the integral (3.8) is countably additive. Moreover we can replace P by a bounded Borel function and the consistency condition, Lemma 3.1, and its proof remain valid. If Y_F is a Borel set in R^n, we define

$$Y = \{q : q(f_1), \ldots, q(f_n) \in Y_F\}. \tag{3.9}$$

Such a set is called a Borel cylinder set, based on F. The Borel cylinder sets form a ring, in the sense of measure theory, and the Borel sets in Q are the elements of the generated σ-ring. The integral defines a measure ν on cylinder sets. If $\{\mu_x^{-\frac{1}{2}} f_j\}$ is orthonormal and if Y and Y_F are related as above, then

$$\begin{aligned} \nu(Y) &= \int \chi_Y(q)\, d_B q = \pi^{-n/2} \int \chi_{Y_F}(\lambda) \exp(-\Sigma \lambda_l^2\, d\lambda) \\ &= \pi^{-n/2} \int_{Y_F} \exp(-\Sigma \lambda_l^2)\, d\lambda, \end{aligned} \tag{3.10}$$

where χ_Y is the characteristic function of Y.

Proposition 3.2. *The measure ν, defined on the ring of Borel cylinder sets, is countably additive.*

We postpone the proof to the end of this Section. As an immediate corollary of Proposition 3.2 and general results of measure theory, we have

Theorem 3.3. *The measure ν extends uniquely to define a countably additive measure $\nu = d_B q$ on the σ-ring of Borel sets.*

Let $\mathscr{H} = L_2(Q, d_B q)$. We call $\mathscr{H}$ Schrödinger space, or more precisely, the Schrödinger space of the free field because the (free) field operators $\phi(f)$ defined below are represented on $\mathscr{H}$ as multiplication operators, as in the Schrödinger representation of ordinary quantum mechanics. Let $\mathscr{D} \subset \mathscr{H}$ be the set of complex polynomial functions on Q.

PROPOSITION 3.4. *$\mathscr{D}$ is dense in $\mathscr{H}$.*

Proof. Let $\mathscr{D}_F$ be the set of polynomials based on F and let $\mathscr{H}_F$ be the set of L_2 functions based on F. Then $\mathscr{D}_F$ is dense in $\mathscr{H}_F$ since the Hermite functions are complete in L_2. (See Chapter 2.) By general measure theory, $\bigcup_F \mathscr{H}_F$ is dense in $\mathscr{H}$, and this completes the proof. ∎

We define $\phi(f)$ to be the multiplication operator

$$\phi(f) : \Phi(q) \to q(f)\Phi(q). \tag{3.11}$$

Then $\phi(f)$ contains $\mathscr{D}$ in its domain and $\phi(f)\mathscr{D} \subset \mathscr{D}$. $\phi(f)$ is interpreted as a position operator, where position refers not to the location of a point x in physical space, R, but rather to the location of the classical time-zero field $\phi(x)$ in its configuration space $Q = \mathscr{S}_{\mathbf{R}}'(\mathbf{R})$. In fact $\phi(f)$ is the position operator which measures the distance from the origin of Q in the coordinate direction f. It is not hard to show that $\phi(f) \in L_p(Q, d_B q)$ for all $p < \infty$, and so $\phi(f)$ defines a self-adjoint operator on $L_2(Q, d_B q)$.

On the domain $\mathscr{D}$, we define

$$\pi(f) = -i \frac{\partial}{\partial \phi(f)} + i\phi(\mu_x f). \tag{3.12}$$

The second term is required to make $\pi(f)$ symmetric, and occurs because of our use of a Gaussian, rather than Euclidean measure (cf. (2.15)). Let $\Omega_0 \in \mathscr{D}$ be the function identically equal to one. On a vector

$$\theta = \phi(f_1) \ldots \phi(f_n)\Omega_0 \in \mathscr{D} \tag{3.13}$$

we have

$$\pi(f)\theta = -i \sum_{l=1}^{n} \langle f, f_l \rangle \phi(f_1) \ldots \phi(f_{l-1})\phi(f_{l+1}) \ldots \phi(f_n)\Omega_0 + i\phi(\mu_x f)\theta.$$

The inner product $\langle\, , \rangle$ is the Euclidean inner product,

$$\langle f, f_l \rangle = \int f(x) f_l(x)\, dx.$$

The commutation relations

$$[\phi(f), \pi(g)] = i\langle f, g\rangle I \tag{3.14}$$

$$[\phi(f), \phi(g)] = 0 = [\pi(f), \pi(g)] \tag{3.15}$$

may be verified directly on $\mathscr{D}$.

In analogy with (2.2) we introduce the operators

$$b(f) = 2^{-\frac{1}{2}}(\phi(\mu_x^{\frac{1}{2}}f) + i\,\pi(\mu_x^{-\frac{1}{2}}f)) \tag{3.16}$$

$$b^*(f) = 2^{-\frac{1}{2}}(\phi(\mu_x^{\frac{1}{2}}f) - i\,\pi(\mu_x^{-\frac{1}{2}}f)). \tag{3.17}$$

THEOREM 3.5. *The representation of b and b^* defined by* (3.16–17) *is the Fock representation for the canonical commutation relations over Q'.*

Proof. The commutation relations for b and b^* follow from (3.14–17). Since ϕ and π are defined on $\mathscr{D}$ and map $\mathscr{D}$ into $\mathscr{D}$, the same is true for b and b^*. Since $\Omega_0 \equiv 1$, we have

$$-i\frac{\partial}{\partial\phi(f)}\Omega_0 = 0$$

$$i\,\pi(\mu_x^{-\frac{1}{2}}f)\Omega_0 = -\,\phi(\mu_x^{\frac{1}{2}}f)\Omega_0$$

and

$$b(f)\Omega_0 = 0.$$

Since $\mathscr{D}$ is spanned by polynomials in ϕ, acting on Ω_0, it is also spanned by polynomials in b^*, acting on Ω_0. The fact that $b^*(f)$ is contained in the adjoint of $b(f)$ follows from (3.16–17) and the fact that $\pi(f)$ is symmetric, proved below.

LEMMA 3.6. *$\pi(f)$ is a symmetric operator on the domain $\mathscr{D}$.*

Proof. It is sufficient to show that

$$\langle \pi(f)\theta_1, \theta_2\rangle = \langle \theta_1, \pi(f)\theta_2\rangle$$

where θ_1 and θ_2 are vectors of the form (3.13). We reason by induction on the number n_1 of factors $\phi(f)$ in the definition of θ_1. Assuming the identity for n_1, we have

$$\begin{aligned}\langle \pi(f)\phi(g)\theta_1, \theta_2\rangle &= i\langle f, g\rangle\langle\theta_1, \theta_2\rangle + \langle\phi(g)\pi(f)\theta_1, \theta_2\rangle\\ &= i\langle f, g\rangle\langle\theta_1, \theta_2\rangle + \langle\theta_1, \pi(f)\phi(g)\theta_2\rangle\\ &= \langle\theta_1, \phi(g)\pi(f)\theta_2\rangle\\ &= \langle\phi(g)\theta_1, \pi(f)\theta_2\rangle\end{aligned}$$

and so the identity holds for $n_1 + 1$. Thus it is sufficient to assume $n_1 = 0$ and establish

$$\langle \pi(f)\Omega_0, \theta\rangle = \langle \Omega_0, \pi(f)\theta\rangle.$$

This is equivalent to the identity

$$\langle \Omega_0, \phi(f)\phi(f_1) \dots \phi(f_n)\Omega_0\rangle$$

$$= \frac{1}{2}\sum_{l=1}^{n} \langle \mu_x^{-1}f, f_l\rangle\langle \Omega_0, \phi(f_1) \dots \phi(f_{l-1})\phi(f_{l+1}) \dots \phi(f_n)\Omega_0\rangle \quad (3.18)$$

once we substitute $\mu_x^{-1}f$ for f.

To establish (3.18), we choose a finite dimensional subspace F of Q' containing all the f's, and we choose a basis $\{g_1, \dots, g_m\}$ for F with the set $\{\mu_x^{-\frac{1}{2}}g_1, \dots, \mu_x^{-\frac{1}{2}}g_m\}$ orthonormal. Since (3.18) is linear in the f's, we may suppose that each of the f's is one of the basis elements, and set $f = g_1$. Then (3.18) is equivalent to the identity

$$\int_{R^m} \lambda_1\lambda_{l_1} \dots \lambda_{l_n} \exp\left(-\sum_{i=1}^{m} \lambda_i^2\right) d\lambda$$

$$= \frac{1}{2}\sum_{l=1}^{n} \delta_{1l_l}\int_{R^m} \lambda_{l_1} \dots \lambda_{l_{l-1}}\lambda_{l_{l+1}} \dots \lambda_{l_n} \exp\left(-\sum_{i=1}^{m} \lambda_i^2\right) d\lambda.$$

The integral over $\lambda_2, \dots, \lambda_m$ contributes the same factor to each side of the identity, and so we are reduced to proving

$$\int_R \lambda^j e^{-\lambda^2}\, d\lambda = \frac{j-1}{2}\int_R \lambda^{j-2} e^{-\lambda^2}\, d\lambda.$$

The latter is proved by integration by parts. This completes the proof of the lemma and the theorem. ∎

Inverting the relations between b, b^* and ϕ, π, we have

$$\left.\begin{aligned} \phi(f) &= 2^{-\frac{1}{2}}(b^*(\mu_x^{-\frac{1}{2}}f) + b(\mu_x^{-\frac{1}{2}}f)) \\ \pi(f) &= i2^{-\frac{1}{2}}(b^*(\mu_x^{\frac{1}{2}}f) - b(\mu_x^{\frac{1}{2}}f)), \end{aligned}\right\} \quad (3.19)$$

which generalizes (2.10).

We now turn to the proof of Proposition 3.2. The measure is finitely additive and it is regular in the sense that

$$\nu(Y) = \inf_{Y' \supset Y} \nu(Y')$$

where Y' ranges over open cylinder sets and Y is a Borel cylinder set. Let

$$h = -(d/dx)^2 + x^2$$

be the Hermite operator as studied in Chapter 2, and let

$$\|q\|_j = \|h^{j/2} q\|_{L_2}.$$

The norms $\|\cdot\|_j$ determine the topology in Q and Q'. Let Q_j be the subspace of Q on which $\|\cdot\|_j$ is finite. Then

$$\ldots \supset Q_j \supset Q_{j+1} \supset \ldots$$

and

$$Q = \bigcup_j Q_j, \qquad Q' = \bigcap_j Q_j.$$

Let

$$S(r, j) = \{q : \|q\|_j \leqslant r\}.$$

PROPOSITION 3.7. *Let ν be a finitely additive regular measure defined on Borel cylinder sets in Q. Suppose that for any $\varepsilon > 0$ there is a $j = j(\varepsilon)$ and an $r = r(\varepsilon)$ such that for any cylinder set Y disjoint from $S(r, j)$, we have $\nu(Y) \leqslant \varepsilon$. Then ν is countably additive.*

Remark. The condition on j and r can be reformulated by saying that the inner measure of $\sim S(r, j)$ is bounded by ε. The converse to the proposition is easily established, i.e. if ν is countably additive then such $j(\varepsilon)$ and $r(\varepsilon)$ exist.

Proof. Let $Y = \bigcup_{k=1}^{\infty} Y_k$ be a Borel cylinder set, expressed as a disjoint union of Borel cylinder sets. Let $Y_0 = Q \sim Y$, and then we must show that

$$\sum_{k=0}^{\infty} \nu(Y_k) = 1.$$

By finite additivity,

$$\sum_{k=0}^{\infty} \nu(Y_k) = \lim_{L \to \infty} \sum_{k=0}^{L} \nu(Y_k) = \lim_{L \to \infty} \nu\left(\bigcup_{k=0}^{L} Y_k\right) \leqslant 1.$$

Because ν is regular, the proof is completed by showing that

$$\sum_{k=0}^{\infty} \nu(Z_k) \geqslant 1$$

whenever Z_k is an open cylinder set containing Y_k.

Let $\varepsilon > 0$ be given. We use the fact that $S(r, j)$ is a ball in a Hilbert space, and so is weakly compact. By weak compactness, there is a finite number of these sets, $Z_0, \ldots, Z_l$, which form a cover for $S(r, j)$. Let

$$Z = Q \sim \bigcup_{k=0}^{l} Z_k.$$

Then Z is a cylinder set, disjoint from $S(r, j)$ and so

$$\varepsilon \geqslant \nu(Z) = \nu\left(Q \sim \bigcup_{k=0}^{l} Z_k\right) \geqslant 1 - \sum_{k=0}^{l} \nu(Z_k) \geqslant 1 - \sum_{k=0}^{\infty} \nu(Z_k).$$

Hence

$$\sum_{k=0}^{\infty} \nu(Z_k) \geqslant 1 - \varepsilon,$$

and since this is true for all $\varepsilon > 0$, the proof is complete. ∎

Proof of Proposition 3.2. We choose j sufficiently negative, so that the operator

$$C = h^{j/2} \mu_x^{-\frac{1}{2}}$$

is Hilbert Schmidt on L_2, and we apply the criterion of Proposition 3.7. We choose

$$r = r(\varepsilon) = \left[\operatorname{Trace} C^*C \, \varepsilon^{-1} \pi^{-\frac{1}{2}} \int_R \lambda^2 e^{-\lambda^2} \, d\lambda\right]^{\frac{1}{2}}.$$

Let Z be a Borel cylinder set disjoint from $S(r, j)$ and let Z be based on a finite dimensional subspace F of Q'. Let $G = \mu_x^{-\frac{1}{2}} F$ and let P_G be the orthogonal projection in L_2 onto G. With

$$D = h^{j/2} \mu_x^{-\frac{1}{2}} P_G,$$

we have

$$\operatorname{Trace} D^*D \leqslant \operatorname{Trace} C^*C.$$

We choose a basis $\{f_1, \ldots, f_n\}$ for F so that the vectors $g_i = \mu_x^{-\frac{1}{2}} f_i$ are eigenvectors for D^*D. Furthermore we can require that the set $\{g_i\}$ is orthonormal. Let δ_i be the corresponding eigenvalue.

Using these coordinates, we define

$$Z_F = \{\lambda \in \mathbf{R}^n : \lambda_i = q(f_i), \; q \in Z\}$$

$$S(r, j)_F = \{\lambda \in \mathbf{R}^n : \lambda_i = q(f_i), \; q \in S(r, j)\}.$$

Then

$$\nu(Z) = \pi^{-n/2} \int_{Z_F} \exp\left(-\Sigma \lambda_i^2\right) d\lambda.$$

Since Z is a cylinder set, and since Z and $S(r,j)$ are disjoint, the projections Z_F and $S(r,j)_F$ are disjoint also. Hence

$$\nu(Z) \leqslant \pi^{-n/2} \int_{\sim S(r,j)_F} \exp(-\Sigma\lambda_i^2)\, d\lambda.$$

However by a change of variables, $q \to \mu_x^{-\frac{1}{2}} q$, we see that

$$\begin{aligned} S(r,j)_F &= \{\lambda \in \mathbf{R}^n : \lambda_i = q(g_i),\ \|Cq\|_{L_2} \leqslant r,\ q \in Q\} \\ \supset S'(r,j)_F &= \{\lambda \in \mathbf{R}^n : \lambda_i = q(g_i),\ \|Dq\|_{L_2} \leqslant r,\ q \in G\} \\ &= \{\lambda \in \mathbf{R}^n : \textstyle\sum_i \delta_i \lambda_i^2 \leqslant r^2\}. \end{aligned}$$

Thus

$$\begin{aligned} \nu(Z) &\leqslant \pi^{-n/2} \int_{\Sigma_i \delta_i \lambda_i^2 \geqslant r^2} \exp(-\Sigma\lambda_i^2)\, d\lambda \\ &\leqslant \pi^{-n/2} \int_{R^n} \frac{\sum_i \delta_i \lambda_i^2}{r^2} \exp(-\textstyle\sum_i \lambda_i^2)\, d\lambda \\ &\leqslant \operatorname{Trace} C^*C\, r^{-2} \pi^{-n/2} \sup_k \int_{R^n} \lambda_k^2 \exp(-\textstyle\sum_i \lambda_i^2)\, d\lambda \\ &\leqslant \varepsilon \end{aligned}$$

and the proof is complete. ∎

4. HERMITE EXPANSIONS AND FOCK SPACE

In Fock space the free Hamiltonian H_0 is diagonalized. This fact is one of the merits of Fock space and the other is that some calculations or estimates are easier to do in Fock space than in Schrödinger space. In order to diagonalize H_0, we must utilize a momentum space (rather than a configuration space) representation for Fock space. First we define Fock space directly and then we identify it as the Hermite expansion of the Schrödinger space $L_2(Q, d_B q)$ using Theorem 2.3.

Let $\mathscr{F}_n$ be the space of symmetric L_2 functions defined on R^n; $\mathscr{F}_0$ is the complex numbers. Let

$$\mathscr{F} = \sum_{n=0}^{\infty} \mathscr{F}_n \tag{4.1}$$

$$\Omega_0 = 1 \in \mathscr{F}_0 \subset \mathscr{F}. \tag{4.2}$$

$\mathcal{F}_n$ is called the n particle subspace of $\mathcal{F}$ and $\mathcal{F}$ is Fock space. With $\theta = \{\theta_0, \theta_1, \ldots\} \in \mathcal{F}$, θ_n is the n particle component of θ. The number (of particles) operator N is defined by the formula

$$N\theta = \{0, \theta_1, 2\theta_2, \ldots, n\theta_n, \ldots\}. \tag{4.3}$$

Ω_0 is the vacuum or no particle state. These definitions refer to particles whose dynamics is governed by the free Hamiltonian H_0 (cf. Corollary 4.5). These particles have a provisional role in our construction and they are not the physical particles, which emerge from the theory as $|t| \to \infty$.

Let S_n be the projection of $L_2(\mathbf{R}^n)$ onto $\mathcal{F}_n$ and let $\mathcal{D}$ be the dense domain in $\mathcal{F}$ spanned algebraically by Ω_0 and vectors of the form

$$S_n f_1(k_1) \ldots f_n(k_n) \tag{4.4}$$

where $f_i \in \mathcal{S}_C(R)$ and $n = 1, 2, \ldots$.

The annihilation operator $a(k)$, $k \in \mathbf{R}$, is defined on $\mathcal{D}$ by the formula

$$(a(k)\theta)_n(k_1, \ldots, k_n) = (n+1)^{\frac{1}{2}}\theta_{n+1}(k, k_1, \ldots, k_n).$$

Then $a(k)\mathcal{D} \subset \mathcal{D}$, and so for $k = k_1, \ldots, k_n \in \mathbf{R}^n$, we define the product

$$a(k) = a(k_1) \ldots a(k_n), \tag{4.5}$$

also mapping $\mathcal{D}$ into $\mathcal{D}$. $a(k)$ is not closable and its adjoint, considered as an operator, has domain $\{0\}$. Nonetheless $a^*(k)$ is a bilinear form on $\mathcal{D} \times \mathcal{D}$. Similarly $a^*(k)a(k') = a^*$ is a bilinear form on $\mathcal{D} \times \mathcal{D}$.

For $\theta_1, \theta_2 \in \mathcal{D}, k \in \mathbf{R}^m, k' \in \mathbf{R}^n$

$$\langle \theta_1, a^*(k)a(k')\theta_2 \rangle \in \mathcal{S}_C(\mathbf{R}^{m+n})$$

and so for any distribution $w \in \mathcal{S}_C'(\mathbf{R}^{m+n})$, the weak integral

$$W = \int_{\mathbf{R}^{m+n}} a^*(k)w(k; k')a(k)\, dk\, dk' = a^{\#}(w) \tag{4.6}$$

is also a bilinear form on $\mathcal{D} \times \mathcal{D}$. W is called a Wick monomial of degree m, n, and any linear combination of Wick monomials is a Wick polynomial. Let $\mathcal{W}(\mathcal{S}')$ be the class of Wick polynomials and for any subspace $\mathcal{X} \subset \mathcal{S}'$, let $\mathcal{W}(\mathcal{X})$ be the class of Wick polynomials whose kernels w are restricted to lie in $\mathcal{X}$. $\mathcal{W}$ is a general class of forms. The important operators of field theory have natural expressions in terms of Wick monomials. Moreover any bounded operator on $\mathcal{F}$ has an expansion in Wick monomials. The expansion converges in terms of bilinear forms on $\mathcal{D} \times \mathcal{D}$.

We also introduce the more restricted class $\mathscr{V}(\mathscr{S}')$ of bilinear forms on $\mathscr{D} \times \mathscr{D}$ expressible as a linear combination of the forms

$$V = \sum_{j=0}^{n} \binom{n}{j} \int_{R^n} v(k) a^*(k_1) \ldots a^*(k_j) a(-k_{j+1}) \ldots a(-k_n)\, dk \tag{4.7}$$

with symmetric kernels $v \in \mathscr{S}'$ having real Fourier transforms. As before, $\mathscr{V}(\mathscr{X})$ is the subclass of $\mathscr{V}(\mathscr{S}')$ obtained by restricting the kernels v to lie in $\mathscr{X}$. As we will see later, $\mathscr{V}(\mathscr{S}')$ is a "maximal abelian" class of forms, and it is a general class of forms expressible as functions of the ϕ's.

We now define the configuration space annihilation and creation operators, using the Fourier transform with the sign convention (3.3). For real test functions $f \in \mathscr{S}_{\mathbf{R}}(\mathbf{R})$, let

$$\begin{aligned} b^*(f) &= \textstyle\int a^*(-k) f(k)^\sim \, dk \\ b(f) &= \textstyle\int a(k) f(k)^\sim \, dk. \end{aligned} \tag{4.8}$$

The commutation relations

$$[b(f), b^*(g)] = \textstyle\int f(g) g(x)\, dx = \int f(k)^\sim g(-k)^\sim \, dk$$

follow from the corresponding commutation relations for a and a^*,

$$[a(k_1), a^*(k_2)] = \delta(k_1 - k_2), \tag{4.9}$$

and (4.9) can be verified directly.

THEOREM 4.1. *With the above definitions, b and b^* are the Fock representation for the canonical commutation relations over $Q' = \mathscr{S}_{\mathbf{R}}(\mathbf{R})$.*

The proof is routine, and will be omitted. By Theorem 2.3, we identify $\mathscr{F}$ and $\mathscr{H} = L_2(Q, d_B q)$. We now define

$$\begin{aligned} \phi(x) &= (4\pi)^{-\frac{1}{2}} \textstyle\int e^{-ikx}(a^*(k) + a(-k))\mu(k)^{-\frac{1}{2}}\, dk \\ \pi(x) &= i(4\pi)^{-\frac{1}{2}} \textstyle\int e^{-ikx}(a^*(k) - a(-k))\mu(k)^{\frac{1}{2}}\, dk. \end{aligned} \tag{4.10}$$

Since the kernels in (4.10) belong to $\mathscr{S}_C'$, $\phi(x)$ and $\pi(x)$ are bilinear forms on $\mathscr{D} \times \mathscr{D}$. Combining (3.19), (4.8) and (4.10), we have

$$\begin{aligned} \phi(f) &= 2^{-\frac{1}{2}} \textstyle\int f(k)^\sim \mu(k)^{-\frac{1}{2}}(a^*(-k) + a(k))\, dk \\ &= (4\pi)^{-\frac{1}{2}} \textstyle\iint e^{-ikx}(a^*(k) + a(-k))\mu(k)^{-\frac{1}{2}} f(x)\, dk\, dx \\ &= \textstyle\int \phi(x) f(x)\, dx \end{aligned}$$

and similarly $\pi(f) = \int \pi(x) f(x)\,dx$. This calculation justifies the definition (4.10). For the interchange of k and x integration, see the proof of Theorem 4.4 below.

We resume our analysis of the Wick monomial W in (4.6).

THEOREM 4.2. *Let w be the kernel of a bounded operator from $S_n L_2(\mathbf{R}^n)$ to $S_m L_2(\mathbf{R}^m)$, with norm $\|w\|$. Then*

$$(N + I)^{-m/2} W (N + I)^{-n/2} \tag{4.11}$$

is also bounded, with norm at most $\|w\|$.

Proof. Let A be the operator (4.11). Then $A : \mathscr{F}_{r+n} \to \mathscr{F}_{r+m}$ and

$$\|A\| = \sup_r \|A \upharpoonright \mathscr{F}_{r+n}\|.$$

For $\theta_1 \in \mathscr{F}_{r+m} \cap \mathscr{D}$ and $\theta_2 \in \mathscr{F}_{r+n} \cap \mathscr{D}$, we have

$$\langle \theta_1, A\theta_2 \rangle = (r + m + 1)^{-m/2} (r + n + 1)^{-n/2} \langle \theta_1, W\theta_2 \rangle$$

and

$$\langle \theta_1, W\theta_2 \rangle = \left(\frac{(r+m)!}{r!} \frac{(r+n)!}{r!} \right)^{\frac{1}{2}} \int \theta_1(k,p)^- w(k;k')\theta_2(k',p)\,dk\,dk'\,dp,$$

where $p \in \mathbf{R}^r$. Thus

$$\begin{aligned} |\langle \theta_1, A\theta_2 \rangle| &\leqslant \|w\| \int \|\theta_1(\cdot, p)\|_2 \, \|\theta_2(\cdot, p)\|_2 \, dp \\ &\leqslant \|w\| \left(\int \|\theta_1(\cdot, p)\|_2^2 \, dp\right)^{\frac{1}{2}} \left(\int \|\theta_2(\cdot, p)\|_2^2 \, dp\right)^{\frac{1}{2}} \\ &= \|w\| \, \|\theta_1\| \, \|\theta_2\| \end{aligned}$$

and the proof is complete. ∎

COROLLARY 4.3. *Let $a + b \geqslant m + n$. Then*

$$\|(N + I)^{-a/2} W (N + I)^{-b/2}\| \leqslant (1 + |m - n|)^{|a-m|/2} \|w\|.$$

Proof. We transfer $|a - m|$ factors of $(N + I)^{-\frac{1}{2}}$ from one side of W to the other. We use the identity

$$(N + (\alpha + 1)I)^{-\frac{1}{2}} W = W(N + (\alpha + m - n + 1)I)^{-\frac{1}{2}}$$

with $\alpha = 0$ if $m > n$ and $\alpha = n - m$ if $m \leqslant n$, and we use

$$\left\|(N + (\alpha + 1)I)^{\frac{1}{2}} (N + I)^{-\frac{1}{2}}\right\| \leqslant (1 + \alpha)^{\frac{1}{2}}$$

for $\alpha = |m - n|$. ∎

Remark. It follows that the bilinear form defines uniquely an operator on the domain $\mathscr{D}(N^{(m+n)/2})$.

Now let W_1 and W_2 be two Wick monomials, with degrees i, j and m, n respectively. Assuming that the kernels w_1 and w_2 are sufficiently regular (e.g. in $\mathscr{S}$), the product $W_1 W_2$ may be defined. $W_1 W_2$ is not a Wick monomial because the creation operators in W_2 precede the annihilation operators in W_1. However using the commutation relations (4.9), the a's and a^*'s can be ordered as in (4.6). Because of the δ function on the right side of (4.9), the repeated use of (4.9) expresses $W_1 W_2$ as a sum of terms. Letting U_r be the contribution to the sum formed by all terms with r δ functions (r contractions) we have

$$W_1 W_2 = \sum_{r=0}^{\min\{j,m\}} U_r \tag{4.12}$$

and

$$U_r = \int a^*(k_1) \ldots a^*(k_{i+m-r}) u_r(k; k') a(k_1') \ldots a(k'_{j+n-r})\, dk\, dk'. \tag{4.13}$$

If w_1 and w_2 are each symmetric in their creating variables ($k_1, \ldots, k_i$ or $k_1, \ldots, k_m$, resp.) and also in their annihilating variables, then

$$\begin{aligned} u_r &= u_r(k, k'; k'', k''') \\ &= r!\binom{j}{r}\binom{m}{r}\int w_1(k; p, k'_{r+1}, \ldots, k_j')w_2(p, k''_{r+1}, \ldots, k_m''; k''')\, dp \end{aligned} \tag{4.14}$$

where $k \in \mathbf{R}^i$, $p \in \mathbf{R}^r$ and $k''' \in \mathbf{R}^n$. The term U_0 is called the Wick product and is denoted $:W_1 W_2:$. The kernel of $:W_1 W_2:$ is just $w_1 \otimes w_2$, that is the product of w_1 and w_2, considered as functions of distinct variables. For any two tempered distributions w_1 and w_2, $w_1 \otimes w_2$ is again a tempered distribution, and so $:W_1 W_2:$ is always defined, even if the product $W_1 W_2$ is not. Similarly we define the Wick product $:U \ldots VW:$ of an arbitrary number of Wick monomials. If P is a noncommutative polynomial in l indeterminants and if $W_1, \ldots, W_l$ are Wick monomials, then $P(W_1, \ldots, W_l)$ is called a formal expression in the $a^{\#}$'s. We define $:P(W_1, \ldots, W_l):$ by linearity. As bilinear forms we have the identity

$$W_1 W_2 = :W_1 W_2: + :U_1: + :U_2: + \ldots .$$

Since U_1 is in general not zero, the Wick product on formal expressions cannot be regarded as a product on bilinear forms.

As a special case of the above discussion, we have $:\phi^n(x): \in \mathscr{V}(\mathscr{S}')$ with the kernel

$$v(k, x) = (4\pi)^{-n/2} \exp(-ix\textstyle\sum_l k_l) \displaystyle\prod_{l=1}^{n} \mu(k_l)^{-\frac{1}{2}}. \tag{4.15}$$

The kernel belongs to $\mathscr{S}_C'(\mathbf{R}^n)$ as a function of k, and x is a parameter. For $n = 2$, one can check that

$$\phi^2(x) - :\phi^2(x): = (4\pi)^{-1} \int \mu(k)^{-1}\, dk. \tag{4.16}$$

The integral on the right is logarithmically divergent and since $:\phi^2(x):$ exists, we conclude that $\phi^2(x)$ does not exist. The facts that $\phi^2(x)$ does not exist (or is identically infinite), that $\phi(x)$ is a bilinear form but not an operator and that the Gaussian measure $d_B q$ is concentrated on distributions but not on functions all are different facets of the same phenomenon.

We are now ready to define the free Hamiltonian H_0. According to the correspondence principle, we obtain H_0 by substituting the quantized fields ϕ and π in the classical free field Hamiltonian (3.2) at time $t = 0$. We have already seen that $\phi^2(x)$ is identically equal to $+\infty$, and the same is true for π^2 and $\nabla\phi^2$. However we are also allowed to subtract a constant, as was done in (2.1) in the case of one degree of freedom. Thus we define

$$H_0 = \int \tfrac{1}{2} :(\pi^2(x) + \nabla\phi^2(x) + m^2\phi^2(x)): dx. \tag{4.17}$$

This formidable looking expression is actually very simple, thanks to our special choice of the bilinear form B. In analogy with (2.3), we have

THEOREM 4.4. $$H_0 = \int \mu(k) a^*(k) a(k)\, dk.$$

Proof. This is a routine calculation, and so we omit some details. The pure creation part of H_0 is

$$2^{-1}(4\pi)^{-1} \int a^*(k_1)a^*(k_2)\, e^{-ix(k_1+k_2)}(-\mu(k_1)\mu(k_2) + (-k_1k_2 + m^2))$$
$$\times\, \mu(k_1)^{-\frac{1}{2}}\mu(k_2)^{-\frac{1}{2}}\, dk_1\, dk_2\, dx,$$

and the point is to justify the interchange of k and x integration. For θ_1, $\theta_2 \in \mathscr{D}$, let

$$f(k_1, k_2) = \langle \theta_1, a^*(k_1)a^*(k_2)\theta_2 \rangle.$$

Then $f \in \mathscr{S}_C(R^2)$ and the above contribution to $\langle \theta_1, H_0\theta_2 \rangle$ is

$$\begin{aligned} 2^{-1}(4\pi)^{-1} \int f(k_1, k_2)\, e^{-ix(k_1+k_2)}(-\mu(k_1)\mu(k_2) - k_1k_2 + m^2) \\ \times \mu(k_1)^{-\frac{1}{2}}\mu(k_2)^{-\frac{1}{2}}\, dk_1\, dk_2\, dx \\ = 4^{-1} \int f(k_1, k_2)\delta(k_1 + k_2)(-\mu(k_1)\mu(k_2) - k_1k_2 + m^2) \\ \times \mu(k_1)^{-\frac{1}{2}}\mu(k_2)^{-\frac{1}{2}}\, dk_1\, dk_2 = 0. \end{aligned}$$

Similarly the pure annihilation part of H_0 is zero, and the remaining contribution to H_0 gives

$$\begin{aligned} (4\pi)^{-1} \int a^*(k_1)a(-k_2)\, e^{-ix(k_1+k_2)}(\mu(k_1)\mu(k_2) - k_1k_2 + m^2)) \\ \times \mu(k_1)^{-\frac{1}{2}}\mu(k_2)^{-\frac{1}{2}}\, dk_1\, dk_2\, dx. \end{aligned}$$

As above we interchange the x and k integrations to obtain

$$\begin{aligned} 2^{-1} \int a^*(k_1)a(-k_2)\delta(k_1 + k_2)(\mu(k_1)\mu(k_2) - k_1k_2 + m^2) \\ \times \mu(k_1)^{-\frac{1}{2}}\mu(k_2)^{-\frac{1}{2}}\, dk_1\, dk_2 \\ = \int \mu(k)a^*(k)a(k)\, dk, \end{aligned}$$

as bilinear forms on the domain $\mathscr{D} \times \mathscr{D}$. ∎

COROLLARY 4.5. *H_0 leaves each subspace $\mathscr{F}_n$ invariant, and on $\mathscr{F}_n$, H_0 is the multiplication operator*

$$\sum_{l=1}^{n} \mu(k_l).$$

COROLLARY 4.6. *H_0 is essentially self-adjoint as an operator on the domain $\mathscr{D}$.*

The proofs are elementary. From now on, we let H_0 denote the self-adjoint operator, with

$$\mathscr{D}(H_0) = \left\{\theta : \sum_{n=1}^{\infty} \left\| \sum_{l=1}^{n} \mu(k_l)\theta_n \right\|_2 < \infty \right\}. \tag{4.18}$$

III Boson Quantum Field Models: Part II. The Solution of Two-Dimensional Boson Models

PART II. THE SOLUTION OF TWO-DIMENSIONAL BOSON MODELS

5. THE INTERACTION HAMILTONIAN

Two-dimensional boson quantum field models employ three cut-offs for treatment of the Hamiltonian operator. The cut-offs are approximations which simplify the Hamiltonian. After obtaining properties of the cut-off

Hamiltonians, the cut-offs are removed by some limit procedure, and one obtains properties of the full Hamiltonian. The first cut-off is a space cut-off, indexed by a non-negative function $g \in L_\infty(\mathbf{R})$ of compact support. g limits the interaction of particles to the space region, support g. The second cut-off is a momentum cut-off κ, and κ indicates roughly the largest magnitude of the momentum occurring in the interaction Hamiltonian. Finally we have a third auxiliary cut-off, which makes evident the fact that the sum of the free and the cut-off interaction Hamiltonians are essentially self-adjoint.

We fix a positive polynomial P of degree p. The interaction Hamiltonian is

$$H_I = \int :P(\phi(x)): dx \in \mathscr{V}(\mathscr{S}'). \tag{5.1}$$

The kernel of the term of degree n in H_I is proportional to

$$\delta\left(\sum_{l=1}^{n} k_l\right) \prod_{l=1}^{n} \mu(k_l)^{-\frac{1}{2}} \tag{5.2}$$

after an interchange of x and k integration, as in the proof of Theorem 4.4. The kernel belongs to $\mathscr{S}_C'(\mathbf{R}^n)$ but not to L_2, because of the δ-function. The δ-function results from conservation of momentum, or what is the same, the translation invariance of H_I. Because of the δ-function, the sum $H_0 + H_I$ is exceedingly singular. After adding an infinite constant to $H_0 + H_I$, we will realize it as a self-adjoint operator on a new Hilbert space, associated with a (presumably) non-Fock representation of the canonical commutation relations. The space cut-off Hamiltonian $H(g)$ is much less singular. We define

$$H_I(g) = \int :P(\phi(x)): g(x)\, dx, \tag{5.3}$$

and here the kernel is proportional to

$$v = g\left(-\sum_{l=1}^{n} k_l\right)^{\sim} \prod_{l=1}^{n} \mu(k_l)^{-\frac{1}{2}}. \tag{5.4}$$

The momentum cut-off field $\phi_\kappa(x)$ is defined by the formula

$$\phi_\kappa(x) = (4\pi)^{-\frac{1}{2}} \int \xi(k\kappa^{-1})\, e^{-ixk} \mu(k)^{-\frac{1}{2}}(a^*(k) + a(-k))\, dk \tag{5.5}$$

where ξ is some fixed element of $\mathscr{S}_C(\mathbf{R})$ satisfying $\xi(0) = 1$ and $\xi(k) = \xi(-k)^-$, so that ξ is the Fourier transform of a real function. We define

$$H_I(g, \kappa) = \int :P(\phi_\kappa(x)): g(x)\, dx \tag{5.6}$$

and

$$H(g, \kappa) = H_0 + H_I(g, \kappa). \tag{5.7}$$

The kernel of $H_I(g, \kappa)$ is proportional to

$$v_\kappa = g\left(-\sum_{l=1}^{n} k_l\right)^{\sim} \prod_{l=1}^{n} \left(\mu(k_l)^{-\frac{1}{2}}\xi(k_l\kappa^{-1})\right). \tag{5.8}$$

The g cut-off is introduced in order to bring the kernel v into L_2. Although $v_\kappa \in \mathscr{S}$ and $H_I(g, \kappa) \in \mathscr{V}(\mathscr{S})$, the extra regularity does not help. Rather, the importance of the κ cut-off is that $H_I(g, \kappa)$ is semi-bounded, while $H_I(g)$ is not. To prove this statement, we need Wick's theorem.

Let A_j be a set of disjoint unordered pairs, selected from $\{1, \ldots, n\}$, let $|A_j|$ be the subset of $\{1, \ldots, n\}$ covered by the pairs in A_j and let $\mathscr{A}_j$ be the set of all A_j's.

THEOREM 5.1. *The relations*

$$\prod_{l=1}^{n} [a^*(k_l) + a(-k_l)]$$

$$= \sum_{j=0}^{[n/2]} \sum_{A_j \in \mathscr{A}_j} \prod_{(\alpha,\beta) \in A_j} \delta(k_\alpha + k_\beta) : \prod_{l \notin |A_j|} \left(a^*(k_l) + a(-k_l)\right) : \tag{5.9}$$

and

$$: \prod_{l=1}^{n} [a^*(k_l) + a(-k_l)] :$$

$$= \sum_{j=0}^{[n/2]} (-1)^j \sum_{A_j \in \mathscr{A}_j} \prod_{(\alpha,\beta) \in A_j} \delta(k_\alpha + k_\beta) \prod_{l \notin |A_j|} \left(a^*(k_l) + a(-k_l)\right) \tag{5.10}$$

are valid in $\mathscr{W}(\mathscr{S})$, after multiplying by smooth test functions and integrating over k. The identity is then an operator identity on the domain $\mathscr{D}$.

Proof. We use (4.9) to reorder the a^* and a operators on the left side. For each pair k_α, k_β of variables, there is one interchange of order between an a^* and an a operator, and so there are two terms, one with a δ-function $\delta(k_\alpha + k_\beta)$ and the other with the a^* and a in reversed order. Thus the expansion of the left side contains one term for each way of selecting disjoint unordered pairs from $\{1, \ldots, n\}$. This term contains δ-functions for the pairs selected and reordered a^* and a operators for the variables not lying in such pairs. The minus sign in (5.10) comes from the direction in which the reordering takes place, i.e. $a^*a = aa^* - \delta$ in (5.10). This completes the proof. ∎

Let

$$c_\kappa = (4\pi)^{-1} \int \xi(k\kappa^{-1})\mu(k)^{-1}\, dk. \tag{5.11}$$

COROLLARY 5.2. *As an operator identity on the domain $\mathscr{D}$, we have*

$$:\phi_\kappa{}^n(x): = \sum_{j=0}^{[n/2]} (-1)^j c_\kappa{}^j \frac{n!}{(n-2j)!\,j!\,2^j} \phi_\kappa{}^{n-2j}(x).$$

Proof. The number of elements of $\mathscr{A}_j$ is just $n!/(n-2j)!\,j!\,2^j$. ∎

COROLLARY 5.3. *Let $g \in L_1$. As an operator inequality on $\mathscr{D} \times \mathscr{D}$, we have*

$$-O(\ln \kappa)^{p/2} \leqslant H_I(g, \kappa).$$

Proof. $c_\kappa = O(\ln \kappa)$ in (5.11). Since P is a positive polynomial, the inequality follows from Corollary 5.2. ∎

The inequality in Corollary 5.3 extends by closure to an inequality for self-adjoint operators, see Theorem 5.6.

LEMMA 5.4. *Let $f_1, f_2 \in \mathscr{S}_R(\mathbf{R})$. Then $\phi(f_1) + \pi(f_2)$ is essentially self-adjoint on $\mathscr{D}$.*

Proof. We establish the inequality

$$\|(\phi(f_1) + \pi(f_2))^n \theta\| \leqslant O(n!)^{\frac{1}{2}}(\|\mu_x{}^{-\frac{1}{2}} f_1\|_2 + \|\mu_x{}^{\frac{1}{2}} f_2\|_2)^n \tag{5.12}$$

for $\theta \in \mathscr{D}$. Since $\mathscr{D}$ is also invariant, θ is then an analytic vector for $[\phi(f_1) + \pi(f_2)] \restriction \mathscr{D}$. Since $\mathscr{D}$ is dense, the lemma follows from Nelson's theorem. (See Simon's chapter.)

To prove (5.12), we expand the nth power on the left as a sum of 4^n terms, each of the form

$$\|a^\#(h_1) \ldots a^\#(h_j)\theta\| \tag{5.13}$$

where $a^\# = a$ or a^* and h_j is proportional to $\mu_x^{\pm\frac{1}{2}} f_i$. If M is the maximum number of particles in θ, (5.13) is dominated by

$$\prod_{j=1}^n \|a^\#(h_i)(N+I)^{-\frac{1}{2}}\| \left\| \prod_{j=1}^n (N + (j+1)I)^{\frac{1}{2}} \theta \right\|$$

$$\leqslant (M+1)^n n!^{\frac{1}{2}} \prod_{j=1}^n \|h_i\|_2 \, \|\theta\|$$

as in the proof of Corollary 4.3. This completes the proof. ∎

We let $\phi(f)$ now denote the self-adjoint closure, $\phi(f) = (\phi(f) \restriction \mathscr{D})^-$. Let $\mathfrak{M}$ be the von Neumann algebra generated by the operators $\{e^{i\phi(f)} : f \in \mathscr{S}\}$.

By definition, a von Neumann algebra is a *-algebra of bounded operators, closed under strong limits. It is easy to see that $\mathfrak{M}$ is also generated by the joint spectral projections of finite families $\phi(f_1), \ldots, \phi(f_n)$, $f_i \in \mathscr{S}$. Such a spectral projection is a multiplication operator on $L_2(Q, d_B q)$. In fact it is multiplication by the characteristic function of a Borel cylinder set. Conversely, any such a multiplication operator is a spectral projection, and so these multiplication operators generate $\mathfrak{M}$. We now prove a simple density theorem.

PROPOSITION 5.5. $\mathfrak{M} = L_\infty(Q)$.

Proof. $\mathfrak{M} \subset L_\infty(Q)$ by construction. To prove the converse, let $\mathscr{R} = \{Y : \chi_Y \in \mathfrak{M}\}$. Then $\mathscr{R}$ is a ring of subsets of Q, containing the Borel cylinder sets. Since a monotone sequence of projections has a strong limit, $\mathscr{R}$ is closed under sequential monotone limits. Hence $\mathscr{R}$ contains all Borel sets, and the proposition follows from this fact. ∎

THEOREM 5.6. *Let $V \in \mathscr{V}(L_2)$. Then V is essentially self-adjoint on $\mathscr{D}$.*

Proof. Let V be defined as an operator on the domain $\mathscr{D}$. We assert that $\mathfrak{M}\Omega_0 \subset \mathscr{D}(V^-)$ and that $\mathfrak{M}$ commutes with V^-. Since $V\Omega_0 = \theta \in \mathscr{H} = L_2(Q)$, it follows that $V^- \restriction \mathfrak{M}\Omega_0$ is multiplication by a real L_2 function. Since any real L_2 multiplication operator is essentially self-adjoint on $L_\infty(Q) = \mathfrak{M}\Omega_0$, $V^- \restriction \mathfrak{M}\Omega_0$ is essentially self-adjoint and V^- is self-adjoint. To prove the assertion, we compute

$$\lim_n \sum_{j \leq n} V(i\phi(f))^j \Omega_0/j! = \lim_n \sum_{j \leq n} (i\phi(f))^j V\Omega_0/j! = e^{i\phi(f)} V\Omega_0.$$

The series above converges, as in (5.10), since $V\Omega_0$ has a finite number of particles. Thus

$$\exp(i\phi(f))\Omega_0 \in \mathscr{D}(V^-) \qquad \text{and} \qquad V^- \exp(i\phi(f))\Omega_0 = \exp(i\phi(f))V^-\Omega_0.$$

The same argument applies to linear combinations in the exponentials, and to polynomials in the exponentials (since polynomials in the exponentials can be written as linear combinations). An arbitrary element M of $\mathfrak{M}$ is the strong limit of such polynomials, M_j. Hence $M\Omega_0 = \lim M_j\Omega_0$ and

$$\lim V^- M_j\Omega_0 = \lim M_j V^- \Omega_0 = MV^-\Omega_0,$$

which proves our assertion. ∎

THEOREM 5.7. *Let $V \in \mathscr{V}(L_2)$. Then $V \in L_r(Q, d_B q)$ for all $r < \infty$, and if v is the kernel of V,*

$$\|V\|_r \leq \text{const } \|v\|_2.$$

Proof. For $r = 2j$ an even integer, we have

$$\|V\|_r^r = \int_Q V^{2j}\, d_B q = \langle \Omega_0, V^{2j}\Omega_0 \rangle$$

$$= \|V^j\Omega_0\|^2 \leqslant \prod_{l=1}^{j} \|(N+I)^{n(l-1)/2} V (N+I)^{-nl/2}\|^2$$

$$\leqslant \text{const}\, \|v\|_2^{2j}$$

by Corollary 4.3. This proves the theorem for $r = 2j$, and since the total measure of Q is finite, the theorem is also valid for smaller values of r, and hence for all $r < \infty$. ∎

PROPOSITION 5.8. *For $g \in L_2$, v and v_κ belong to $L_2(\mathbf{R}^n)$ and for any $\varepsilon > 0$,*

$$\|v - v_\kappa\| \leqslant O(\kappa^{-\frac{1}{2}+\varepsilon}).$$

For g_1 and g_2 in L_2, the associated kernels v_1 and v_2 of (5.4) *satisfy*

$$\|v_1 - v_2\| \leqslant O(\|g_1 - g_2\|_2).$$

Proof. Let Σ be the sector

$$\mu(k_1) \leqslant \ldots \leqslant \mu(k_n).$$

Since v and v_κ are symmetric functions,

$$\|v - v_\kappa\|_2^2 = n! \int_\Sigma |v - v_\kappa|^2\, dk$$

$$= \text{const} \int_\Sigma \left|1 - \prod_i \xi(k_i\kappa^{-1})\right|^2 \prod_l \mu(k_l)^{-1} \left|g^\sim\left(-\sum_i k_i\right)\right|^2 dk.$$

Now

$$\left|1 - \prod_i \xi(k_i\kappa^{-1})\right|^2 \leqslant \text{const} \sup_i |1 - \xi(k_i\kappa^{-1})|.$$

Since $|1 - \xi(k_i\kappa^{-1})|\, \mu(k_n)^{-1}$ is bounded from above by const κ^{-1} and also by const $\mu(k_n)^{-1}$ on the sector Σ, we have

$$|1 - \xi(k_i\kappa^{-1})|\, \mu(k_n)^{-1} \leqslant \text{const}\, \kappa^{-1+\varepsilon} \mu(k_n)^{-\varepsilon}$$

on Σ. Thus

$$\|v - v_\kappa\|_2^2 \leqslant \text{const}\,\kappa^{-1+\varepsilon} \int_\Sigma \prod_{l<n} \mu(k_l)^{-1} \mu(k_n)^{-\varepsilon} \left| g^\sim\left(-\sum_i k_i \right) \right|^2 dk$$

$$\leqslant \text{const}\,\kappa^{-1+\varepsilon} \int_\Sigma \prod_{l<n} \mu(k_l)^{-1-\varepsilon/n-1} \left| g^\sim\left(-\sum_i k_i \right) \right|^2 dk$$

$$\leqslant O(\kappa^{-1+\varepsilon})\, O(\|g\|_2^2),$$

where we integrate first over k_n. This proves the first estimate, and it also proves the second, since $v_1 - v_2$ is the kernel associated with the cut-off function $g_1 - g_2$. We note that g is not required to be positive in this proposition. ∎

Remark. It follows that $H_I(g) \in \mathscr{V}(L_2)$ and so it is a self-adjoint operator. Corollary 5.3 and Proposition 5.8 express the fact that $H_I(g)$ consists of a semibounded part and a small remainder.

Let B be a bounded open region of space and let $\mathfrak{M}(B)$ be the von Neumann algebra generated by the field operators $e^{i\phi(f)}$ where $f \in \mathscr{S}_{\mathbf{R}}(\mathbf{R})$ and where $\operatorname{suppt} f \subset B$.

THEOREM 5.9. *Let* $g(x) \equiv 0$ *on* $\sim B$. *Then* $\exp(itH_I(g)) \in \mathfrak{M}(B)$ *for all* $t \in \mathbf{R}$.

Proof. The multiplication operators $V \in \mathscr{V}(L_2)$ are each essentially self-adjoint on the dense domain L_∞. If a sequence V_j converges, as L_2 functions, to a limit V, then the sequence also converges strongly as operators on the domain L_∞, and in this case the unitary groups $\exp(itV_j)$ converge strongly also. By Theorem 5.7, the convergence of the kernels v_j in L_2 implies the convergence of V_j in $L_2(Q, d_B q)$. Since $\mathfrak{M}(B)$ is closed under strong limits, it is sufficient to prove the theorem for kernels v_j approximating the kernels v of $H_I(g)$ in L_2.

As our first approximation, we choose

$$g_j(x) = \begin{cases} g(x) & \text{if } \operatorname{dist}(x, \operatorname{Bdry} B) > j^{-1} \\ 0 & \text{otherwise.} \end{cases}$$

Then $g_j \to g$ in L_2, and the kernels v_j converge in L_2 by Proposition 5.8. Next we approximate $H_I(g_j)$ by $H_I(g_j, \kappa_l)$, $\kappa_l \to \infty$. Again the kernels converge in L_2 by Proposition 5.8. As a further assumption, we choose the momentum cut-off function ξ to be the Fourier transform of a function of compact support. Then

$$\phi_\kappa(x) = 2\pi\kappa^{-1} \int \eta(\kappa(x-y))\phi(y)\, dy$$

and for $x \in \operatorname{supp} g_j$ and l sufficiently large,

$$\operatorname{supp} \eta(\kappa_l(x - \cdot)) \subset B.$$

Thus $\exp(it\phi_\kappa(x)) \in \mathfrak{M}(B)$ by definition and so the spectral projections of $\phi_\kappa(x)$ belong to $\mathfrak{M}(B)$. The same is true for $:P(\phi_\kappa(x)):$, since it is a polynomial in $\phi_\kappa(x)$ (Corollary 5.2) and the same is true for $H_I(g, \kappa)$ since the integration over x converges strongly on the core $L_\infty(Q)$ of the operator $H_I(g, \kappa)$ (Theorem 5.7). This completes the proof. ∎

6. THE FREE HAMILTONIAN

We now develop properties of the operator H_0. For the proof of essential self-adjointness of $H(g)$, the key property of H_0 is the fact that e^{-tH_0} has a nonnegative kernel. This fact results from H_0 being a second order elliptic operator with real coefficients, and it implies the existence of a positive measure on the associated path space. For removal of the space cut-off g, the key property of H_0 is its finite propagation speed. The finite propagation speed follows from the same property for the classical field, and is a consequence of the hyperbolic character of the equation (3.1).

Definition 6.1. Let X be a measure space and let dx be a measure on X. Let A be a bounded operator on $L_2(X, dx)$. Then A has a nonnegative kernel if for any two nonnegative functions θ and ψ in $L_2(X, dx)$,

$$0 \leq \langle \theta, A\psi \rangle. \tag{6.1}$$

If B is another bounded operator on $L_2(X, dx)$, and if $B - A$ has a nonnegative kernel, then we say that the kernel of B is greater than the kernel of A.

THEOREM 6.2. *e^{-tH_0} has a nonnegative kernel.*

Proof. We take three reductions and then appeal to the finite dimensional case, Theorem 2.4. The first reduction is to replace e^{-tH_0} by operators converging strongly to it. We choose the approximating operators to have the form e^{-tH_n} where

$$H_n = \sum_{l=1}^{j(n)} \mu_{l,n} a^*(e_{l,n}) a(e_{l,n}), \qquad 0 \leq \mu_{l,n}, \qquad e_{l,n} \in \mathscr{S}(R)$$

and where the family $\{e_{l,n}\}$ for fixed n is orthonormal. Furthermore since the kernel $w(k;p) = \mu(k)\delta(k-p)$ of H_0 satisfies the configuration reality condition, $w(k;p) = w(-k;-p)^-$, we can require that $e_{l,n}(k) = e_{l,n}(-k)^-$ and then $e_{l,n}$ is the Fourier transform of a real function $g_{l,n}$. Hence by (4.8),

$$H_n = \sum_{l=1}^{j(n)} \mu_{l,n} b^*(g_l) b(g_l).$$

The second reduction is to replace θ and ψ by strongly approximating vectors $|\theta_v| \geqslant 0$ and $|\psi_v| \geqslant 0$. We want θ_v and ψ_v to depend on a finite number of degrees of freedom, and we accomplish this by requiring that θ_v and ψ_v belong to $\mathscr{D}$. $\mathscr{D}$ is dense in $L_2(Q)$ and by the triangle inequality

$$-\,|\theta - \theta_v| \leqslant |\theta| - |\theta_v| \leqslant |\theta - \theta_v|$$

we see that absolute values of elements of $\mathscr{D}$ are dense in the positive elements of $L_2(Q)$.

Let G be a finite dimensional subspace of $\mathscr{S}_{\mathbf{R}}(\mathbf{R})$ such that $g_l \in G$ and such that θ_v and ψ_v are polynomials based on $F = \mu_x^{\pm}G$. Since $\mu_l = 0$ is permitted, we may suppose that the set $\{g_l\}$ is a basis for G.

The third reduction is to appeal to the unitary equivalence of Theorem 2.3. Let $\mathscr{D}_G$ be the linear subspace of $\mathscr{D}$ spanned algebraically by polynomials based on $\mu_x^{\pm}G$ and let $\mathscr{H}_G = \mathscr{D}_G^- \subset L_2(Q)$. Then $\mathscr{H}_G$ is also the cyclic subspace generated by the operators $b^*(g_l)$ applied to Ω_0, and we check easily that

$$G \ni g \to b^*(g), b(g)$$

is the Fock representation for the canonical commutation relations over G. Since

$$e^{-tH_n} b^*(g_l) = b^*(\exp(-t\mu_{l,n})g_l)\, e^{-tH_n},$$

e^{-tH_n} leaves $\mathscr{H}_G$ invariant, and so the unitary equivalence U between $\mathscr{H}_G$ and $L_2(\mathbf{R}^{2j(n)})$ given by Theorem 2.3 preserves the inner product (6.1). Let $\mathfrak{M}(F)$ be the von Neumann algebra generated by $\{e^{i\phi(f)} : f \in F\}$. Then $\mathfrak{M}(F)$ also leaves $\mathscr{H}_G$ invariant, and one checks that $(\mathfrak{M}(F)\Omega_0)^- = \mathscr{H}_G$. For $M \in \mathfrak{M}(F)$, $UM \cdot 1 = M' \in L_\infty(\mathbf{R}^{2j(n)})$. The map $M \to M'$ preserves products, hence square roots and hence absolute values. Thus U preserves absolute values on $\mathscr{H}_G$. The theorem now follows from Theorem 2.4. ∎

COROLLARY 6.3. *Let $V(q) \in L_r(Q, d_B q)$ for some $r \geqslant 1$. Suppose that $H_0 + V$ is essentially self-adjoint on the domain*

$$\mathscr{D}(H_0) \cap \mathscr{D}(V)$$

and suppose that for some constant M,

$$-M \leqslant V,$$

Then

$$0 \leqslant \text{kernel}\, e^{-t(H_0+V)} \leqslant \text{kernel}\, e^{tM}\, e^{-tH_0}.$$

Proof. We use the Trotter formula to approximate $\exp(-t(H_0 + V))$. Let θ and ψ be nonnegative, as elements of $L_2(Q, d_B q)$, and let

$$\theta_j = e^{-(jt/n)H_0}\,\theta$$
$$\psi_j = (e^{-(t/n)H_0}\, e^{-(t/n)V})^{n-j}\,\psi.$$

Then θ_j and ψ_j are nonnegative by Theorem 6.2 and

$$\langle\theta, (e^{-(t/n)H_0}\, e^{-(t/n)V})^n\psi\rangle = \langle\theta_0, \psi_0\rangle = \langle\theta_1, e^{-(t/n)V}\psi_1\rangle$$
$$\leqslant e^{(t/n)M}\langle\theta_1, \psi_1\rangle \leqslant \ldots \leqslant e^{tM}\langle\theta_n, \psi_n\rangle = \langle\theta, e^{tM}\, e^{-tH_0}\psi\rangle.$$

The proof is completed by the limit $n \to \infty$. ∎

Let

$$\begin{aligned} \phi_0(x, t) &= e^{itH_0}\,\phi(x)\,e^{-itH_0} \\ \pi_0(x, t) &= e^{itH_0}\pi(x)\,e^{-itH_0} \end{aligned} \tag{6.2}$$

$$\begin{aligned} \phi_0(f, t) &= \int \phi_0(x, t) f(x)\,dx \\ \pi_0(f, t) &= \int \pi_0(x, t) f(x)\,dx. \end{aligned} \tag{6.3}$$

Because of the commutation relations

$$\begin{aligned} e^{itH_0}a^*(f) &= a^*(e^{it\mu(\cdot)}f)\,e^{itH_0} \\ [H_0, a^*(f)] &= a^*(\mu(\cdot)f) \\ [H_0, a(f)] &= -\,a(\mu(\cdot)f), \end{aligned} \tag{6.4}$$

H_0 and e^{itH_0} map $\mathscr{D}$ into $\mathscr{D}$. Thus $\phi_0(x, t)$ and $\pi_0(x, t)$ are bilinear forms defined on $\mathscr{D} \times \mathscr{D}$.

THEOREM 6.4. *Let θ and ψ be in $\mathscr{D}$. Then*

$$F(x, t) = \langle\theta, \phi_0(x, t)\psi\rangle$$

is a solution of the Klein–Gordon equation (3.1), *with Cauchy data*

$$F(x, 0) = \langle\theta, \phi(x)\psi\rangle$$
$$(d/dt)\,F(x, 0) = \langle\theta, \pi(x)\psi\rangle.$$

Proof. The formal calculation

$$\frac{d}{dt} F(x, t) = (4\pi)^{-1} \int e^{-ikx} \mu(k)^{-\frac{1}{2}}$$

$$\times \langle e^{-itH_0}\theta, [iH_0, a^*(k) + a(-k)]\, e^{-itH_0}\psi\rangle\, dk$$

is valid on $\mathscr{D} \times \mathscr{D}$, since H_0 and e^{itH_0} map $\mathscr{D}$ into $\mathscr{D}$. By (6.4),

$$\frac{d}{dt} F(x, t) = (4\pi)^{-1}\, i \int e^{-ikx} \mu(k)^{\frac{1}{2}}$$

$$\times \langle e^{-itH_0}\theta, \big(a^*(k) - a(-k)\big)\, e^{-itH_0}\psi\rangle\, dk$$

$$= \langle \theta, \pi_0(x, t)\psi\rangle.$$

Similarly,

$$\frac{d^2}{dt^2} F(x, t) = -\,(4\pi)^{-1} \int e^{-ikx} \mu(k)^{3/2}$$

$$\times \langle e^{-itH_0}\theta, \big(a^*(k) + a(-k)\big)\, e^{-itH_0}\psi\rangle\, dk,$$

and the proof is completed by writing

$$\mu(k)^{3/2} = \mu(k)^2 \mu(k)^{-\frac{1}{2}} = (k^2 + m^2)\mu(k)^{-\frac{1}{2}}. \ \blacksquare$$

The fundamental solution E for the Klein–Gordon equation with data

$$E(x, 0) = 0, \qquad E_t(x, 0) = \delta(x)$$

is denoted $E = \Delta(x, t)$ in the physics literature. As is well known, $\Delta(x, t)$ has support in the double light-cone $|x| \leqslant |t|$.

COROLLARY 6.5. *As bilinear forms on* $\mathscr{D} \times \mathscr{D}$,

$$\phi_0(x, t) = \int \Delta(x - y, t)\pi(y)\, dy + \int \Delta_t(x - y, t)\phi(y)\, dy$$

$$\pi_0(x, t) = \int \Delta_t(x - y, t)\pi(y)\, dy + \int \Delta_{tt}(x - y, t)\phi(y)\, dy.$$

THEOREM 6.6. *Let* f_1 *and* f_2 *belong to* $\mathscr{S}_{\mathbf{R}}(\mathbf{R})$. *The operator*

$$\phi_0(f_1, t) + \pi_0(f_2, t)$$

is essentially self-adjoint on the domain $\mathscr{D}$.

Proof. e^{itH_0} is a unitary operator which leaves $\mathscr{D}$ invariant and transforms $\phi_0(f_1, t) + \pi_0(f_2, t)$ onto $\phi(f_1) + \pi(f_2)$, which is essentially self-adjoint by Lemma 5.4. ■

Let B be a bounded open region of space and let $\mathfrak{A}(B)$ be the von Neumann algebra generated by the operators

$$\exp\left(i(\phi(f_1) + \pi(f_2))\right)$$

where $f_1, f_2 \in \mathscr{S}_{\mathbf{R}}(\mathbf{R})$ and $\operatorname{suppt} f_1$, $\operatorname{suppt} f_2 \subset B$. For any real number t, let B_t be the set of points with distance less than $|t|$ to B.

THEOREM 6.7. $e^{itH_0}\mathfrak{A}(B)\,e^{-itH_0} \subset \mathfrak{A}(B_t)$.

Remark. We reformulate the theorem by saying that H_0 has propagation speed at most one.

Proof. By Corollary 6.5,

$$\phi_0(f_1, t) + \pi_0(f_2, t) = \int_{y\in B_t} h_1(y)\phi(y) + h_2(y)\pi(y)\,dy$$

where $h_1 = f_1{}^*\Delta_t + f_2{}^*\Delta_{tt}$ and $h_2 = f_1{}^*\Delta + f_2{}^*\Delta_t$. The identity is valid as bilinear forms on $\mathscr{D} \times \mathscr{D}$, but by Theorem 6.6, it extends to an identity between essentially self-adjoint operators. Since the unitary group generated by the right side belongs to $\mathfrak{A}(B_t)$ by definition, the unitary group generated by the left side also belongs to $\mathfrak{A}(B_t)$, and the proof is complete. ∎

THEOREM 6.8. *If B and C are disjoint bounded open regions of space then the operators in $\mathfrak{A}(B)$ commute with the operators in $\mathfrak{A}(C)$.*

Proof. Let $f_{i,B} \in \mathscr{S}_{\mathbf{R}}(\mathbf{R})$ be supported in B and let $f_{i,C} \in \mathscr{S}_{\mathbf{R}}(\mathbf{R})$ be supported in C, $i = 1, 2$, and let

$$B = \exp\left(i[\phi(f_{1,B}) + \pi(f_{2,B})]\right)$$
$$C = \exp\left(i[\phi(f_{1,C}) + \pi(f_{2,C})]\right).$$

From the commutation relations (3.13), valid on the domain $\mathscr{D}$, and the fact that vectors in $\mathscr{D}$ are analytic for operators of the form $\phi(f_1) + \pi(f_2)$, we see that B and C commute. Since such operators generate $\mathfrak{A}(B)$ and $\mathfrak{A}(C)$ respectively, the theorem follows. ∎

COROLLARY 6.9. *Let $g \in L_2$, and let $g(x) \equiv 0$ on B. Then $\exp\left(itH_I(g)\right) \in \mathfrak{A}(B)'$ for all $t \in \mathbf{R}$.*

Proof. As in the proof of Theorem 5.9, we approximate $H_I(g)$ by polynomials in the field $\phi(x)$, with x localized in bounded open sets $C_j \subset \sim B$.

7. SELF-ADJOINTNESS OF $H(g)$

PROPOSITION 7.1. *Let H_j be a sequence of self-adjoint operators. The resolvents $R_j(\zeta) = (H_j - \zeta)^{-1}$ converge strongly to the resolvent of a self-adjoint operator H provided the following three conditions are satisfied:*

(1) *The operators H_j are bounded from below, uniformly in j.*

(2) *The operators H_j converge strongly on some dense domain $\mathscr{D}$.*

(3) *The operators e^{-tH_j} converge strongly, uniformly in t, for t bounded away from zero and infinity.*

Proof. By (1) and (3), the resolvents

$$R_j(\zeta) = \int_0^\infty e^{-t(H_j - \zeta)}\, dt$$

converge for ζ sufficiently negative. From Kato, (1966b p.428, Theorem 1.3), $R(\zeta) = \lim_j R_j(\zeta)$ is the resolvent of a closed operator H once we show that the null space of $R(\zeta)$ is zero. Moreover for negative ζ, $R(\zeta)$ is self-adjoint, and thus so is H.

Let $R(\zeta)\theta = 0$. Then for $\psi \in \mathscr{D}$,

$$\langle \psi, \theta \rangle = \langle (H_j - \zeta)\psi,\ R_j(\zeta)\theta \rangle \to \langle (H - \zeta)\psi,\ R(\zeta)\theta \rangle = 0$$

using (2), and so $\theta = 0$ as required. This completes the proof. ∎

Remark. The conclusion of the proposition remains valid if we replace (1) and (3) by the assumption that the sequences $R_j(\zeta)$ and $R_j(\zeta)^*$ converge strongly for some ζ.

PROPOSITION 7.2. *Let $H_j = (H_0 + V_j)^-$ where H_0 and V_j are self-adjoint. Assume the hypothesis of Proposition 7.1 and the conditions*

(4) *V_j converges to a self-adjoint limit V on a core $\mathscr{D}_0$ for V*

(5) *Range $e^{-tH_j} \subset \mathscr{D}(H_0) \cap \mathscr{D}(V_j)$*

(6) *The sequence $V_j e^{-tH_j}$ converges strongly, $0 < t$. Then $H = (H_0 + V)^-$ is essentially self-adjoint on $\mathscr{D}(H_0) \cap \mathscr{D}(V)$ and for $t > 0$*

$$\text{Range}\, e^{-tH} \subset \mathscr{D}(H_0) \cap \mathscr{D}(V).$$

Remark. $H_0 + V$ is taken with domain $\mathscr{D}(H_0) \cap \mathscr{D}(V)$.

Proof. From the functional calculus and Proposition 7.1, $H_j e^{-tH_j}$ converges to He^{-tH} for $t > 0$. By (6),

$$H_0 e^{-tH_j}\theta = H_j e^{-tH_j}\theta - V_j e^{-tH_j}\theta$$

converges for any vector θ. Since H_0 is closed, $e^{-tH}\theta \in \mathscr{D}(H_0)$ and

$$H_0 e^{-tH}\theta = H e^{-tH}\theta - \lim V_j e^{-tH_j}\theta.$$

Let $\psi \in \mathscr{D}_0$. By (4), $V_j\psi \to V\psi$, so

$$\left\langle \psi, \lim_j V_j e^{-tH_j}\theta \right\rangle = \lim_j \langle V_j\psi, e^{-tH_j}\theta\rangle = \langle V\psi, e^{-tH}\theta\rangle$$
$$= \langle \psi, (V \restriction \mathscr{D}_0)^* e^{-tH}\theta\rangle.$$

Since $\mathscr{D}_0$ is a core for V and since V is self-adjoint, $(V \restriction \mathscr{D}_0)^* = V$ and so Range $e^{-tH} \subset \mathscr{D}(V)$. Thus

$$H \restriction (\text{Range}\, e^{-tH}) = (H_0 + V) \restriction (\text{Range}\, e^{-tH}).$$

Since H is essentially self-adjoint on the range of e^{-tH},

$$H = [(H_0 + V) \restriction \text{Range}\, e^{-tH}]^- \subset (H_0 + V)^-.$$

Thus $(H_0 + V)^-$ is a symmetric extension of the self-adjoint operator H, and so $H = (H_0 + V)^-$. ∎

THEOREM 7.3. *Let $V \in \mathscr{V}(L_2)$ be bounded from below. Then $H = (H_0 + V)^-$ is essentially self-adjoint on $\mathscr{D}(H_0) \cap \mathscr{D}(V)$ and for $t > 0$ and $V' \in \mathscr{V}(L_2)$,*

$$\text{Range}\, e^{-tH} \subset \mathscr{D}(H_0) \cap \mathscr{D}(V').$$

Proof. Let

$$F_j(\lambda) = \begin{cases} \lambda & \text{for} \quad |\lambda| \leqslant j \\ 0 & \text{otherwise} \end{cases}$$

and let $V_j = F_j(V)$. Then $H_j = H_0 + V_j$ is self-adjoint and bounded from below, uniformly in j. $V_j \to V$ on $L_r(Q)$ by the Lebesgue bounded convergence theorem. We choose the domain $\mathscr{D}$ as in (4.4). Since $\mathscr{D} \subset L_r(Q)$ also, for all $r < \infty$, V_j and H_j converge on $\mathscr{D}$. This verifies (1) and (2). In (4) we take $\mathscr{D}_0 = L_\infty(Q)$, while (5) follows from the inclusions

$$\text{Range}\, e^{-tH_j} \subset \mathscr{D}(H_0) = \mathscr{D}(H_0) \cap \mathscr{D}(V_j).$$

LEMMA 7.4. *The sequence $F_j(V')e^{-tH_j}$ converges in norm, for $t > 0$ and $V' \in \mathscr{V}(L_2)$.*

The lemma yields (3) and (6) and moreover for $\psi \in L_\infty(Q)$

$$\left\langle \psi, \lim_j F_j(V')e^{-tH_j}\theta \right\rangle = \langle \psi, (V' \restriction \mathscr{D}_0)^* e^{-tH}\theta\rangle = \langle \psi, V'e^{-tH}\theta\rangle$$

as in the proof of Proposition 7.2. Thus Range e^{-tH} is contained in $\mathscr{D}(V')$ and the theorem is proved. ∎

Proof of Lemma 7.4. The Duhamel formula states that

$$e^{-tH_j} - e^{-tH_l} = -\int_0^t e^{-sH_j}\delta V e^{-(t-s)H_l}\,ds \tag{7.1}$$

where

$$\delta V = H_j - H_l = V_j - V_l.$$

Let $j \leqslant l$. For vectors θ and $\psi \in L_2(Q, d_Bq)$,

$$\begin{aligned}|\langle\theta, (e^{-tH_j} - e^{-tH_l})\psi\rangle| &\leqslant t \sup_{0\leqslant s\leqslant t} \langle e^{-sH_j}|\theta|, |\delta V|e^{-(t-s)H_l}|\psi|\rangle\\ &\leqslant \text{const} \sup_s \langle e^{-sH_0}|\theta|, |\delta V|e^{-(t-s)H_0}|\psi|\rangle\\ &\leqslant j^{-1}\,\text{const} \sup_s \langle |\theta|, e^{-sH_0}V^2 e^{-(t-s)H_0}|\psi|\rangle\\ &\leqslant j^{-1}\,\text{const} \sup_s \|\theta\|\;\|\psi\|\;\|e^{-sH_0}V^2 e^{-(t-s)H_0}\|,\end{aligned}$$

using Corollary 6.3 and the inequality $|\delta V| \leqslant j^{-1}V^2$. Thus

$$\|e^{-tH_j} - e^{-tH_l}\| \leqslant j^{-1}\,\text{const} \sup_s \|e^{-sH_0}V^2 e^{-(t-s)H_0}\|.$$

Similarly

$$\begin{aligned}&\|F_j(V')e^{-tH_j} - F_l(V')e^{-tH_l}\|\\ &\qquad\leqslant j^{-1}\,\text{const} \sup_s \|(1 + V'^2)e^{-sH_0}V^2 e^{-(t-s)H_0}\|\\ &\qquad\quad + j^{-1}\,\text{const}\,\|(V')^2 e^{-tH_0}\|.\end{aligned}$$

N and H_0 are commuting self-adjoint operators and $N \leqslant m^{-1}H_0$ where m is the mass in (3.1). If deg V, deg $V' \leqslant n$,

$$\|(V')^2 e^{-tH_0}\| \leqslant \|(V')^2(N + I)^{-n}\|\;\|(N + I)^n e^{-tH_0}\| < \infty$$

by these properties of N and H_0 and by Theorem 4.2. If $t - s > t/2$, then

$$\|(1 + V'^2)e^{-sH_0}V^2 e^{-(t-s)H_0}\|$$
$$\leqslant \|(1 + V'^2)(N + I)^{-n}\| \, \|e^{-sH_0}\|$$
$$\times \|(N + I)^n V^2 (N + I)^{-2n}\| \, \|(N + I)^{2n} e^{-tH_0/2}\| < \infty$$

by Corollary 4.3. If $s > t/2$, we use the factor e^{-sH_0} to dominate the powers of $N + I$, and so the proof is complete. ∎

THEOREM 7.4. *$H(g) = (H_0 + H_I(g))^-$ is essentially self-adjoint on $\mathscr{D}(H_0) \cap \mathscr{D}(H_I(g))$. $H(g)$ is bounded from below and for $t > 0$ and $V' \in \mathscr{V}(L_2)$, we have*

$$\text{Range}\, e^{-tH} \subset \mathscr{D}(H_0) \cap \mathscr{D}(V').$$

Proof. Let $j = \kappa$, $H_j = H(g, \kappa)$. We verify conditions (1)–(6) of Propositions 7.1 and 7.2. Conditions (2), (4) and (5) are trivial. We prove (1). The modifications required for (3) and (6) are routine, and similarly for the convergence of $V'e^{-tH_j}$. As before, these facts complete the proof.

Our proof of (1) is a specialization to two dimensions of methods developed for the three-dimensional ϕ^4 model. We use the nth order Duhamel expansion, and to simplify the formulas, we use time ordered products. For example the integrand in (7.1) is expressed as

$$\left(\exp\left[-\int_0^t H(\sigma, s)\, d\sigma\right] \delta V(s)\right)_+$$

where

$$H(\sigma, s) = \begin{cases} H_j & \text{for } \sigma \leqslant s \\ H_l & \text{for } \sigma \geqslant s \end{cases}$$

and $\delta V(s)$ inserts δV at the time s. The subscript $+$ denotes antitime ordering (earlier times to the left in the product).

Let $\kappa_\nu = \exp(\nu^{2/p})$ where p is the degree of the polynomial P in H_I. Let

$$h_\nu = \begin{cases} H(g, \kappa_\nu) & \text{if } \kappa_\nu \leqslant \kappa \\ H(g, \kappa) & \text{if } \kappa \leqslant \kappa_\nu \end{cases}$$

and let $\delta h_\nu = H(g, \kappa) - H(g, \kappa_\nu)$. Iterating (7.1) we obtain

$$\exp(-tH(g, \kappa)) = \sum_{n=0}^{\infty} (-1)^n \int \left(\exp\left[-\int_0^t H(\sigma, s)\, d\sigma\right] \prod_{\nu=1}^{n} \delta h_\nu(s_\nu)\right)_+ ds \tag{7.2}$$

where the time integration ds extends over the domain

$$0 \leqslant s_1 \leqslant \ldots \leqslant s_n \leqslant t$$

and $H(\sigma, s) = h_\nu$ if $s_{\nu-1} \leqslant \sigma \leqslant s_\nu$. The series terminates for $\kappa_n \geqslant \kappa$.

To dominate $|\delta h_\nu|$, we use the upper bound

$$|\delta h_\nu| \leqslant \kappa_\nu^{-\frac{1}{4}} + \kappa_\nu^{\frac{1}{4}}(\delta h_\nu)^2 \equiv f_\nu. \tag{7.3}$$

Then

$$\left|\left\langle \theta, \left(\exp\left[-\int_0^t H(\sigma, s)\, d\sigma\right] \prod_{\nu=1}^n \delta h_\nu(s_\nu)\right)_+ \psi\right\rangle\right|$$

$$\leqslant \|\theta\| \, \|\psi\| \, e^{O(n)t} \left\| \left(\exp\left[-\int_0^t H_0\, d\sigma\right] \prod_{\nu=1}^n f_\nu(s_\nu)\right)_+ \right\| \tag{7.4}$$

by Corollary 6.3, since $-O(n) \leqslant H_I(g, \kappa_\nu)$ for $\nu \leqslant n$.

Each f_ν is a sum of a bounded number of Wick monomials,

$$f_\nu = \sum_{l=1}^L d_{\nu,l}.$$

The L_2 norm of the kernel of $d_{\nu,l}$ can be bounded in terms of the L_2 norms of the kernels of δh_ν, by the definition (7.3) and the Schwarz inequality. By Proposition 5.8 we get the bound $O(\kappa_\nu^{-1/4})$ as the bound on the kernel of $d_{\nu,l}$, since the kernel of $(\delta h_\nu)^2$ in (7.3) has an L_2 norm bounded by $O(\kappa_\nu^{-1+\varepsilon}) \leqslant O(\kappa_\nu^{-\frac{1}{2}})$. Thus by Corollary 4.3

$$\|d_{\nu,l}(N + I)^{-p}\| \leqslant O(\kappa_\nu^{-1/4}) = \exp(-\tfrac{1}{4}\nu^{2/p}). \tag{7.5}$$

The maximum length $s_j - s_{j-1}$ of the time intervals in the time ordered product is $t/(n+1)$ at least, since there are $(n+1)$ intervals of total length t. We insert the product

$$(N + I)^p(N + (1 + 2p)I)^p \ldots \left(N + (1 + (n-1)2p)I\right)^p$$
$$\leqslant (2p)^{np}(n-1)!\,(N + I)^{np}$$

and its inverse in the maximum time interval in (7.4). We use $(N + \alpha I)^{-p} W = W(N + (\alpha + m - n)I)^{-p}$ to move the inverse powers and obtain a factor $(N + (1 + \alpha)I)^{-p}$, $\alpha \geqslant 0$, in each time interval. This yields the bound

$$\left\| \left(\exp\left[-\int_0^t H_0 d\sigma\right] \prod_{\nu=1}^n f_\nu(s_\nu)\right)_+ \right\|$$

$$\leqslant (4np)^{np}(n!)^p \prod_{\nu=1}^n \kappa_\nu^{-\frac{1}{4}} \|N^{pn} e^{-tH_0/n+1}\|$$

$$\leqslant M^n(n!)^{3p} \prod_{\nu=1}^n \kappa_\nu^{-\frac{1}{4}},$$

for some constant M. The product

$$\prod_{\nu=1}^{n} \kappa_\nu^{-\frac{1}{4}} = \exp\left((-\tfrac{1}{4}) \sum_{\nu=1}^{n} \nu^{2/p}\right) = \exp\left(-O(n^{1+2/p})\right)$$

is rapidly converging to zero, and it dominates the other factors above. Substituting in (7.4) and (7.2), we obtain

$$\|e^{-tH(g,\kappa)}\| \leqslant \sum_{n=0}^{\infty} e^{O(tn)} \exp\left(-O(n^{1+2/p})\right),$$

which is bounded uniformly in κ. This completes the proof.

8. THE LOCAL ALGEBRAS AND THE LORENTZ GROUP AUTOMORPHISMS

Haag and Kastler have formulated the following axioms for quantum field theory. Their axioms are independent of the underlying representation of the field operators and so they can be verified at the present stage of the construction, with the field operators acting on $\mathscr{F} = L_2(Q, d_B q)$, the Fock space of the bare particles, rather than the Hilbert space of physical states (i.e. the Fock space of the physical, or in and out particles, see Section 10). The axioms are:

(a) To each bounded open region B of space time there is associated a C^*-algebra $\mathfrak{A}(B)$ containing the identity.

(b) Isotony: If $B_1 \subset B_2$ then $\mathfrak{A}(B_1) \subset \mathfrak{A}(B_2)$.

(c) Locality: If B_1 and B_2 are space like separated then $\mathfrak{A}(B_1)$ commutes with $\mathfrak{A}(B_2)$.

(d) The algebra $\mathfrak{A}$ of local observables is defined as the norm closure of $\bigcup_B \mathfrak{A}(B)$. $\mathfrak{A}$ is *primitive*; in other words $\mathfrak{A}$ has a faithful irreducible representation.

(e) Poincaré covariance: Let $\{a, \Lambda\}$ be an element of the inhomogeneous Lorentz group G. There is a representation $\sigma_{\{a,\Lambda\}}$ of G by a group of * automorphisms of $\mathfrak{A}$ such that

$$\sigma_{\{a,\Lambda\}}(\mathfrak{A}(B)) = \mathfrak{A}(\{a, \Lambda\}B). \tag{8.1}$$

In the two-dimensional boson models we are considering the time translations (i.e. the dynamics) and the Lorentz rotations are the only non-trivial axioms. The use of automorphisms is equivalent to working in the Heisenberg picture, in which the observables, represented by Hilbert space operators, change with time, while the states, represented by Hilbert space vectors, remain fixed. In a quantum system with a Hamiltonian H, the Heisenberg picture dynamics is given by the formula

$$A(t) = e^{itH} A(0)\, e^{-itH}$$

and then $A(t)$ is the observable at time t corresponding to the time zero observable $A(0)$. In our model we have locally correct Hamiltonians $H(g)$ but no global Hamiltonian, and we construct the Heisenberg picture dynamics nonetheless. We do this by restricting the observables to lie in the local algebras $\mathfrak{A}(B)$ and by using the finite propagation speed implicit in (8.1).

THEOREM 8.1. *Let B be a bounded open region of space, let $t \in R$, let g be a nonnegative function in $L_1 \cap L_2$ and let g be incidentally equal to one on B_t. For $A \in \mathfrak{A}(B)$,*

$$\sigma_t(A) \equiv e^{itH(g)} A\, e^{-itH(g)}$$

is independent of g and $\sigma_t(A) \in \mathfrak{A}(B_t)$.

Proof. Let

$$\sigma_t^{(0)}(A) = e^{itH_0} A\, e^{-itH_0}$$
$$\sigma_t^I(A) = e^{itH_I} A\, e^{-itH_I}.$$

Trotter's product formula is valid for the unitary group $\exp\big(it(H_0 + H_I(g))\big)$. This fact yields the product formula for the associated automorphism group:

$$\sigma_t(A) = \lim_{n \to \infty} (\sigma_{t/n}^{(0)} \sigma_{t/n}^I)^n (A). \tag{8.2}$$

Each interaction automorphism σ^I maps each $\mathfrak{A}(B_s)$ into itself and is independent of g on $\mathfrak{A}(B_s)$ for $|s| \leqslant |t|$. To see this, let $\chi(B_s)$ be the characteristic function of B_s. We assert that

$$\sigma_{t/n}^I(C) = \exp\big(i(t/n)H_I(\chi(B_s))\big) C \exp\big(-i(t/n)H_I(\chi(B_s))\big) \tag{8.3}$$

for $C \in \mathfrak{A}(B_s)$ and that $\sigma_{t/n}^I(C) \in \mathfrak{A}(B_s)$ also. In other words the interaction automorphism has propagation speed zero and is independent of g on $\mathfrak{A}(B_s)$ for $|s| \leqslant |t|$. The theorem follows from (8.2), (8.3) and Theorem 6.7. To prove (8.3), we write

$$H_I(g) = H_I(\chi(B_s)) + H_I(g[1 - \chi(B_s)]).$$

as a sum of commuting self-adjoint operators. By Theorem 5.6, $\exp\big(itH_I(\chi(B_s))\big) \in \mathfrak{A}(B_s)$ and so the right side of (8.3) belongs to $\mathfrak{A}(B_s)$. By Corollary 6.8,

$$\exp\big(itH_I(g[1 - \chi(B_s)])\big) \in \mathfrak{A}(B_s)'$$

and (8.3) follows. ∎

Definition 8.2. Let B be a bounded open region of space time and for any time t, let $B(t) = \{x : x, t \in B\}$ be the time t time slice of B. We define $\mathfrak{A}(B)$ to be the von Neumann algebra generated by

$$\bigcup_s \sigma_s\Big(\mathfrak{A}(B(s))\Big). \tag{8.4}$$

THEOREM 8.3. *The Haag–Kastler axioms* (a)–(e) *are valid for these local algebras* $\mathfrak{A}(B)$.

Proof. (except Lorentz rotations). The axioms (a) and (b) are obvious, while (c) follows easily from the finite propagation speed, Theorem 8.1, together with the time zero locality, Theorem 6.8. (d) is a well known property of the free field (Proposition 9.8). Because the time zero fields coincide with the time zero free fields, and because the time zero fields generate $\mathfrak{A}$ by Theorem 8.1 and the definition of the local algebras, the free field result carries over to our model with interaction $H_I \neq 0$.

In the Poincaré covariance axiom (e), the time translation is given by σ_t. Let $B + t$ be the time translate of the space time region B. Then $(B + t)(s) = B(s - t)$ and so

$$\begin{aligned}\sigma_t \bigcup_s \sigma_s\Big(\mathfrak{A}(B(s))\Big) &= \bigcup_s \sigma_{s+t}\Big(\mathfrak{A}(B(s))\Big)\\ &= \bigcup_s \sigma_s\Big(\mathfrak{A}(B(s-t))\Big) = \bigcup_s \sigma_s\Big(\mathfrak{A}((B+t)(s))\Big).\end{aligned}$$

Thus $\sigma_t(\mathfrak{A}(B)) = \mathfrak{A}(B + t)$, and (8.1) is verified for time translations. Since the local algebras are norm dense in $\mathfrak{A}$ and since automorphisms of C^*-algebras preserve the norm, σ_t extends to an automorphism of $\mathfrak{A}$.

To define the space translation automorphism σ_x, let

$$P = \int k\, a^*(k) a(k)\, dk$$

and define

$$\sigma_x(A) = e^{-ixP} A\, e^{ixP}. \tag{8.5}$$

Then

$$e^{-ixP}\phi(y)\, e^{ixP} = \phi(x + y)$$

$$e^{-ixP}\pi(y)\, e^{ixP} = \pi(x + y).$$

We leave as an exercise the proof of the following lemma, which completes the proof of Theorem 8.5 except for Lorentz rotations. ∎

LEMMA 8.4. $\sigma_x(\mathfrak{A}(B)) = \mathfrak{A}(B + x)$, σ_x *extends to a* $*$ *automorphism of* $\mathfrak{A}$, *and* $x, t \to \sigma_x\sigma_t = \sigma_t\sigma_x$ *defines a two parameter abelian automorphism group of* $\mathfrak{A}$.

To construct automorphisms for the full Lorentz group and to complete the proof of Theorem 8.1, there are four separate steps. The first is to construct a self-adjoint locally correct generator for Lorentz rotations, along the lines of Theorem 7.4. This generator then defines a locally correct unitary group and automorphism group. The second step is to prove this statement for the fields, by showing that the field $\phi(x, t)$, considered as a bilinear form on a suitable domain, is transformed locally correctly by our unitary group. The third step is to show that the local algebras $\mathfrak{A}(B)$ are also transformed correctly. The final step is to reconstruct the Lorentz group automorphisms from the locally correct pieces given by the first three steps. This final step is not difficult.

We now sketch the first step. Let $H_0(x)$ denote the integrand in (4.17), so that $H_0 = \int H_0(x)\,dx$. The formal generator of Lorentz rotations is

$$M = M_0 + M_I = \int xH_0(x) + \int x{:}P(\phi(x)){:}\,dx.$$

The local Lorentzian with which we are concerned is

$$M(g_1, g_2) = \varepsilon H_0 + H_0(g_1) + H_I(g_2)$$

where $H_0(g_1) = \int H_0(x)g_1(x)\,dx$. We require that $0 < \varepsilon$ and that g_1 and g_2 be nonnegative C_0^∞ functions. In the second step one requires more, for example that $\varepsilon + g_1(x) = x$ and $g_2(x) = x$ in some local space region. This region is necessarily contained in the interval $[\varepsilon, \infty)$, and general space time regions are considered only in the fourth step.

Although the Lorentzians M, M_0 and M_I are each unbounded from below, $M(g_1, g_2)$ is semibounded, as we shall see, and this fact means that the methods of Section 7 can be used to study it. We begin by decomposing $H_0(g_1)$ as a sum of a diagonal and an off-diagonal term:

$$\begin{aligned} H_0(g_1) &= \int v_D(k,l)a^*(k)a(l)\,dk\,dl \\ &\quad + \int v_{OD}(k,l)[a^*(k)a^*(l) + a(-k)a(-l)]\,dk\,dl \\ &= H_0{}^D(g_1) + H_0{}^{OD}(g_1), \end{aligned}$$

where

$$\begin{aligned} v_D &= 2^{-1}(2\pi)^{-1}(\mu(k)\mu(l) + kl + m^2)\mu(k)^{-\frac{1}{2}}\mu(l)^{-\frac{1}{2}}g_1{}^{\sim}(-k+l) \\ v_{OD} &= 4^{-1}(2\pi)^{-1}(-\mu(k)\mu(l) - kl + m^2)\mu(k)^{-\frac{1}{2}}\mu(l)^{-\frac{1}{2}}g_1{}^{\sim}(-k-l), \end{aligned}$$

as in the proof of Theorem 4.4.

We require the following information about this decomposition.

LEMMA 8.5. (a) $v_{OD} \in L_2(\mathbf{R}^2)$.

(b) v_D *is the kernel of a nonnegative operator and* $\varepsilon\mu(k)\delta(k-l) + \beta v_D$ *is the kernel of a positive self-adjoint operator, for* $\beta \geqslant 0$. *These operators are real in configuration space.*

Proof. (a) is a routine calculation based on the facts that $|g_1^{\sim}(k+l)| \leqslant$ const $\mu(k+l)^{-n}$ for any n and that $\mu(k) - |k| = O(|k|^{-1})$ as $k \to \infty$. We omit further details. (b) is proved using a finite sequence of Kato perturbations. Let $v_\beta = \varepsilon\mu(k)\delta(k-l) + \beta v_D$ and let V_β and V_D denote the operators with kernels v_β and v_D respectively. V_D is a sum of three terms having the form

$$A^* M_{g_1} A$$

in configuration space, where M_{g_1} is multiplication by $g_1 \geqslant 0$. Hence $0 \leqslant V_D$. Moreover for γ sufficiently small, but chosen independently of β,

$$\gamma V_D \leqslant \tfrac{1}{2} V_0 \leqslant \tfrac{1}{2}(V_0 + \beta V_D) = \tfrac{1}{2} V_\beta$$

and so

$$V_{\beta+\gamma} = V_\beta + \gamma V_D$$

is a Kato perturbation, in the sense of bilinear forms. Consequently if V_β is self-adjoint, so is $V_{\beta+\gamma}$, and $\mathscr{D}(V_{\beta+\gamma}^{\frac{1}{2}}) = \mathscr{D}(V_\beta^{\frac{1}{2}})$. A finite induction starting with $V_0 = V_0^*$ now shows that V_β is self-adjoint, for all $\beta \geqslant 0$. ∎

LEMMA 8.6. $H_0{}^D(g_1)$ *is nonnegative and* $\varepsilon H_0 + \beta H_0{}^D(g_1)$ *is self-adjoint, for all* $\beta > 0$.

Proof. The sum is taken as a sum of bilinear forms (in order to avoid the slightly more complicated estimates required to define the sum of operators). One appeals to Lemma 8.5 and general properties of the biquantization map, e.g. $V_D \to H_0{}^D(g_1)$ or one repeats the arguments used in the proof of Lemma 8.5.

Let

$$M(\beta) = \varepsilon H_0 + H_0{}^D(g_1) + \beta H_0{}^{OD}(g_1) + H_I(g_2).$$

Then $M(0)$ is self-adjoint and semibounded by the same construction as was applied to $H(g)$ in Chapter 7. Again we go from $M(0)$ to $M(1) = M(g_1, g_2)$ be a finite sequence of Kato bilinear form perturbations. By Lemma 8.5 and Corollary 4.3,

$$\pm \gamma H_0{}^{OD}(g_1) \leqslant (\varepsilon m_0/2)N + I \leqslant (\varepsilon/2)H_0 + I$$

for γ sufficiently small. We assert that

$$(\varepsilon/2)H_0 \leqslant \tfrac{3}{4}M(\beta) + c(\beta)$$

where $c(\beta)$ is a constant depending on β, $0 \leqslant \beta \leqslant 1$. Using the assertion, we show by induction that $M(\gamma), M(2\gamma), \ldots$ and $M(1) = M(g_1, g_2)$ are all self-adjoint. Moreover

$$\mathscr{D}(M(g_1, g_2)^{\frac{1}{2}}) = \mathscr{D}(M(0)^{\frac{1}{2}}) = \mathscr{D}(H_D^{\frac{1}{2}}).$$

The assertion is equivalent to showing that

$$(\varepsilon/3)H_0 + H_0{}^D(g_1) + \beta H_0{}^{OD}(g_1) + H_I(g_2).$$

is semibounded. Again by the methods of Section 7,

$$(\varepsilon/6)H_0 + (1 - \beta)H_0{}^D(g_1) + H_I(g_2)$$

is semibounded. Thus it is sufficient to consider the remainder

$$(\varepsilon/6)H_0 + \beta H_0(g_1).$$

To show that this operator is semibounded, we introduce a momentum cut-off in $H_0(g_1)$. The contribution from the low momentum region is semibounded by undoing the Wick order (cf. Section 5). The contribution from the high momentum region is a sum of a diagonal part, which is nonnegative (cf. Lemma 8.5) and an off-diagonal part, which has an L_2 kernel. This kernel has a small norm if the momentum cut-off is sufficiently large, and in this case the off-diagonal term is dominated by $(\varepsilon/6)H_0$, by Corollary 4.3. This completes the construction of the locally correct self-adjoint generator for Lorentz rotations. See Cannon and Jaffe (1970), Rosen (1972a) and Klein (1971).

The second step is the most difficult. One verifies the equation

$$(tD_x + xD_t)\phi = [M, \phi]. \tag{8.6}$$

The field $\phi(x, t) = e^{itH(g)}\phi(x)\,e^{-itH(g)}$ is not taken at time zero, and before the commutator can be evaluated, M must be commuted with the unitary operator $e^{itH(g)}$. The commutator $[M, e^{itH(g)}]$ cannot be evaluated explicitly in terms of Wick monomials. However, the part of $[M, e^{itH(g)}]$ which cannot be evaluated does not contribute to $[M, \phi]$, by virtue of its localization. The calculations to justify (8.6) are extensive, in order to deal with questions of operator domains. See Cannon and Jaffe (1970) and Rosen (1972a).

The key to the third step is to give a covariant definition of the local algebras $\mathfrak{A}(B)$. Let $f \in \mathscr{D}(B)$, the class of real C^∞ functions with support in B, let $\alpha_1, \ldots, \alpha_n$ be real numbers and consider

$$\phi(f) = \int \phi(x, t) f(x, t)\, dx\, dt \tag{8.7}$$

$$\phi(f, t) = \int \phi(x, t) f(x, t)\, dx \tag{8.8}$$

$$\alpha_1 \phi(f, t_1) + \ldots + \alpha_n \phi(f, t_n) \tag{8.9}$$

$$\pi(f, t) = \int \pi(x, t) f(x, t)\, dx. \tag{8.10}$$

For $g \equiv 1$ on a sufficiently large set (the domain of dependence of the region B), the time integration in (8.7) converges strongly, and all four operators above are symmetric and defined on $\mathscr{D}(H(g))$.

THEOREM 8.7. *The operators* (8.7)–(8.10) *are essentially self-adjoint on any core for* $H(g)^{\frac{1}{2}}$.

If a sequence $\{A_n\}$ of self-adjoint operators converges strongly to a self-adjoint limit A on a core for A then the unitary operators e^{itA_n} converge strongly to e^{itA} (Kato, 1966b). Using this fact, one can show that the operators (8.7) and (8.9) generate the same von Neumann algebra, $\mathfrak{A}_1(B)$ and that $\mathfrak{A}_1(B) \supset \mathfrak{A}(B)$. To show that $\mathfrak{A}_1(B) \subset \mathfrak{A}(B)$, recall that a self-adjoint operator A commutes with a bounded operator C provided $C\mathscr{D} \subset \mathscr{D}(A)$ and $CA = AC$ on $\mathscr{D}$, for some core $\mathscr{D}$ of A. Equivalent is the condition that C commute with all bounded functions of A. Also equivalent is the relation $CA = AC$ on $\mathscr{D}(A)$. We choose $\mathscr{D} = \mathscr{D}(H(g))$. If C commutes with all operators of the form (8.8), it also commutes on $\mathscr{D}(H(g))$ with all operators of the form (8.9). Hence $\mathfrak{A}(B)' \subset \mathfrak{A}_1(B)'$ and so

$$\mathfrak{A}_1(B) = \mathfrak{A}_1(B)'' \subset \mathfrak{A}(B)'' = \mathfrak{A}(B).$$

This proves the following result, which gives the covariant definition of $\mathfrak{A}(B)$.

THEOREM 8.8. $\mathfrak{A}(B)$ *is the von Neumann algebra generated by bounded functions of operators of the form* (8.7).

Proof of Theorem 8.7. We consider first the sharp time operators (8.8) and (8.10). Using the unitary operator $U(t) = \exp(itH(g))$, we transform $\phi(f, t)$ and $\pi(f, t)$ into the time zero operators $\phi(f, 0)$ and $\pi(f, 0)$. Since

$U(t)$ maps a core for $H(g)^{\frac{1}{2}}$ into another core for $H(g)^{\frac{1}{2}}$, we may suppose that $t = 0$. For $A = \phi(f, 0)$ or $A = \pi(f, 0)$, we have

$$\|A(H(g) - \zeta)^{-\frac{1}{2}}\| \leqslant \|A(N + I)^{-\frac{1}{2}}\| \, \|(N + I)^{\frac{1}{2}}(H_0 + I)^{-\frac{1}{2}}\| \times \|(H_0 + I)^{\frac{1}{2}}(H(g) - \zeta)^{-\frac{1}{2}}\|. \tag{8.11}$$

The first factor on the right is bounded by Corollary 4.3. The second factor is bounded since $N \leqslant m^{-1}H_0$ and since N and H_0 commute. To bound the third factor, we appeal to the semiboundedness in Theorem 7.4. Let

$$\mathscr{D}_1 = \mathscr{D}(H_0) \cap \mathscr{D}(H_I(g))$$
$$\mathscr{D}_{\frac{1}{2}} = (H(g) - \zeta)^{\frac{1}{2}}\mathscr{D}_1.$$

Since $\mathscr{D}_1$ is a core for $H(g)$ (Theorem 7.4), $\mathscr{D}_1$ is also a core for $H(g)^{\frac{1}{2}}$ and hence $\mathscr{D}_{\frac{1}{2}}$ is dense in $\mathscr{F}$. By Theorem 7.4, with g replaced by $2g$, we have

$$0 \leqslant \tfrac{1}{2}(H_0 + H_I(2g) + \text{const})$$

on $\mathscr{D}_1 \times \mathscr{D}_1$. Hence

$$\tfrac{1}{2}H_0 \leqslant H_0 + H_I(g) + \text{const}$$

and

$$0 \leqslant (H(g) - \zeta)^{-\frac{1}{2}} H_0 (H(g) - \zeta)^{-\frac{1}{2}} \leqslant \text{const}$$

for ζ real and sufficiently negative. The latter inequality holds on $\mathscr{D}_{\frac{1}{2}} \times \mathscr{D}_{\frac{1}{2}}$. Since $\mathscr{D}_{\frac{1}{2}}$ is dense, the third factor in (8.11) is a bounded operator.

Now let $\mathscr{D}_0$ be any core for $H(g)^{\frac{1}{2}}$. From (8.11), we see that

$$(A \restriction \mathscr{D}_0)^- = \left(A \restriction \mathscr{D}(H(g)^{\frac{1}{2}})\right)^- \supset (A \restriction \mathscr{D})^-,$$

where $\mathscr{D}$ is the domain introduced in Chapters 3 and 4. By Lemma 5.4, A is essentially self-adjoint on $\mathscr{D}$ and hence on $\mathscr{D}_0$.

Next we consider the time averaged operators (8.7) and (8.9), which we again denote by A. Integrating (8.11) above over t, we obtain the bound

$$\|A(H(g) - \zeta)^{-\frac{1}{2}}\| + \|\dot{A}(H(g) - \zeta)^{-\frac{1}{2}}\| \leqslant \text{const} \tag{8.12}$$

where $\dot{A} = i[H(g), A]$ is defined as a bilinear form on $\mathscr{D}(H(g)) \times \mathscr{D}(H(g))$. Theorem 8.7 is now a consequence of the following general result. ∎

THEOREM 8.9. *Let H be a positive self-adjoint operator and let A be a symmetric operator defined on $\mathscr{D}(H^{\frac{1}{2}})$. Suppose that*

$$\|AR^{\frac{1}{2}}\| + \|\dot{A}R^{\frac{1}{2}}\| \leqslant \text{const} \tag{8.13}$$

where $R = (H + I)^{-1}$ and $\dot{A} = i[H, A]$. Then A is essentially self-adjoint on any core for $H^{\frac{1}{2}}$.

Proof. First we construct a self-adjoint extension C for $A_0 = A \upharpoonright \mathscr{D}(H)$, using the remark following Proposition 7.1. Let $H = \int \lambda \, dE_\lambda$ be the spectral resolution for H and let $C_n = E_n A_0 E_n$. C_n is bounded (by (8.13)) and symmetric, and hence self-adjoint. We assert that $\|(C_n - A)R\| \to 0$. Then $C_n \to A_0$ strongly on the dense domain $\mathscr{D}(H)$, verifying condition (2) of Proposition 7.1. To prove the assertion, we use the identity

$$RA = AR + i R\dot{A}R,$$

valid on the domain $\mathscr{D}(H)$. We have

$$\begin{aligned}\|(C_n - A)R\| &\leqslant \|E_n A(1 - E_n)R\| + \|(I - E_n)AR\| \\ &\leqslant \|AR^{\frac{1}{2}}\| \, \|(I - E_n)R^{\frac{1}{2}}\| + \|RA(I - E_n)AR\|^{\frac{1}{2}} \\ &\leqslant O(n^{-\frac{1}{2}}) + (\|RA(I - E_n)RA\| + \|RA(I - E_n)R\dot{A}R\|)^{\frac{1}{2}} \\ &\leqslant O(n^{-\frac{1}{2}}) + O(n^{-1/4}) \to 0.\end{aligned}$$

Let $K_n = (C_n \pm iy)^{-1}$. Next we show that $\|(H + I)K_n R\| \leqslant 1$ for y sufficiently large. The operators C_n and K_n leave the domain of H invariant and so for any vector $\theta \in \mathscr{F}$, $K_n R\theta = \psi \in \mathscr{D}(H)$. Thus

$$\begin{aligned}&\langle \psi, (C_n \mp iy)(H + I)^2 (C_n \pm iy)\psi \rangle \\ &\quad \geqslant y^2 \langle \psi, (H + I)^2 \psi \rangle - y \, |\langle \psi, (H + I)\dot{C} + \dot{C}(H + I)\psi \rangle| \\ &\quad \geqslant (y^2 - \text{const}\, y)\langle \psi, (H + I)^2 \psi \rangle.\end{aligned}$$

For y^2 sufficiently large, $y^2 - \text{const}\, y \geqslant 1$, and

$$\langle \theta, RK_n^*(H + I)^2 K_n R\theta \rangle \leqslant \|\theta\|^2$$

as asserted.

The next step is to prove the convergence of the resolvents K_n. For $\psi \in D(H)$,

$$\begin{aligned}\|(K_n - K_m)\psi\| &= \|K_n(C_m - C_n)K_n\psi\| \\ &\leqslant \|K_n\| \, \|(C_n - C_m)R\| \, \|(H + I)K_m R\| \, \|(H + I)\psi\|.\end{aligned}$$

The second factor tends to zero as $n \to \infty$ and the remaining factors are bounded as $n \to \infty$. Thus the resolvents converge strongly. By Proposition 7.1 and the remark following it, $K = \lim_{n \to \infty} K_n$ is the resolvent of a self-adjoint operator C. It is easy to see that $C \supset A_0$ since for $\theta \in \mathscr{D}(H)$ and $\psi \in \mathscr{F}$,

$$\langle K\psi, (C \pm iy)\theta \rangle = \langle \psi, \theta \rangle = \lim_n \langle K_n \psi, (C_n + iy)\theta \rangle = \langle K\psi, (A_0 \pm iy)\theta \rangle.$$

Since the range of K is dense, $(C \pm iy)\theta = (A_0 \pm iy)\theta$.

Let $\theta \in \mathscr{D}(C)$. We show that $\theta \in \mathscr{D}(A_0^-)$ and that $C\theta = A_0^-\theta$. Thus A is essentially self-adjoint on $\mathscr{D}(H)$, and since $AR^{\frac{1}{2}}$ is a bounded operator, we see that A is also essentially self-adjoint on any core for $H^{\frac{1}{2}}$.

Let $\chi = (C_n - iy)\theta$ and let $P_j = (1 + H/j)^{-1}$. Then $\theta_j = P_j\theta \in \mathscr{D}(H) = \mathscr{D}(A_0)$ and $\theta_j \to \theta$. Furthermore

$$(A_0 - iy)\theta_j = \lim_n (C_n - iy)P_j K_n \chi$$

since $K_n\chi \to K\chi = \theta$, since $\|(C_n - A_0)R\| \to 0$ and since $(H + I)P_j$ is bounded. However

$$\begin{aligned}\|(C_n - iy)P_jK_n\chi - P_j\chi\| &= \|(C_n - iy)[P_j, K_n]\chi\| \\ &= j^{-1}\|(C_n - iy)K_nP_j\dot{C}_nP_jK_n\chi\| \\ &\leqslant j^{-1}\|\dot{C}_nP_j\| \, \|K_n\chi\| \\ &= O(j^{-1})\|\dot{C}_nR^{\frac{1}{2}}\| \, \|(H + I)^{\frac{1}{2}}P_j\| \\ &= O(j^{-1})\,O(j^{\frac{1}{2}}) = O(j^{-\frac{1}{2}}).\end{aligned}$$

Thus

$$\begin{aligned}\|(A_0 - iy)\theta_j - \chi\| &\leqslant \limsup_n \|(C_n - iy)\theta_jK_n\chi - \chi\| \\ &\leqslant \|(P_j - I)\chi\| + O(j^{-\frac{1}{2}}).\end{aligned}$$

Hence $\theta \in \mathscr{D}(A_0^-)$ and

$$A_0^-\theta = \chi + iy\theta = C\theta,$$

which completes the proof. ∎

IV Boson Quantum Field Models: Part III. Further Developments

PART III. FURTHER DEVELOPMENTS

9. LOCALLY NORMAL REPRESENTATIONS OF THE OBSERVABLES

The local algebras, as constructed in Section 8, are not acting on the Hilbert space of physical particles. On the physical Hilbert space, as in the laboratory, the states have a simple asymptotic description for large values of $|t|$. One observes isolated particles or clusters formed as bound states of several elementary particles. Because they are widely separated, the elementary particles or bound states do not interact, and they behave asymptotically like free particles. We present here the functional analysis preparation for the construction of the physical Hilbert space $\mathscr{F}_{ren}$ in Section 10. On $\mathscr{F}_{ren}$, the above asymptotic description of the states should be valid. We begin by listing without proof three general results. A state on a C^*-algebra is by definition a positive linear functional ω which satisfies the normalization condition $\omega(I) = 1$.

THEOREM 9.1. *Let $\mathfrak{A}$ be a C^*-algebra and let ω be a state on $\mathfrak{A}$. There is a Hilbert space $\mathscr{H}_\omega$, a representation π_ω of $\mathfrak{A}$ by operators on $\mathscr{H}_\omega$ and a vector $\Omega \in \mathscr{H}_\omega$ such that*

$$\omega(A) = \langle \Omega, \pi_\omega(A)\Omega \rangle.$$

The vectors $\pi_\omega(\mathfrak{A})\Omega$ are dense in $\mathscr{H}_\omega$.

THEOREM 9.2. *With $\mathfrak{A}$ and ω as above, let $\{\sigma_\lambda : \lambda \in \Lambda\}$ be a group of * automorphisms of $\mathfrak{A}$, and suppose that $\omega \,.\, \sigma_\lambda(A) = \omega(A)$ for all A in $\mathfrak{A}$ and all λ in Λ. Then there is a unique unitary operator U_λ such that $U_\lambda \Omega = \Omega$ and*

$$U_\lambda \pi_\omega(A) U_\lambda^* = \pi_\omega(\sigma_\lambda(A)).$$

Furthermore, $\lambda \to U_\lambda$ is a representation of the group Λ.

THEOREM 9.3. *Let ω be a state on a von Neumann algebra $\mathfrak{A}$. The following four conditions are equivalent.*

(i) *There is a trace class matrix T such that $\omega(A) = \mathrm{Tr}(TA)$ for all $A \in \mathfrak{A}$.*

(ii) *ω is a (possibly infinite) convex combination of the vector states $\omega_\theta(A) = \langle \theta, A\theta \rangle$.*

(iii) *ω is ultraweakly continuous.*

(iv) *If $\{A_\alpha\}$ is a monotone increasing net of operators, bounded from above, then $\sup_\alpha \omega(A_\alpha) = \omega(\sup_\alpha A_\alpha)$.*

Definition. Such an ω is called a normal state. The operator T in (i) is a density matrix for ω.

COROLLARY 9.4. *Let $\{\omega_n\}$ be a norm convergent net of normal states defined on a von Neumann algebra $\mathfrak{A}$. Then the limit state is normal also.*

Proof. Let ω be the limit state, let $\{A_\alpha\}$ be a monotone net of operators as in (iv), and let $A = \sup_\alpha A_\alpha$. Then

$$\omega(A - A_\alpha) = (\omega - \omega_n)(A) + \omega_n(A - A_\alpha) + (\omega_n - \omega)(A_\alpha).$$

We are given a bound on $\|A\|$ and $\sup_\alpha \|A_\alpha\|$, and so if n is chosen large enough, the first and third terms are bounded by $\varepsilon/3$. For fixed n and large α, the second term is also bounded by $\varepsilon/3$, and this completes the proof. ∎

THEOREM 9.5. *If ω is a normal state on $\mathfrak{A}$, then π_ω is normal (preserves the ultraweak topology).*

Proof. Let θ_n be a sequence of vectors in $\mathscr{H}_\omega$. If $\theta_n = \pi_\omega(C_n)\Omega$, $C_n \in \mathfrak{A}$, then the ultraweak continuity of $\mathfrak{A}(B) \ni A \to \langle \theta_n \pi_\omega(A)\theta_n\rangle = \omega(C_n{}^*AC_n)$ follows from that of $\omega \upharpoonright \mathfrak{A}$. Since the vectors $\pi_\omega(C_n)\Omega$ are dense in $\mathscr{H}_\omega$, normality of the above functional for arbitrary θ_n in $\mathscr{H}_\omega$ is a consequence of Corollary 9.4. Thus the functional

$$A \to \sum_{n=1}^{j} \langle \theta_n, \pi_\omega(A)\theta_n\rangle$$

is normal and providing $\Sigma\|\theta_n\|^2 < \infty$, the same is true for the infinite sum by Corollary 9.4. Hence π_ω is normal. ∎

We now specialize to the C^*-algebra $\mathfrak{A}$ of local observables. A state ω of $\mathfrak{A}$ is called locally normal if the restriction $\omega \upharpoonright \mathfrak{A}(B)$ is normal for each of the local subalgebras $\mathfrak{A}(B)$. Similarly a representation π_ω of $\mathfrak{A}$ is locally normal if each of the restrictions $\pi_\omega \upharpoonright \mathfrak{A}(B)$ is normal. A sequence (or net) $\{\omega_n\}$ of states is locally norm convergent if the sequence $\{\omega_n \upharpoonright \mathfrak{A}(B)\}$ is norm convergent for each B. In this case $\{\omega_n\}$ is w^* convergent to a state ω of $\mathfrak{A}$. The w^* convergence follows from the facts that $\|\omega_n\| = 1$ and that $\{\omega_n\}$ is w^* convergent on the norm dense subset $\cup_B \mathfrak{A}(B)$.

Remark. A local norm limit of locally normal states is locally normal. A locally normal state ω yields a locally normal representation. The first statement follows from Corollary 9.4. The second is a corollary to the proof of Theorem 9.5.

We note that the algebra $\mathfrak{A}$ has the following two properties:

(P1) $\mathfrak{A}$ is the norm closure of an increasing sequence of separable factors $\mathfrak{A}(B)$.

(P2) For $B^- \subset \operatorname{Int} C$, $\mathfrak{A}(C) \cap \mathfrak{A}(B)'$ contains a type I_∞ factor.

Haag, *et al.* (1970) have studied similar properties of C^*-algebras. In their terminology, a C^*-algebra is a sequential funnel if it is the norm closure of an increasing sequence of factors.

We work with the time zero algebras $\mathfrak{A}(B)$ of Chapter 6 and the space region $B_n = (-n, n)$. Araki (1963) proved (P1).

THEOREM 9.6. *Let B be an interval. The time zero algebra $\mathfrak{A}(B)$ is a factor.*

Proof. We prove below that

$$(\mathfrak{A}(B) \cup \mathfrak{A}(\sim B))'' = \mathscr{B}(\mathscr{F}). \tag{9.1}$$

Then $\mathfrak{A}(B)' \cap \mathfrak{A}(\sim B)' = \{\lambda I\}$. However $\mathfrak{A}(B) \subset \mathfrak{A}(\sim B)'$, and consequently

$$\mathfrak{A}(B)' \cap \mathfrak{A}(B) = \{\lambda I\},$$

so $\mathfrak{A}(B)$ is a factor. ∎

The proof of (9.1) is contained in the following two propositions.

PROPOSITION 9.7. $(\mathfrak{A}(B) \cup \mathfrak{A}(\sim B))'' = \mathfrak{A}''$.

Proof. Let χ_B be the characteristic function of the interval B. For any test function f,

$$f = \chi_B f + (1 - \chi_B)f$$

and each of the terms $\chi_B f$ and $(1 - \chi_B)f$ can be approximated in L_2 by a sequence of localized test functions:

$$g_n \to \chi_B f; \qquad \operatorname{supp} g_n \subset B$$
$$h_n \to (1 - \chi_B)f; \qquad \operatorname{supp} h_n \subset \operatorname{Int}(\sim B).$$

It follows that

$$\exp(i\phi(g_n)) \exp(i\phi(h_n)) \to \exp(i\phi(f))$$

and so

$$\exp(i\phi(f)) \in (\mathfrak{A}(B) \cup \mathfrak{A}(\sim B))''.$$

Because of the extra singularity of the kernel of π, this argument does not apply to $\exp(i\pi(f))$ in general, but it does apply if f is zero on the boundary points a and b of $B = (a, b)$. Let

$$g_{a,n}(x) = g(n(x - a)) f(a)$$

where $g \in C_0^\infty$ and g is chosen so that $g(0) = 1$, and let $g_{b,n}$ be similarly defined. Then

$$\begin{aligned} \|\mu_x^{\frac{1}{2}} g_{a,n}\|^2 &= |f(a)|^2 \int \mu(k)\, n^{-2}\, |g(k/n)^\sim|^2\, dk \\ &= |f(a)|^2\, n^{-1} \int \mu(nk)\, |g(k)^\sim|^2\, dk \\ &\to |f(a)|^2 \int k\, |g(k)^\sim|^2\, dk. \end{aligned}$$

Let L_a denote the above limit. The sequence $\|\mu_x^{\frac{1}{2}} g_{a,n}\|$ converges weakly to zero. Using these facts, one can show that

$$\exp\left(i\pi(g_{a,n})\right) \to \exp\left(-L_a/2\right) \neq 0 \tag{9.2}$$

in the weak operator topology, using a Taylor series expansion to define the exponential. Consequently

$$\exp\left(i\pi(f - g_{a,n} - g_{b,n})\right) \in \left(\mathfrak{A}(B) \cup \mathfrak{A}(\sim B)\right)'',$$

by the argument given for $\phi(f)$, and using the convergence (9.2),

$$\exp\left(i\pi(f)\right) \in \left(\mathfrak{A}(B) \cup \mathfrak{A}(\sim B)\right)''.$$

This completes the proof. ∎

PROPOSITION 9.8. $\mathfrak{A}'' = \mathscr{B}(\mathscr{F})$.

Proof. We use the fact that Ω_0 is a cyclic vector for $\mathfrak{A}$ and we show that $\mathfrak{A}' = \{\lambda I\}$.

If $\{e_i\}$ is an orthonormal basis in the space of test functions, one can show that

$$N = \sum_{i=1}^{\infty} a^*(e_i^-)a(e_i).$$

Let

$$N(n) = \sum_{i=1}^{n} a^*(e_i^-)a(e_i).$$

Then $N(n) \to N$ strongly on a core for N, and so $e^{itN(n)} \to e^{itN}$ strongly. One can show that $e^{itN(n)} \in \mathfrak{A}''$, and then $e^{itN} \in \mathfrak{A}''$ also. We use the fact that Ω_0 is a simple eigenvector for N for the eigenvalue zero.

Since $\mathfrak{A}'$ is a von Neumann algebra, $\mathfrak{A}'$ is generated by its orthogonal projections. Let E be a projection in $\mathfrak{A}'$. If $E\Omega_0 = 0$ then $E\mathfrak{A}\Omega_0 = 0$ and since $\mathfrak{A}\Omega_0$ is dense in $\mathscr{F}$, $E = 0$. If $E\Omega_0 \neq 0$ then $E\Omega_0$ is an eigenvector for N with eigenvalue zero, since

$$e^{itN}E\Omega_0 = Ee^{itN}\Omega_0 = E\Omega_0$$

for all $t \in R$. Since zero is a simple eigenvalue, $E\Omega_0$ is proportional to Ω_0. Consequently Ω_0 is in the range of E, and so $E\Omega_0 = \Omega_0$. Thus $EA\Omega_0 = AE\Omega_0 = A\Omega_0$ for all $A \in \mathfrak{A}$, and since Ω_0 is cyclic, $E = I$. Hence $\mathfrak{A}' = \{\lambda I\}$ and $\mathfrak{A}'' = \mathscr{B}(\mathscr{F})$. ∎

To verify (P2) we choose $f \neq 0$ in $\mathcal{S}_{\mathbf{R}}(\mathbf{R})$ with support in $C \sim B$ and let $\mathfrak{M}$ be the von Neumann algebra generated by the operators $\exp(it\phi(f) + is\pi(f))$. By the von Neumann uniqueness theorem in the case of one degree of freedom, $\mathfrak{M}$ is a factor of type I_∞. (See Simon's Chapter.) ∎

THEOREM 9.9. *Locally normal representations of* $\mathfrak{A}$ *are faithful.*

Proof. Let π be a locally normal representation. Then $(\text{kernel}\, \pi) \cap \mathfrak{A}(B)$ is an ultraweakly closed two-sided ideal in $\mathfrak{A}(B)$. Since a factor contains no ultraweakly closed two-sided ideals (Dixmier, 1957, p. 43, Cor. 3), $\pi \restriction \mathfrak{A}(B)$ is one-to-one. A one–one representation is isometric, and this is true for each B. By (P1), π is thus isometric on a dense subalgebra of $\mathfrak{A}$, and hence isometric on $\mathfrak{A}$. This completes the proof. ∎

Remark. It is not difficult to show that $\mathfrak{A}$ is simple, using (P1) and (P2). In fact (P2) has the consequence that any nonzero projection $E \in \mathfrak{A}(B)$ is infinite in $\mathfrak{A}(C)$, and hence not in the kernel of an arbitrary representation of $\mathfrak{A}$.

LEMMA 9.10. *The Hilbert space* $\mathscr{H}_\omega$ *constructed from a locally normal state* ω *of* $\mathfrak{A}$ *is separable.*

Proof. Since $\mathfrak{A}(B)$ is separable in the ultraweak topology and $\pi_\omega \restriction \mathfrak{A}(B)$ is normal, $\pi_\omega(\mathfrak{A}(B))\Omega^-$ is a separable subspace of $\mathscr{H}_\omega$. Since a countable family of such subspaces span $\mathscr{H}_\omega$ by (P1), $\mathscr{H}_\omega$ is separable also. ∎

THEOREM 9.11. *Let* ω *be a locally normal state on* $\mathfrak{A}$. *Then* $\pi_\omega \restriction \mathfrak{A}(B)$ *is unitarily implemented, for each bounded region* B.

Proof. We only use properties (P1) and (P2) of $\mathfrak{A}$. By Theorem 9.9, $\mathfrak{A}(B)$ and $\pi_\omega(\mathfrak{A}(B))$ are isomorphic von Neumann algebras and by (P2) $\mathfrak{A}(B)'$ and $\pi_\omega(\mathfrak{A}(B))'$ contain infinite families of orthogonal equivalent projections, constructed within $\mathfrak{A}(C)$ and $\pi_\omega(\mathfrak{A}(C))$ respectively. Since $\mathscr{H}_\omega$ is separable, $\mathfrak{A}(B)'$ and $\pi_\omega(\mathfrak{A}(B))'$ are of countable type. Consequently the theorem follows from Dixmier (1957). ∎

COROLLARY 9.12. *Let* ω *be a locally normal state of* $\mathfrak{A}$ *and let* Λ *be the subgroup of the Lorentz group leaving* ω *fixed*: $\omega \circ \sigma_\lambda = \omega$, $\lambda \in \Lambda$. *Then the unitary representation* U_λ *of Theorem 9.2 is strongly continuous on* Λ.

Proof. σ_λ is locally implemented by continuous unitary groups, by the proof of Theorem 8.1. Thus the same is true for $\pi_\omega \circ \sigma_\lambda$ by Theorem 9.10. Hence

$U_\lambda \pi_\omega(A) U^* \Omega = U_\lambda \pi_\omega(A)\Omega$ depends continuously on λ, for A in some local algebra. Since the vectors $\bigcup_B \pi_\omega(\mathfrak{A}(B))\Omega$ are dense in $\mathscr{H}_\omega$, the proof is complete. ∎

10. THE CONSTRUCTION OF THE PHYSICAL VACUUM

We apply the theory of Section 9 to the physical vacuum state ω. The vacuum state ω which we construct is locally normal and translation invariant. We conjecture that it is invariant under the full Lorentz group. On the renormalized Hilbert space $\mathscr{F}_{\text{ren}} = \mathscr{H}_\omega$, the space time automorphism $\sigma_{t,x}$ is implemented by a unitary operator $U(t, x) = \exp(itH - ixP)$ and $H = \lim_{g \to 1} H(g)$. All but two of the Wightman axioms have been verified for the field operators acting on $\mathscr{F}_{\text{ren}}$ (Glimm and Jaffe, In press). The missing axioms are the uniqueness and the Lorentz invariance of the vacuum. (The spectrum condition, also a Wightman axiom, is known to be equivalent to the Lorentz invariance of the vacuum for $P(\phi)_2$ (Streater, 1972)).

The construction of ω begins with vacuum state $\Omega_g \in \mathscr{F}$ for the Hamiltonian $H(g)$. Let

$$E(g) = \inf \text{spectrum}\,(H_0 + H_I(g)). \tag{10.1}$$

$E(g)$ is the vacuum energy of $H(g)$. We renormalize, by subtracting this vacuum energy from $H(g)$, and it is also convenient to change notation, and call the renormalized operator $H(g)$, so that now

$$H(g) = H_0 + H_I(g) - E(g). \tag{10.2}$$

The spectrum of $H(g)$ now begins at zero.

THEOREM 10.1. *Zero is a simple eigenvalue of* $H(g)$.

The proof of this theorem is not difficult, but it depends on a periodic approximation $H(g, V)$ to $H(g)$, and since we have not set up the notation to handle $H(g, V)$, we refer the reader to Glimm and Jaffe (1972).

Let Ω_g be the eigenvector of $H(g)$ for the eigenvalue zero. We call Ω_g the vacuum state of $H(g)$. As $g \to 1$, the vacuum states Ω_g converge weakly to zero, according to formal perturbation theory. In order to obtain the vacuum state ω, we take limits in the sense of C^*-algebra states, rather than in the sense of Hilbert space vectors. Let

$$\omega_g(A) = \langle \Omega_g, A\Omega_g \rangle. \tag{10.3}$$

We select a sequence $g_n(\cdot) \to 1$ and we average the state ω_{g_n} over space translations, over a distance depending on n and increasing to infinity as $n \to \infty$, see Glimm and Jaffe (1970a). (For the periodic space cut-off approximation, this spatial averaging of the vacuum is unnecessary, see Glimm and Jaffe (1971a). Let ω_n be the resulting averaged state. Then ω_n is normal, and hence locally normal, by construction. The most difficult step in the construction of ω is the next theorem.

THEOREM 10.2. *Let B be a bounded region. The sequence $\{\omega_n \restriction \mathfrak{A}(B)\}$ of states of $\mathfrak{A}(B)$ lies in a norm compact set.*

Given the theorem, we select a subsequence $\omega_{n(j)}$, which converges in norm on each local subalgebra $\mathfrak{A}(B)$. Let ω be the limit. Then ω is locally normal by the remark to Theorem 9.5. Because of the space translation invariance built into the sequence ω_n by the space averaging, it is easy to show that ω is invariant under space translation. It is even easier to show that ω is invariant under time translations, since each state ω_g is invariant:

$$\omega_g(A) = \omega_g(e^{itH(g)} A\, e^{-itH(g)}).$$

Thus Theorems 9.1, 9.2, 9.9, 9.11 and Corollary 9.12 are applicable.

We call $\mathscr{H}_\omega = \mathscr{F}_{\text{ren}}$ the physical Hilbert space, and $\Omega \in \mathscr{F}_{\text{ren}}$ is the physical vacuum. Let H and P be the infinitesimal generators of the time and space translation unitary groups. H is the Hamiltonian operator, and all cut-offs are now removed. P is the momentum operator. Since the unitary operators in the translation group leave Ω fixed (cf. Theorem 9.2),

$$H\Omega = 0 = P\Omega.$$

It is not difficult to relate the spectrum of H to that of $H(g)$:

$$\sigma(H) \subset \limsup_{g \to 1} \sigma(H(g))$$

and consequently

$$0 \leqslant H.$$

Beyond the missing Wightman axioms, the major open problem for the $P(\phi)_2$ model concerns the spectrum of the mass operator $(H^2 - P^2)^{\frac{1}{2}}$ and the construction of scattering states.

We now sketch some of the ideas in the proof of Theorem 10.2. The first step is an estimate on $E(g)$ in (9.1).

THEOREM 10.3. *Let $0 \leqslant g \leqslant 1$ and let g have compact support. There is a constant M independent of g such that*

$$-M \times (\operatorname{diam} \operatorname{supp} g + 1) \leqslant E(g).$$

Proof. Let $b(x)$ be the configuration space annihilation operator, so that $b(f) = \int b(x)f(x)$. Let

$$N_i = \int_{i-\frac{1}{2}}^{i+\frac{1}{2}} b^*(x)b(x)\,dx$$

$$N_{\text{loc}} = \sum_{i=-\infty}^{\infty} N_i e^{-m|i|/2}.$$

With g supported in a fixed interval and $d_{v,l}$ defined as in the proof of Theorem 7.4, we assert that

$$\|d_{v,l}(N_{\text{loc}} + I)^{-p}\| \leqslant O(\kappa_v^{-\frac{1}{4}}). \tag{10.4}$$

Our proof of Theorem 7.4 is valid with $\varepsilon N_{\text{loc}}$ replacing N_0 and (10.4) replacing (7.5). Thus

$$0 \leqslant \varepsilon N^{\text{loc}} + H_I(g) + O(1).$$

Summing this inequality over translates and using the bound $N \leqslant m^{-1}H_0$ yields Theorem 10.3. The proof of (10.4) is not difficult, see Glimm and Jaffe (1971b). ■

The next step in the proof of Theorem 10.2 is to pass from an estimate on $E(g)$ to an estimate on a local free energy operator $H_{0,\zeta}$. Let ζ be a non-negative C_0^{∞} function and let

$$H_{0,\zeta} = \int b^*(x)\zeta(x)k(x-y)\zeta(y)b(y)\,dx\,dy$$

where k is the kernel of the operator μ_x. ■

THEOREM 10.4. *There is a constant $M = M(\zeta)$, independent of n, such that*

$$\omega_n(H_{0,\zeta}) \leqslant M(\zeta) \tag{10.5}$$

and $M(\zeta)$ is invariant under space translations.

Formal proof. We use the fact that $\operatorname{diam} \operatorname{supp} g_n = O(n)$. Thus $E(g_n) = O(n)$ and also $E(2g_n) = O(n)$. Hence

$$0 \leqslant H_0 + 2H_I(g_n) + O(n),$$

and so

$$0 \leqslant \tfrac{1}{2}H_0 + H_I(g_n) + O(n),$$

$$\tfrac{1}{2}H_0 \leqslant H(g_n) + O(n),$$

and

$$\omega_{g_n}(H_0) = \langle \Omega_{g_n}, H_0\Omega_{g_n}\rangle \leqslant 2\langle \Omega_{g_n}, (H(g_n) + O(n))\Omega_{g_n}\rangle \leqslant O(n).$$

Let $\zeta_j(x) = \zeta(x - j)$ be the translate of ζ, and let

$$H_0' = \Sigma' H_{0,\zeta_j}$$

where the sum Σ' runs over a subset of the integers. Then $H_0' \leqslant \text{const}\, H_0$ and so $\omega_{g_n}(H_0') \leqslant O(n)$. Recall that ω_n was obtained from ω_{g_n} by an average over space translates over a distance $O(n)$. We choose Σ' to run over $O(n)$ integers, namely the integers lying in the above set of space translates. Then one sees that

$$\omega_n(H_{0,\zeta}) = O(n^{-1}) \int_0^1 \omega_{g_n}(\sigma_a(H_0'))\, da$$

where σ_a is the space translation automorphism. The above bounds on $\omega_{g_n}(H_0')$ apply equally well to $\omega_{g_n}(\sigma_a(H_0'))$, and so $\omega_n(H_{0,\zeta})$ is bounded, as required. ∎

The inequality (10.5) defines the set of states in Theorem 10.2. The most difficult component of Theorem 10.2 is to show that the set of states satisfying inequalities of the form (10.5) is norm compact, on restriction to $\mathfrak{A}(B)$. We now explain one ingredient of the proof, namely Rellich's theorem. Rellich's theorem asserts that the set of functions having support in a bounded region C, and having L_2 derivatives with norm at most one is compact in L_2. Rellich's theorem is also valid for L_2 derivatives of fractional order. Now the inequality in Theorem 10.4 asserts that the wave functions making up the state ω_n have L_2 derivatives of order $\frac{1}{2}$, and locally there is a bound on these L_2 derivatives. Thus a subsequence ω_{n_j} can be selected so that the sequence of wave functions occurring in ω_{n_j} converges in the Fock space norm, on restriction to any local region.

However this analysis does not complete the proof. The localization of states is expressed by the Newton–Wigner localization, that is by the operator $b^*(x)$, while the localization of operators is expressed through the field operator

$$\phi(x) = 2^{-\frac{1}{2}}\mu_x^{-\frac{1}{2}}(b^*(x) + b(x)).$$

If $k^{\pm}(x, y)$ is the kernel of $\mu_x^{\pm\frac{1}{2}}$, then

$$k^{\pm}(x, y) = O \exp\left(-(m - \varepsilon)|x - y|\right)$$

as $|x - y| \to \infty$. Thus the difference between the relativistic localization of the operators $\phi(x)$ and $\pi(x)$ and the Newton Wigner localization of states is small.

Let $\mathfrak{A}_{NW}(B)$ be the von Neumann algebra generated by the operators

$$\exp\left(i\phi(\mu_x^{\frac{1}{2}}f) + i\pi(\mu_x^{-\frac{1}{2}}g)\right)$$

where f and g range over test functions with support in B. The localization of operators $A \in \mathfrak{A}_{NW}(B)$ agrees with the localization of states, and one uses Rellich's theorem to show that the sequence $\{\omega_n \restriction \mathfrak{A}_{NW}(B)\}$ has a norm convergent subsequence, for each bounded region B.

Any bounded operator on Fock space can be expanded as an infinite series in the monomials

$$\int :\phi(x_1) \ldots \phi(x_m)\pi(y_1) \ldots \pi(y_n): w(x, y)\, dx\, dy.$$

Given the region B and given $\delta > 0$, we choose a large region $B_1 = B_1(\delta, B) \supset B$, and we use the $\phi - \pi$ expansion to express certain $A \in \mathfrak{A}(B)$ as a sum,

$$A = A_1 + A_2, \tag{10.7}$$

where

$$A_1 \in \mathfrak{A}_{NW}(B_1), \qquad |\omega_n(A_2)| \leqslant \delta \|A\|. \tag{10.8}$$

It is the small tail (10.6) which leads to the small tail A_2 in (10.7). Theorem 10.2 follows without difficulty from the expansion (10.7–8) and the norm compactness property for the Newton–Wigner localized sequences $\{\omega_n \restriction \mathfrak{A}(B_1)\}$. See Glimm and Jaffe (1970a) for details. ∎

11. FORMAL PERTURBATION THEORY AND MODELS IN THREE SPACE–TIME DIMENSIONS

In three or more space–time dimensions the degree of ultraviolet divergence depends on the degree of the polynomial P. Thus we put aside general polynomial interactions and study the superrenormalizable case ${\phi_3}^4$. The model ${\phi_3}^6$ is more difficult; it is renormalizable but not super renormalizable, and it is equal in difficulty to ${\phi_4}^4$. To say that a model is super renormalizable means that the ultraviolet, or large k, infinities can be written as a polynomial in the coupling constant whose coefficients are divergent integrals. The coupling constant is the coefficient of H_I in the Hamiltonian. The problem of Section 9 and 10, the limit $g \to 1$, involves a divergence proportional to diam supp g in each order of the coupling constant, and hence displays in the infinite volume limit a nonsuperrenormalizable phenomenon.

The $P(\phi)_2$ model is also superrenormalizable. To see this, let $H(g) = H_0 + \lambda H_I(g)$, $0 < \lambda$. The only infinities in $H(g)$ are those removed by the Wick ordering in H_I, and they have the form

$$\lambda c^j = \lambda \left[(4\pi)^{-1} \int \frac{dk}{\mu(k)} \right]^j, \qquad j \leqslant \tfrac{1}{2} \deg P, \tag{11.1}$$

as in (5.11). The $P(\phi)_2$ model is the only model with such simple ultraviolet renormalizations. The next most difficult model is the two-dimensional Yukawa interaction of bosons and fermions. This model involves ultraviolet divergent shifts in the vacuum energy and in the Boson mass. Nonetheless, the Yukawa$_2$ model is now understood practically at the same level of completeness as $P(\phi)_2$, see Glimm and Jaffe (1972). The Yukawa$_2$ model is intermediate in difficulty between $P(\phi)_2$ and $\phi_3{}^4$, about which only preliminary technical results are known.

In this section we give a simple picture on a formal level of renormalization in the $\phi_3{}^4$ model. This formal picture is justified by a rigorous mathematical analysis. Glimm (1968) established that $H(g)$ is a densely defined symmetric operator. Hepp (1970) and Fabrey (1971) constructed the time zero algebra $\mathfrak{A}$ and its local subalgebras $\mathfrak{A}(B)$. While these steps may sound easy, and are trivial for the $P(\phi)_2$ model, they are in fact difficult because of the renormalizations involved. Eventually, our program is to construct a $\phi_3{}^4$ model by proving the results known in $P(\phi)_2$. See also Eckmann and Osterwalder (1971) and Eckmann (1972). Thus some main steps yet to be established are:

(a) Positivity of $H(g)$: $0 \leqslant H(g)$

(b) Self-adjointness of $H(g)$: $H(g) = H(g)^*$

(c) To show that $H(g)$ implements locally a * automorphism σ of $\mathfrak{A}$ with finite propagation speed.

In dealing with $\phi_3{}^4$, we are motivated by perturbation theory which comprises the formal calculations mentioned above. It is convenient to use Friedrichs graphs to describe Wick monomials in our calculations. The Friedrichs graph of the Wick monomial W of (4.6) has a single vertex and $m + n$ legs, the m creating legs pointing to the left and the n annihilating legs pointing to the right.

$W =$ 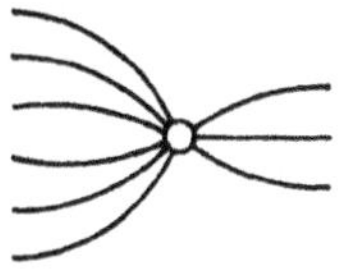 (11.2)

m creating legs $\qquad$ n annihilating legs.

Thus in the language of Friedrichs graphs,

$\phi(x) =$ 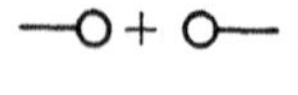 (11.3)

$:\phi^2(x): =$ 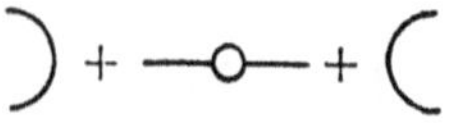 (11.4)

$:\phi^4(x): =$  (11.5)

The commutation relations $[a, a^*] = \delta I$ can also be expressed in Friedrichs graphs:

$a\,a^* =$ 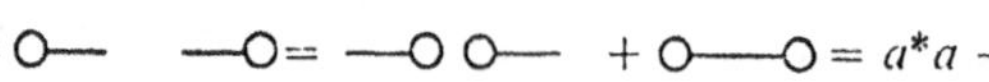 $= a^*a + \delta I.$

Here the contracted term, δI, has two vertices and a line joining them. More generally in the Wick expansion (4.12), we introduce a graph for U_r by contracting r creating legs from the vertex of W_1 to r annihilating legs from the vertex of W_2. We note in passing that the factor $r!\binom{j}{r}\binom{m}{r}$ in (4.14) is just the number of ways of performing these contractions. As a special case of (4.12), we have

 (11.6)

The first term is the Wick ordered term, and it has the least singular kernel. The last term is the fully contracted term. It is a multiple of the identity operator and its kernel is the most singular of the terms above.

The Friedrichs graph, for example (11.2), together with the kernel w uniquely determines the operator. For multivertex graphs, as in (11.6), there is a kernel for each vertex. Generally the choice of the kernel is implied by the context, and then the graph itself uniquely determines the operators. In our discussion of perturbation theory, the kernels come from H_I, and are proportional to (5.2), (5.4) or (5.8), depending on which cut-offs are employed in H_I.

For a kernel w of degree m, n, let

$$\gamma w(k, k') = \frac{w(k, k')}{\sum_{i=1}^{m} \mu(k_i) - \sum_{j=1}^{n} \mu(k_j')} \tag{11.7}$$

and let ΓW be the Wick monomial of degree m, n with kernel γw. The proper interpretation of the singularity in the denominator is important for some aspects of perturbation theory, for example in distinguishing between incoming and outgoing states, but it is unrelated to the ultraviolet divergences we are studying, and so we take the simplest prescription, which is the principal value. The merit of the Γ operation is that

$$[H_0, \Gamma W] = W, \tag{11.8}$$

or in other words, Γ is an inverse to $\operatorname{ad} H_0$,

$$\operatorname{ad} H_0 : X \to [H_0, X] \tag{11.9}$$

$$\operatorname{ad} H_0 \circ \Gamma : X \to X. \tag{11.10}$$

In graph notation, we write a Γ near the vertex to indicate the Γ operation, so that

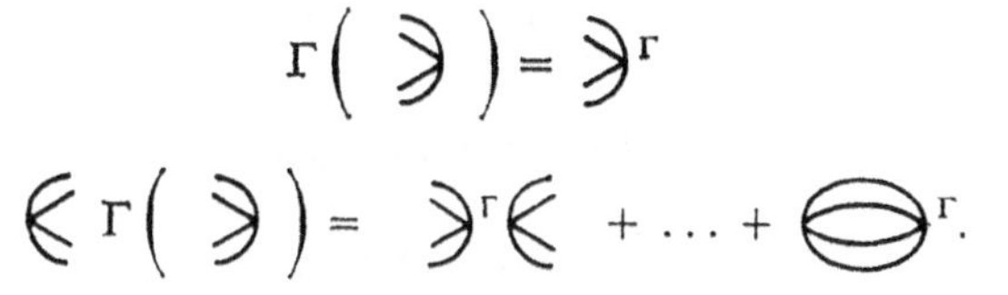

(11.11)

Motivated by the wave operator from potential scattering, we seek an operator T which intertwines H and H_0,

$$HT = TH_0. \tag{11.12}$$

For T to exist, H and H_0 must have the same spectrum. We adjust the spectrum by adding counterterms permitted by the physical interpretation of H. With

$$H = H_0 + \lambda H_I(g) - \tfrac{1}{2}\delta m^2 \int :\phi^2(x): g^2(x)\, dx - E(g), \tag{11.13}$$

we seek to determine the renormalization constants $E(g)$ and δm^2 as well as T. Let us expand formally

$$T = I + \lambda T_1 + O(\lambda^2)$$
$$E(g) = \lambda E_1 + \lambda^2 E_2 + O(\lambda^3)$$
$$\delta m^2 = \lambda \delta m_1{}^2 + \lambda^2 \delta m_2{}^2 + O(\lambda^3)$$

so that

$$E(g) = \inf \operatorname{spectrum} (H_0 + \lambda H_I(g) - \tfrac{1}{2}\delta m^2 :\phi^2(g^2):).$$

Let

$$\Omega(g) = \Omega_0 + \lambda\Omega_1 + O(\lambda^2).$$

We show that $E_1 = 0$. Since $\langle \Omega(g), \Omega_0 \rangle \neq 0$,

$$E(g) = \frac{\langle \Omega_0, (H_0 + \lambda H_I(g) - \frac{1}{2}\delta m^2 \colon \phi^2(g^2) \colon)\Omega(g) \rangle}{\langle \Omega(g), \Omega_0 \rangle}$$

$$= \frac{\lambda \langle \Omega_0, (H_I(g) - \frac{1}{2}\delta m_1{}^2 \colon \phi^2(g^2) \colon)\Omega_0 \rangle + O(\lambda^2)}{1 + O(\lambda)}. \tag{11.14}$$

Since H_I and $:\phi^2:$ are Wick ordered,

$$E_1 = \langle \Omega_0, (H_I(g) - \tfrac{1}{2}\delta m_1{}^2 \colon \phi^2(g) \colon)\Omega_0 \rangle = 0.$$

A similar analysis shows that $\delta m_1 = 0$, assuming that the coefficient of $:\phi^2:$ in H_I is zero. Substituting these values in (11.12), we have

$$(H_0 + \lambda H_I(g))(I + \lambda T_1) = (1 + \lambda T_1)H_0 + O(\lambda^2),$$

or

$$\lambda H_I(g) = \lambda[T_1, H_0].$$

Thus we set

$$T = I - \lambda \Gamma H_I(g) + O(\lambda^2).$$

Then

$$\Omega_g = T\Omega_0 / \| T\Omega_0 \| = \frac{\Omega_0 - \lambda \Gamma H_I(g)\Omega_0 + O(\lambda^2)}{\| \Omega_0 - \lambda \Gamma H_I(g)\Omega_0 + O(\lambda^2) \|}$$

$$= \frac{\Omega_0 - \lambda \Gamma H_I(g)\Omega_0 + O(\lambda^2)}{\left(1 + \lambda^2 \left(\Gamma \quad \Gamma\right) + O(\lambda^2)\right)^{\frac{1}{2}}}$$ 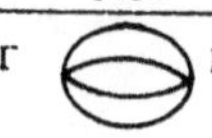

$$= \Omega_0 - \lambda \quad {}^{\Gamma}\Omega_0 + O(\lambda^2).$$ 

Substituting in (11.14), we have

$$E(g) = \frac{-\frac{1}{2}\delta m^2 \lambda^2 \langle \Omega_0 \colon \phi^2(g^2) \colon \Omega_0 \rangle - \lambda^2 \left\langle \Omega_0, H_I(g) \quad {}^{\Gamma}\Omega_0 \right\rangle + O(\lambda^3)}{1 + O(\lambda^2)}$$ 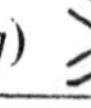

$$= -\lambda^2 \left\langle \Omega_0, H_I(g) \quad {}^{\Gamma}\Omega_0 \right\rangle + O(\lambda^3)$$ 

$$= -\lambda^2 \quad {}^{\Gamma} + O(\lambda^3)$$ 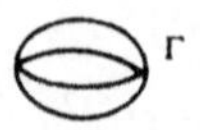

and so

$$E_2 = - \quad {}^{\Gamma}.$$ 

Let

$$-\tfrac{1}{2}\delta m^2 \colon \phi^2(g^2) \colon = \quad + \quad + $$ 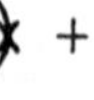 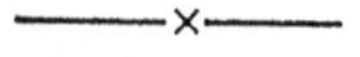 

Similarly it can be shown that $\delta m_2{}^2$ is chosen so that the kernel of the second of the following three operators

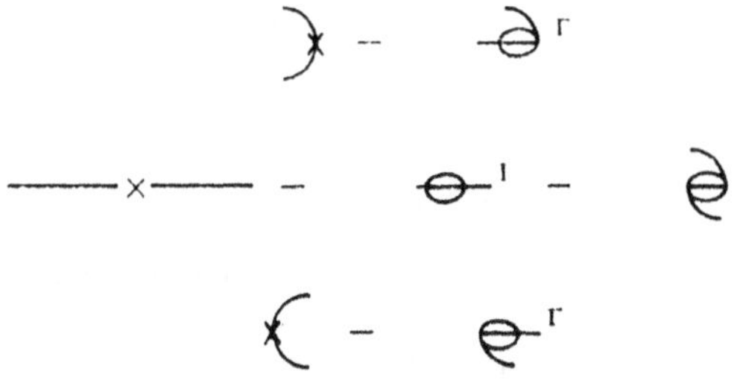

vanish at zero momentum. With this choice of $\delta m_2{}^2$, all three operators above have finite valued kernels, in the limit $\kappa = \infty$. We now determine T to second order:

$$H(g)T - TH_0 = -\lambda^2(H_I\Gamma H_I + E_2 + \tfrac{1}{2}\delta m_2{}^2 :\phi^2(g^2):) + [H_0, T_2].$$

Thus

$$T_2 = \Gamma(H_I\Gamma H_I + E_2 + \tfrac{1}{2}\delta m^2 :\phi^2(g^2):). \tag{11.15}$$

Now $H_I\Gamma H_I$ is not defined. The terms with three and four contractions have kernels which are divergent integrals. However in (11.15), these infinities are cancelled. Thus E_2 cancels the fully contracted term and $\frac{1}{2}\delta m^2 :\phi^2(g^2):$ cancels the infinite part of the terms with three contractions. The calculation of E and δm^2 to higher order reveals one further infinite counterterm,

$E_3 = \Gamma$  Γ .

In addition to the infinities E and δm^2, the $\phi_3{}^4$ model has an infinite wave function renormalization, Λ. From (11.15) and the choice of E_2, we see that $\langle T_2\Omega_0, \Omega_0\rangle = \langle\Omega_2, \Omega_0\rangle = 0$. Thus

$\|T\Omega_0\|^2 = 1 + \lambda^2 \Gamma$ 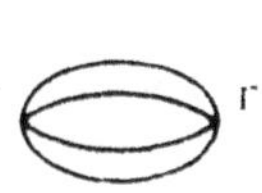 $\Gamma + O(\lambda^3) = 1 + \lambda^2\Lambda_2 + O(\lambda^3)$

and Λ_2 is logarithmically infinite. When computed to higher order, one finds

$$\|T\Omega_0\|^2 = \exp(\lambda^2\Lambda_2 + \dots)$$

and $\Lambda_3, \dots$ are all finite. The fact that Λ_2 is infinite means that T maps out of Fock space. The range of T lies in the renormalized Fock space

$\mathscr{F}_{\text{ren}}$. The change of Hilbert space can be described mathematically as a change to an inequivalent representation of the canonical commutation relations. Eckmann (1972) proved that the representations of a fixed local algebra $\mathfrak{A}(B)$ on $\mathscr{F}_{\text{ren}}$ are independent (in the sense of unitary equivalence) of the space cut-off g, provided $g \equiv 1$ on a neighborhood of B.

TWO NOTES

1. In general, we use real valued functions when working with the position variable x and we use complex valued functions when working with the Fourier transform, or momentum variable k. However the Hilbert spaces on which the canonical commutation relations are represented are always complex.

2. Minlos (1959) proved that a measure is countably additive on the dual of a nuclear space, provided the measure satisfies a weak continuity condition. For Gaussian measures, the proof is simpler. We follow the presentation of Gel'fand and Vilenkin (1961). The existence of a Schrödinger representation, diagonalizing the time zero field operators is possible only in two or three space time dimensions, according to formal considerations, for fields with conventional nonlinear interactions. For free fields or fields with linear interactions there is no such restriction. The Schrödinger representation is implicit in many discussions of free fields, see for example Heisenberg and Pauli (1930) and von Neumann (1932). von Neumann (ca. 1934–1935) constructed a Gaussian measure over an abstract (pre-) Hilbert space $\mathscr{H}$, and showed that the measure was countably additive, when defined on the algebraic dual $\mathscr{H}^* \supset \mathscr{H}$. This result defined the Hilbert space on which the Schrödinger representation exists; the result was not published and was reviewed posthumously in 1961. Friedrichs (1953) outlined the construction of the Schrödinger representation in detail. He used a Gaussian measure and a Hermite expansion to define the inner product in $L_2(d_B q)$; the commutation relations are given in unbounded form and the unitary equivalence with the Fock representation can be read from Friedrichs' formulas. The connection with Kolmogoroff's theorem and the construction of a measure space with a countably additive measure was promised in a later publication. (Cf. Friedrichs and Shapiro 1957). In 1956, Segal gave a self contained and mathematically complete account of the Schrödinger representation from an abstract point of view. He established the canonical commutation relations in the integrated, or Weyl form. We note finally that fields with nonlinear interaction lead to the consideration of non-Gaussian measures on Q.

THE PARTICLE STRUCTURE OF THE WEAKLY COUPLED $P(\varphi)_2$ MODEL AND OTHER APPLICATIONS OF HIGH TEMPERATURE EXPANSIONS, PART I:

PHYSICS OF QUANTUM FIELD MODELS

James Glimm*
Courant Institute
New York, N. Y.

Arthur Jaffe†
Harvard University
Cambridge, Mass.

Thomas Spencer‡
Courant Institute
New York, N. Y.

1. FIVE YEARS OF MODELS

1.1. An Overview

Constructive quantum field theory has moved rapidly in the past few years. A continuation of this progress clearly requires the introduction of new formal ideas, methods and/or points of view. In these lectures, we will review past progress, present some new results and point out possible directions for future work. In this Part I, we emphasize the basic ideas and concepts, while in Part II we present in full detail some core results from our program. In particular, we present the vacuum cluster expansion and estimates to establish its convergence.

For the two dimensional $P(\varphi)$ model, a fairly detailed structure is known, and this structure is in qualitative agreement with basic ideas of physics. From the Haag-Ruelle axioms (verified, for instance, for small coupling), we know that fields describe discrete mass particles. An isometric scattering operator expresses the interactions between these particles. We know that the vacuum state is unique for certain coupling constants (e.g., small coupling, or a large - or in some cases, a nonzero - external field). Dobrushin and Minlos have announced that there are multiple phase solutions for even $P(\varphi) + m_0^2 \varphi^2$ models with $\lambda >> m_0^2$. Symmetry breaking plays a key role in current theories of weak interactions, hence the interest in this phenomenon. There is no direct experimental evidence for or against occurence of broken symmetries in elementary particle physics, since the interparticle coupling constants cannot be varied experimentally (in distinction to the case of statistical mechanics where we can, for example, turn off a magnetic field). Consequently the definitive argument in favor of broken symmetries may come from constructive quantum field theory.

The Yukawa$_2$ (Y_2) and φ_3^4 models are less highly developed. Yet many of the formal ideas developed for $P(\varphi)_2$ models appear to apply to superrenormalizable models in general. Clearly then, one set of problems is to develop stronger techniques, to make these ideas applicable to Y_2, φ_3^4 and Y_3. We propose, in fact, four groups of problems.

I. Physical Properties. One important direction for future work is to develop further the physics of existing quantum field models. The particle structure program, bound states, resonances and scattering present interesting problems. Likewise, the long distance and infrared behavior of our models contains much physics. The general particle structure program is: Which interaction polynomials and which coupling constants give rise to which particles, bound states and resonances? How do the masses and half lives depend on the coupling constants? How do cross sections behave asymptotically? We discuss these problems further in Section 1.5 and Chapter 3.

The long distance behavior of our models pertains to the existence of multiple phases, to the existence of a critical point and to the scaling behavior of the models at a critical point. We ask: Does the φ_2^4 model have a critical point? Does it admit scaling properties with anomalous dimensions? What parameters describe the critical point? We discuss these questions further below and in Section 1.5.

II. Four Dimensions (Renormalizable Models). A second important direction is the question of four space-time dimensions, or in other words how to deal with renormalizable interactions, since there are no super-renormalizable interactions in four dimensions. Clearly this is our most challenging goal, to prove the existence of, for example, φ_4^4. Our present methods have been tied to superrenormalizability ($4 - \epsilon$ dimensions) and for $\epsilon = 0$ new ideas are required. We ask: Can an understanding of the renormalization group be an aid to removing the $\epsilon = 0$ ultraviolet cutoff? Do the ideas in the lectures of Symanzik yield insight into charge renormalization? We discuss these questions further in Section 1.5.

III. Simplification. Aside from these two major directions, there is the question of simplifying the present methods. Clearly the major need for simplification concerns problems with ultraviolet divergences, and a major goal of such a program would be to improve the techniques and isolate their essential elements in order to make tractable more complicated superrenormalizable models, such as Y_3, or even Y_2.

IV. Esthetic Questions. Furthermore, there are esthetic or foundational questions. For example, the Schrödinger representation $\mathcal{H} = \mathcal{L}_2(dq)$ exists for $P(\varphi)_2$ models ; what is the fermion representation corresponding to this non-Gaussian boson measure on $\mathcal{S}'$? What are the properties of the path space measures in models with interaction? Related are interesting, but purely mathematical questions motivated by field theory, which we do not pursue here.

In this connection, we remark that the drive toward simplicity and elegance is important and also has been quite successful in the $P(\varphi)_2$ model. However, we emphasize here those methods that admit (or appear to admit) generalization to other more singular interactions. The reason for this emphasis is two-fold. First, we believe that, in the long run, our ability to handle more singular problems will determine the extent to which the model program has succeeded. Second, we believe that a premature emphasis on the simplicity and elegance of the details can divert energy away from central issues, and thereby delay or obstruct progress.

1.2. Survey of Results

To begin, we review the status of the φ_2^4, Y_2 and φ_3^4 models. We give a chronological summary in Figure 1, plotting models of increasing complexity versus results of increasing complexity. In this chart, we enter the years in which these results were proved.

In Figure 2 we give details and references for various φ_2^4 models. The results quoted for $\lambda/m_0^2 \ll 1$, also hold for $\lambda P(\varphi)_2$ models with $\lambda/m_0^2 \ll 1$. We make several comments: The Wightman axioms require a unique ground state (vacuum), namely the existence of a single vector, invariant under inhomogeneous Lorentz transformations. Alternatively, we consider the C^* algebra vacuum state of a finite volume theory, and its infinite volume limits. Each infinite volume state yields a representation, and a Hilbert space vacuum vector. Uniqueness of the vacuum, as required for the Wightman axioms, refers to vectors in this Hilbert space, and is equivalent to irreducibility of the representation. The infinite volume vacuum state is determined as a limit of finite volume states. The latter are determined by parameters in the energy density $\mathcal{H}(x)$ and the boundary conditions. If the parameters in $\mathcal{H}(x)$ alone are sufficient to specify a unique vacuum, independent of the boundary conditions, then there is said to be a unique phase, and otherwise there are multiple phases. Convergent cluster expansions [Gl Ja Sp 1, 2] yield for certain couplings both a unique vacuum and a single phase.

In a $P(\varphi)_2$ theory satisfying the Wightman axioms, except for the uniqueness of the vacuum, the decomposition theorem of Bratteli [Br 1] allows us to decompose the observables and recover a unique vacuum. The local perturbation estimate [Gl Ja IV] and a result of Streater [St] ensure the spectral condition for the decomposed theory. In this manner we arrive at a Wightman theory for an arbitrary $\lambda\varphi^4 + \frac{1}{2}m_0^2\varphi^2 - \mu\varphi$ interaction. In the case $\mu \neq 0$, the Lee-Yang theorem shows that the decomposition is unnecessary [Gr Si, Si II].

In Figure 3, we have details and references for Y_2, and clearly much work needs to be done to bring it to the level of φ_2^4.

	$P(\varphi)_2$	Y_2	φ_3^4	φ_4^3	Y_3
Critical Point					
Asymptotic Completeness					
Resonances					
Bound States	3				
Broken Symmetries	3				
Analyticity in Coupling	3				
Perturbation Theory Asymptotic	3				
Single Particle States	3				
Mass Gap	2				
Wightman Axioms, $V \to$ Convergence	2				
Euclidean Formulation	1	2			
Wightman Functions	1				
Physical Representation	9	1			
Haag-Kastler Axioms	9				
Equations of Motion	9	1			
Space Time Covariance	8	0			
$H_V = H_V^*$	8	9			
$0 \leq H_V$	5	8	2		
Local Observables	4	7	1		
H_V	4	7	8	0	

3 = 1973
0 = 1970
9 = 1969
4 $\leq$ 1964

Figure 1. Main Results and Year Established

Renormalized Charge, Anomalous Dimensions, Renormalization Group				
Critical Points				
S-Unitary (Asymptotic Completeness)				
Bound States Resonances		Chapter 3		
Multiple Phases (Broken Symmetries)	No: Gl Ja Sp 1	Yes: $\lambda \gg m_0^2$ (Dob Min)	No: Gr Si, Si II	→
Analyticity of the Schwinger Fns.	$\mathrm{Re}\,\lambda > 0$, $0 < \lvert\lambda\rvert < \lambda_0$, $\lvert\mu\rvert < \mu_0$ Gl Ja Sp 2		$\mathrm{Re}\,\lambda > 0$ Sp 2	→
One Particle States and S-Matrix	Gl Ja Sp 1			
Measure dq	New 1, Fr 2	Fr 2	→	→
Schrödinger Repn.	Fr 2			Fr 2
Perturbation Theory is Asymptotic	Di 3			Di 3
m	$m > 0$ Gl Ja Sp 1	monotonic in m_0 Gu Ro Si 3, Si I	monotonic in $\lvert\mu\rvert$ Gr Si	$m > 0$ Sp 2
Verify Wightman Axioms	Gl Ja Sp 1	Nel 5, Gu Ro Si 3, Gl Ja IV, Os Sch 3, Br 1, St	→ Gr Si, Si II	→ → Sp 2
Formulate Euclidean Axioms	Sy 2, Nel 3, Os Sch 3	→	→	→
Physical Representation	Gl Ja III-IV, 5	→	→	→
Haag-Kastler Axions	Gl Ja I-II, Ca Ja Ro 1-3	→	→	→
Preliminary	Ja 1, 2, Nel 1, Gl 1 Se 1	→	→	→
	$\frac{\lambda}{m_0^2} \ll 1$, $\frac{\lvert\mu\rvert}{m_0^2} \ll 1$. Also $\lambda P(\varphi)_2$.	$\mu = 0$	$\mu \neq 0$	$\lvert\mu\rvert$ large

Figure 2. Details for $\left(\lambda\varphi^4 + \frac{1}{2} m_0^2 \varphi^2 - \mu\varphi\right)_2$

Bound States	
Particle Structure	
Schrödinger representation	
Wightman axioms	
Euclidean Fields and Feynman-Kac formula	Os Sch 4
Properties of Currents	McB 1
Equations of Motion	Di 1
Physical Representation Exists	Sch 2
Space - Time Covariance	Gl Ja 10
$0 \leq H_V = H_V^*$	Gl 2,3,Gl Ja 9, 11

Figure 3. Details for Y_2

We conjecture that the Euclidean fields given by Osterwalder and Schrader [Os Sch 4], and by Ozkaynak [Oz] for higher spin, can be used as a starting point for a cluster expansion, yielding for small coupling: the $V \to \infty$ limit, the Wightman axioms, a unique vacuum, and single particle states. We also hope that Euclidean gauge fields will lead to an understanding of function space integrals for higher spin. In particular, we mention as problems the factorization of the integration over the symmetry (gauge) group, and the renormalization scheme of Faddeev and Popov [Fa Po, Sa St].

The φ_3^4 interaction is at an even more primitive stage. The φ_3^4 renormalized Hamiltonian is bounded from below [Gl Ja 8]. Similar estimates [Fe 2] show that in a finite space time volume, approximate Schwinger functions have $\varkappa \to \infty$ limits, see Chapter 4. We hope that the Wightman axioms will soon be verified for φ_3^4.

There are several recent developments which do not fit into the above description and we mention in particular: Federbush's study of the generalized Yukawa$_2$ model, $P(\varphi) + \bar{\psi}\psi Q(\varphi)$, see [Fed] the work of Albeverio and Hoegh-Krohn on bounded nonpolynomial interactions [AHK 1 - 4], and recent work on the polaron model by Gross[Gro 1] and by Fröhlich [Fr 1]. Furthermore Fröhlich's recent results on $P(\varphi)_2$ models [Fr 2] yield the Markov property, DLR equations and cyclicity of the vacuum (for time zero fields) when $\lambda \ll m_0^2$ or $|\mu| \gg m_0^2, \lambda$.

There are three main unifying techniques, yielding the results charted above:

1. Euclidean Formulation. The use of imaginary time in constructive quantum field theory dates to Symanzik's Euclidean field formulation and to Nelson's positivity proof for the $\varphi^4_{2,V}$ Hamiltonian. This approach was used in most of the key technical estimates of the $P(\varphi)_2$ and φ^4_3 boson models, and in proving finite propagation speed for the Y_2 model. The covariant formulation in terms of Euclidean fields or Green's functions is important as a conceptual advance yielding Euclidean axioms, the relativistic Feynman-Kac formula and the Markov property [Nel 4, Os Sch 3, Sy 1, Nel 3, Fel 1]. The covariant formulation is important as a technical advance in proving estimates [Gu 1,Gl Ja 8], in simplifying estimates [Gu Ro Si 1, Si], and in reducing the number of properties which must be verified in order to construct a Wightman-Haag-Ruelle theory [Nel 3, Os Sch 2]. Furthermore, the Euclidean framework makes clear the connection of field theory with statistical mechanics, a central tool in the heuristic approach of Fisher and Wilson and in the work of Guerra, Rosen and Simon. The lattice approximation [Ko Wi, Gu Ro Si 3] then becomes natural, and it relates boson interactions to an Ising model with continuous spin and nearest neighbor interaction.

2. Correlation Inequalities [Gu Ro Si 3, Nel 5, Gr Si, SiII]. The convergence of the lattice approximation and the existence of the thermodynamic limit leads to the generalization of correlation inequalities and the Lee-Yang theorem from statistical mechanics to quantum field theory. These methods appear especially suited to boson interactions and are discussed in the lectures of Guerra, Rosen, Simon, Nelson and in Chapter 3 of these lectures.

3. Cluster Expansions [Gl Ja Sp 1, 2]. These high temperature expansions provide (within their region of convergence - see Figure 6) the most detailed information about $P(\varphi)_2$ models. In combination with low temperature expansions (which we believe to exist), the expansions should converge for all interactions

sufficiently far from a critical point. In fact the inductive expansions for $P(\varphi)_2$ yield an analysis of an arbitrary interval [0, a] of the spectrum of the Hamiltonian H. Related are the inductive bounds for $P(\varphi)_2$ [Gl Ja IV] and for φ_3^4 [Gl Ja 8]. The bounds for φ_3^4 have not yet been refined to yield such detailed results as are known for $P(\varphi)_2$. In contrast to the expansions, the bounds hold without restriction on the magnitude of the coupling constants.

1.3. The Goldstone Structure of $P(\varphi)_2$

We now return to the $P(\varphi)_2$ model, and give a more detailed discussion. There is a very simple picture, partly due to Goldstone, and now part of the physics folklore, which accounts for the fact that some interaction polynomials P produce degenerate vacuums, while others do not. See [Go, Go Sa We, Jo L], and for recent discussions see [Co 1, Wi]. Classically, the (Euclidean) ground state of the $P(\varphi)$ model is the function $\Phi \in \mathcal{S}'(R^2)$ which minimizes the classical (Euclidean) energy

$$\int [\nabla \Phi(x)^2 + P(\Phi)]dx \quad .$$

This function Φ is a constant, equal to the value ξ which minimizes $P(\xi)$. Quantum mechanically, the (Euclidean) ground state is a probability distribution on $\mathcal{S}'(R^2)$, centered (in some sense) about the classical minimum. The quantum mechanical picture requires corrections, including Wick ordering and the higher order mass counterterms, see Coleman and Weinberg [Co We]. In $P(\Phi)_2$ models, Hepp obtained the classical limit $\hbar \to 0$ explicitly, see [He 3]. The Goldstone picture also states that (i) the physical mass m is the curvature of P at its minimum and (ii) in the case of a double (or multiple) minimum, multiple phases exist. In our discussion, we include in P the mass in the free Hamiltonian H_0, but because we ignore quantum corrections beyond (bare) Wick order, our discussion is approximate. Let $\frac{1}{2}m_0^2$ be the coefficient of Φ^2 in P. We consider the polynomials

(a) $$P(\xi) = \sum_{j=1}^{2n} c_j \xi^j \; ; \; c_{2n} > 0 \, , \; \frac{1}{2}m_0^2 = c_2 \gg c_j \quad , \; j \neq 2 \quad .$$

(b) $$P(\xi) = \lambda\xi^4 + \frac{1}{2}m_0^2\xi^2 - \mu\xi \; , \; \lambda > 0 \quad , \; \mu \neq 0 \quad .$$

(c) $$P(\xi) = P_0(\xi) - \mu\xi \; ; \; P_0 \text{ bounded below, } \mu \gg 1 \quad .$$

(d) $$P(\xi) = \lambda P_0(\xi) + \frac{1}{2}m_0^2\xi^2 \; ; \; \lambda > 0 \; , \; m_0^2 < 0 \; , \; P_0 \text{ even.}$$

(e) $$P(\xi) = \lambda\xi^4 \; ; \quad \lambda > 0 \quad .$$

See Figure 4.

In case (b) a cubic term may be added through the trivial transformation $\xi \to \xi + \text{const.}$ In (c), ξ may be replaced by any odd power ξ^{2j+1}, $2j+1 < \deg P_0$. In (d), a polynomial $\xi^6 + \cdots$ of higher degree may have more than two minima.

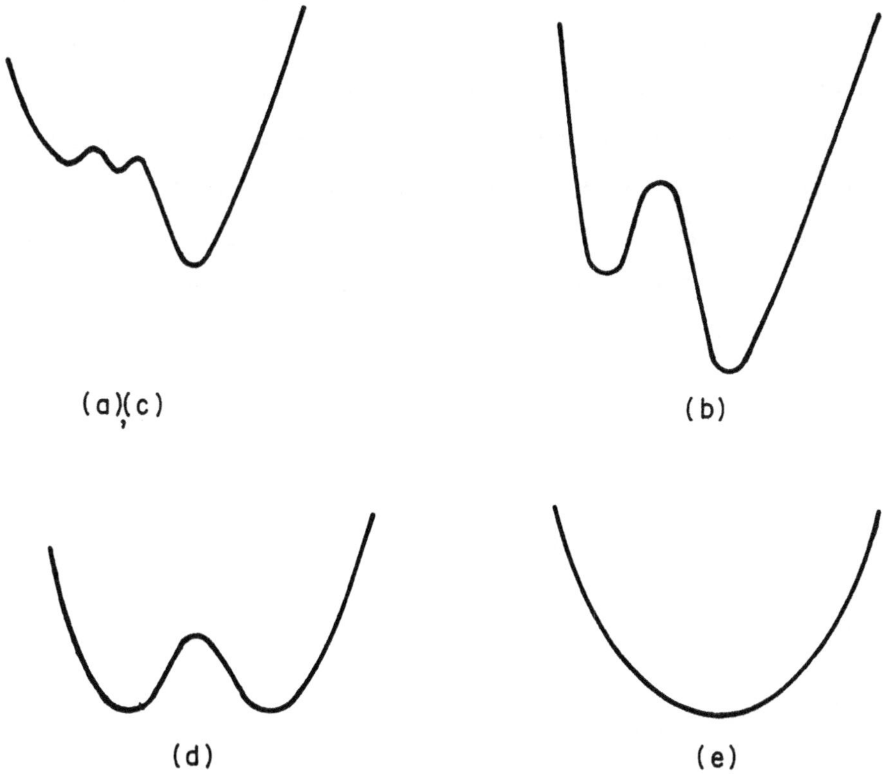

Figure 4. Various interaction polynomials $\mathcal{P}(\xi)$.

In cases (a) - (c) the polynomial $P(\xi)$ has a unique minimum with positive curvature. This suggests the existence of a single phase, and the existence of massive particles. Case (d) has a double minimum, corresponding to the symmetry $\varphi \to -\varphi$, with positive curvature at each minimum. Thus we expect two pure phases, with positive mass particles in each pure phase. The two vacuum states for the two pure phases are localized approximately at minima $\xi = \pm a$. Thus we expect in the vacuum states for these two phases that the expectation value of the field, $\langle\varphi\rangle$, approximately equals $+a$ or $-a$. Case (e) is the limiting case between Case (d) and a particular Case (a). Here $P(\xi)$ has a unique minimum at $\xi = 0$, with curvature zero. The Goldstone picture suggests the presence of zero mass particles, and is the Goldstone picture of a critical point. These figures yield the mean field approximation, which we believe gives the correct qualitative, rather than quantitative, picture. Thus the existence of the limiting case (e) suggests the existence of critical values of the coupling constants for which $m = 0$. In this connection, Figure 5 represents the Goldstone conjecture for even $\lambda P_0(\Phi) + \frac{1}{2}m_0^2\Phi^2$ models. R. Baumel [Ba] remarked that changing the definition of Wick ordering (e.g., Wick ordering in a bare vacuum with a different bare mass) permits the same H to be written with different bare parameters m_0, P. Thus the location of the critical point, when written in these parameters, can be changed, so the sign of m_0^2 at the critical point is not significant.

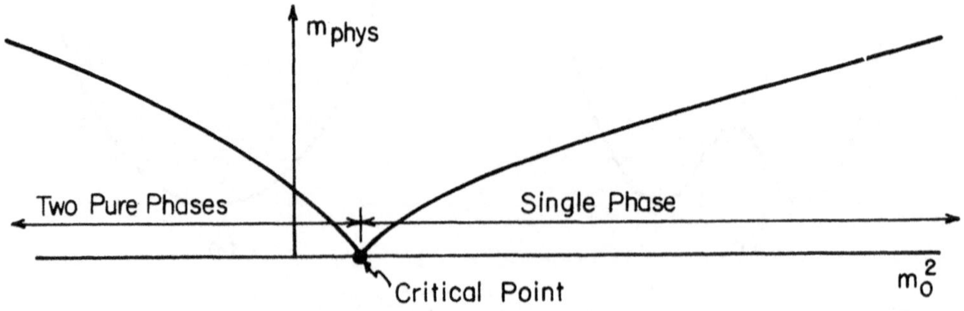

Figure 5. Conjectured symmetry breaking in even $\lambda P_0(\phi) + \frac{1}{2}m_0^2\phi^2$

Results. We give the rigorous results proved in cases (a) - (e). In case (a), convergent cluster expansions show that the mass operator $M = (H^2 - P^2)^{1/2}$ has isolated eigenvalues at 0 and $m > 0$, [Gl Ja Sp 1]. These results have been generalized to case (c) [Sp 2]. A single phase exists, the vacuum is unique and the mass $M = m$ eigenvalue is nondegenerate in the sense that the Lorentz group acts irreducibly on the one particle space. Any additional mass spectrum in the interval $(m, 2m)$ should be discrete and would describe bound states.

In case (b), Griffiths and Simon [Gr Si] extend the Lee-Yang theorem to φ_2^4 and use this result to show [SiII] that the ground state is unique. For λ large and μ small, the structure of the mass spectrum above zero is not known.

In case (d), Dobrushin and Minlos [Dob Min] have announced the existence of at least two phases.

Continuous Symmetry Breaking. Another form of the classical picture [Go Sa We] concerns a continuous symmetry group, rather than the discrete reflection group $\varphi \to -\varphi$ that we discussed above. In the continuous case, the minimum of $P(\xi)$ occurs on a manifold of dimension greater than zero, and translation along this manifold leaves $P(\xi)$ constant. The Goldstone picture now states that in the case of broken symmetry ($\delta \neq 0$ as defined below) particles of mass zero occur. Thus these particles have a mass given by the minimum curvature of $P(\xi)$ at its minimum.

In the physics literature, this broken symmetry is defined in terms of the conserved current, $\partial^\mu j_\mu = 0$, associated with the symmetry group. For classical field theories, a standard variational argument establishes the existence of the conserved current. The generator of the symmetry group is $Q = \int j_0 d\vec{x}$, and the Goldstone picture concerns the vacuum expectation value

$$\delta = \langle [iQ, \varphi] \rangle \quad .$$

If $\delta \neq 0$, discrete zero mass particles (Goldstone bosons) are believed to exist. In case that the symmetry group of automorphisms can not be unitarily implemented,

Kastler, Robinson and Swieca have shown that the mass spectrum extends down to zero [Ka Ro Sw], and Swieca [Sw] has shown that zero mass particles exist.

Two simple examples of such conserved currents are (1) a zero mass free field, with energy density $\pi^2 + (\nabla\varphi)^2$, invariant under translations $\varphi \to \varphi$ + const., and (2) a multicomponent field invariant under orthogonal transformations in the space of components.

In case (1), $j_\mu = \partial_\mu \varphi$, the parameter $\delta = 1$ and the Goldstone bosons are just the zero mass free particles. (We note this argument is not applicable in two dimensions where zero mass free scalar fields do not exist.)

Case (2) corresponds to Figure 4(d) with a rotational symmetry about the $\xi = 0$ axis. If our field has two components φ_i , then $\delta_i = \pm \langle\varphi_j\rangle$ is the vacuum expectation value of the field components. The Goldstone bosons occur when the vector, formed of the components δ_i , does not vanish i.e., $\vec{\delta} \neq 0$. In terms of $\vec{\varphi}$, the broken symmetry condition requires that the vacuum expectation $\langle\vec{\varphi}\rangle \neq 0$. Let us suppose that $\langle\vec{\varphi}\rangle \neq 0$, and let $\vec{n}$ be the unit vector in the $\langle\vec{\varphi}\rangle$ direction. We decompose $\vec{\varphi}$ into longitudinal and transverse parts, $\vec{\varphi} = \vec{\varphi}_L + \vec{\varphi}_T$, where

$$\vec{\varphi}_L = (\vec{\varphi} \cdot \vec{n})\,\vec{n} \quad , \quad \vec{\varphi}_T \cdot \vec{n} = 0 \quad .$$

The conventional picture is that along its trough, $P(\cdot)$ has zero curvature and displays zero mass excitations, while in the orthogonal direction the curvature of $P(\cdot)$ is positive and one expects massive particles. Thus one expects exponential clustering for $\vec{\varphi}_L$,

$$\langle\vec{\varphi}_L(\vec{x})\vec{\varphi}_L(\vec{y})\rangle \leq O(e^{-md}) \quad , \quad m > 0 \; ,$$

but polynomial clustering for $\vec{\varphi}_T$.

As we remarked above, two dimensions displays especially singular infra red behavior. In fact, Coleman [Co 2] has proved that for $d = 2$, a continuous symmetry as above always yields $\delta = 0$. Thus the usual Goldstone picture never ensures the existence of zero mass particles. The basic ingredient of Coleman's proof is the distribution character of the two point function. This precludes zero mass particles appearing in a scalar two point function and forces $\delta = 0$. His argument does not, however, prevent the mass spectrum from going down to zero. Therefore we ask: Does the $\lambda(\vec{\varphi}^{\,2})^2$ model with $\lambda >> m_0^2$ have a mass gap? We conjecture that the mass gap for small λ vanishes as λ is increased to the critical point λ_c, and that the mass gap is zero for $\lambda \geq \lambda_c$. In other words, for $\lambda \geq \lambda_c$ we expect that neither Goldstone bosons nor a mass gap occur.

1.4 Field Theory and Statistical Mechanics

The equivalence of relativistic quantum field theory with statistical mechanics has a long history. Older work includes both the Landau-Ginzberg theory and Symanzik's program to construct Euclidean models. Recent work includes that of Fisher, Wilson, Griffiths and a number of lecturers at this conference. We mention here some selected aspects of this correspondence for boson quantum fields. Ideas of this nature in models with fermions have not yet proved fruitful.

The Partition Function. Let dq denote the Euclidean measure for a boson quantum field model. The partition function

$$Z\{J\} = \int e^{\Phi(J)}\, dq$$

is the generating function for Schwinger functions, and has been studied in $P(\varphi)_2$ models by Fröhlich [Fr 2]. As mentioned above, the Euclidean field model has a natural approximation by a continuous spin ferromagnetic Ising lattice with nearest neighbor interaction, see for instance [Ko Wi]. The convergence of the lattice approximation [Gr Ro Si 3] and the approximation of φ^4 by spin 1/2 Ising models [Gr Si] sharpens this correspondence, see also [New 2]. The one point Schwinger function $\int \phi(x)\, dq$, which parameterizes symmetry breaking in the Goldstone picture above, corresponds to spontaneous magnetization in the Landau-Ginzberg theory. The coupling constant λ/m_0^2, which measures the deviation from a free theory, corresponds to the inverse temperature $\beta = (kT)^{-1}$. In this way the Goldstone picture of the vacuum corresponds to a picture of phase transitions in many body systems. The existence of a gap in the spectrum of H, and exponential clustering, corresponds to a finite correlation length $\xi = m^{-1}$ in the many body system.

One Particle Structure. We define G{J} by

$$G\{J\} = \ln Z\{J\} - \int \Phi(J)\, dq \quad ,$$

and then $G\{J\}$ is the generating function for the connected (truncated) Euclidean Green's functions. The one particle structure is displayed by an entropy principle (Legendre transformation)

$$\Gamma\{A\} = \inf_J \, [-J\cdot A + G\{J\}] \quad ,$$

or in differential form,

$$\Gamma\{A\} = -J\cdot A + G\{J\} \quad ,$$

where J is determined by $A(x) = \delta G\{J\}/\delta J(x)$. This transformation was introduced in statistical mechanics by De Dominicis and Martin [De Ma], in quantum field theory by Jona-Lasinio [Jo L], and was developed by Symanzik [Sy 4]. The analysis of $\Gamma\{A\}$ in quantum field models [Gl Ja 13] may complement our study of the spectrum of the Hamiltonian by expansions described below. The functional $\Gamma\{A\}$ generates the (amputated, one particle irreducible) vertex functions. These functions are directly related to the magnitude of interparticle forces, i.e., the physical charge.

Bound States. In Chapter 3 we study the presence and absence of bound states in certain quantum field models. Our results in Section 3.3 about the absence of bound states in pure φ_2^4 models depend on methods both from field theory and from statistical mechanics. We use high temperature expansions from field theory (see below). We also use an idea of Lebowitz from statistical mechanics to obtain two-particle clustering for the four point vertex function.

Conversely, in Section 3.4 we sketch a proof that bound states occur in φ_2^4 models in a strong external field. We remark that in statistical mechanics, bound state excitations appear in the transfer matrix for large values of chemical potential μ.

High Temperature Expansions. The high temperature expansions in statistical mechanics yield the existence of the thermodynamic limit and high temperature analyticity, i.e., the absence of phase transitions. These Kirkwood-Salsburg or

Mayer-Montroll expansions converge for T/T_c sufficiently large, where T_c is the critical temperature. Related to these expansions are the virial expansions which converge for large values of the chemical potential μ, and which also yield analyticity (absence of phase transitions), see [Ru]. In field theory, the cluster expansions play an analagous role. They converge for large inverse coupling m_0^2/λ [Gl Ja Sp 1, 2] (large T) and for large external field [Sp 2] (large μ). As a result, the cluster expansions establish the existence of the infinite volume limit in field theory, and the existence of a single phase with a unique vacuum vector. These high temperature expansions do not, in general, arise from Kirkwood-Salsburg (or other) integral equations, but have a wider range of validity. We have however, obtained Kirkwood-Salsburg equations for the partition function Z, see Chapter 6 of Part II. These integral equations are a useful tool in our proof of analyticity of the Schwinger functions.

In addition to yielding information about the vacuum, the high temperature expansions give us detailed information about the spectrum of the Hamiltonian H, e.g., the particle structure and the presence or absence of bound states, see Chapters 2, 3 of these lectures. We remark that these more detailed field theory techniques may yield insights into statistical mechanics.

Low Temperature Expansions. The Peierls argument [Pe] is the basic proof of the existence of phase transitions at low temperatures. The proof considers the energy associated with boundaries (contours) separating up spins from down spins. For temperatures T sufficiently below T_c, it is energetically favorable to have spins all up or all down. Griffiths, Dobrushin and others have modified and extended these results, see for example [Dob 1-3, Gi, Gr 1, ML, Min Sin 1-2]. In particular, the contour methods yield exponential clustering in pure phases of low temperature spin systems. Some continuous spin systems have been studied [Bo Gr].

We regard these methods as convergent low temperature expansions. We believe that such low temperature contour expansions exist in quantum field models.

They should converge sufficiently far from the critical point. (Such an expansion may have been used in the proof of the announced result [Dob Min].) We believe that low temperature expansions exist independent of whether multiple phases exist. In a pure phase, we believe that they exhibit exponential clustering and thus are useful to investigate particle structure.

In Figure 6 we show our conjectured region of convergence of the high temperature (cluster) expansion and presumed low temperature (contour) expansion in the φ^4 model. For models such as $\lambda\varphi^4 - \mu\varphi$, $\mu \gg \lambda$, in which symmetry breaking does not occur, the regions of convergence of the high and low temperature expansions may overlap.

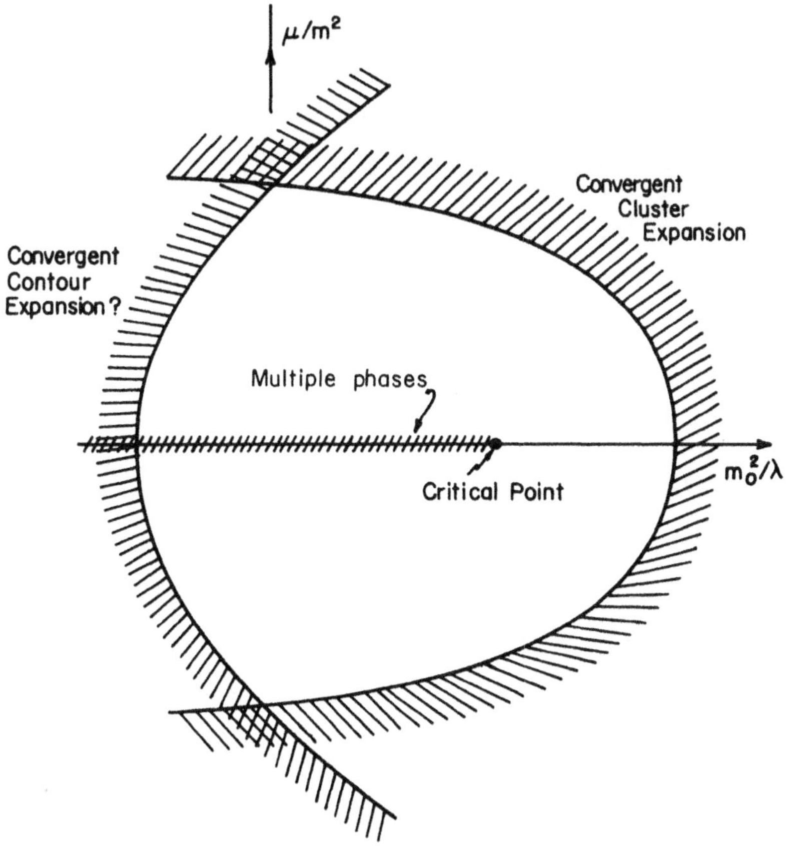

Figure 6. Presumed convergence of cluster and contour expansions.

Correlation Inequalities and the Lee-Yang Theorem. These methods yield the convergence of the Schwinger functions for even $P(\varphi)_2$ models, [Gu Ro Si 3, Nel 5], and a unique phase for $(\varphi^4 + m_0^2\varphi^2 - \mu\varphi)_2$ models, $\mu \neq 0$, [GrSi, Si II]. These methods and related developments are included in the lectures of Guerra, Nelson, Rosen, Simon, to which we refer the reader for further discussion.

1.5. Some Problems

We discuss several open problems for $P(\varphi)_2$. In addition, problems closely related to other sections are mentioned throughout the lectures.

Asymptotic completeness. In a pure φ^4 model with small coupling, is the S-matrix unitary? Can this be proved (see Chapter 3)? In models with bound states, does the inclusion of the bound state in and out fields for mass spectrum below 2m yield asymptotic completeness? Related to these questions is the possibility of performing a cluster expansion in asymptotic fields, as suggested formally by the LSZ expansion of the scattering matrix or the Yang-Feldman equations.

Asymptotic Perturbation Theory. It is known that the Euclidean Green's functions are asymptotic to all orders in the coupling, in the region of convergent cluster expansions, [Di 3]. We conjecture that the S-matrix is asymptotic to its Feynman perturbation series, $S = I + \lambda S_1 + \cdots$. Since $S_1 \neq 0$, this would yield $S \neq I$. We conjecture that the physical mass m is asymptotic in the coupling constant expansion,

$$m = m_0 + \lambda^2 m_2 + \lambda^3 m_3 + \cdots + \lambda^n m_n + O(\lambda^{n+1}) \quad .$$

Cluster Expansions. The high temperature expansions in Part II are based on expanding the Gibbs factor e^{-V_0} in the Gaussian measure $d\Phi$ into $e^{-V_0} = 1 + (e^{-V_0} - 1)$. They yield Kirkwood-Salsburg equations for Z, and related expansions for Z times the Schwinger functions, $ZS(x_1, \cdots, x_n)$. Do these expansions generalize in a natural way to yield particle structure? What is the optimal convergence domain for these expansions?

Contour Expansions. We believe that a low temperature contour expansion exists and converges, independent of whether a $P(\varphi)$ model has an internal symmetry. What is this expansion? Does it yield the existence of the infinite volume limit, of particles and of other properties of the mass spectrum.

Analyticity. In Part II, we show that the $\lambda P(\varphi)_2$ Schwinger functions are analytic in a half circle $0 < |\lambda| < \lambda_0$, Re $\lambda > 0$. What is the complex domain into which the Schwinger functions can be continued? In statistical mechanics, the Lee-Yang theorem is used to extend the analyticity domain of high temperature (small λ/m_0^2 , large m_0^2/λ) expansions and of virial (large μ) expansions. Are the Schwinger functions for $\lambda\varphi^4 - \mu\varphi$ real analytic in λ , μ except for a cut from λ_{crit} to ∞ ? In other words, are the Schwinger functions real analytic in all of Figure 6, except for a cut along the line of multiple phases? What is the domain of complex analyticity? For $\lambda > 0$ and Re $\mu \neq 0$, the pressure is analytic [Sp 2].

Haag-Doplicher-Roberts axioms. Is duality, the missing HDR axiom, valid for $P(\varphi)_2$ models? The HDR analysis of superselection sectors applies only in three and four dimensions, but duality is still presumably true for $P(\varphi)_2$.

Critical Points. If a critical point exists (See Figure 5) how do m , $\langle\varphi\rangle$, etc., behave in a neighborhood of it? Do the mass, spontaneous magnetization, etc., vary with power laws (given by critical exponents)? For $\lambda < \lambda_{crit}$, m is monotone in m_0^2 [Gu Ro Si 3]. Is the mass monotone above the critical point? Since Coleman has shown that $\delta = 0$ for $\lambda(\vec{\varphi}^{\,2})^2$ models, do multiple phases exist for this model? Is there more than one phase at the critical point for $\lambda\varphi_2^4$? Do zero mass particles occur in φ_2^4 at the critical point? (We remark that zero mass particles do not occur in the two point function, since it is a tempered distribution [Gl Ja IV].) What is the locus of multiple phases for a φ^6 or φ^8 model, etc? Do critical manifolds exist for these models?

Structure Analysis. With our control over the particle spectrum, we have the ingredients to carry out the particle structure analysis of Green's functions, as proposed by Symanzik [Sy 1]. It is also of interest to perform a structure analysis of models in statistical mechanics. As a first step, one can prove the existence and analyticity of the generating functional for one particle irreducible (IPI) Green's functions, as given in [Gl Ja 13] . These vertex parts are important in the study of symmetry breaking and of the renormalization group. In the former direction, Jona-Lasinio has an effective potential which one believes gives corrections to the mean field Goldstone picture of Section 1. 3. Such potentials have by studied heuristically in [Co We]. In what sense is the mean field or the effective potential model a limit of quantum field theory?

Anomalous Dimensions. An extremely interesting circle of problems concerns the more refined aspects of $P(\varphi)_2$ models at the critical point. These ideas also make close contact with ideas of high energy theorists. The short distance behavior of $P(\varphi)_2$ and $\varphi_3^4(g)$ models is canonical, and a rigorous proof should follow from the local perturbation estimates [Gl Ja IV, Fel 2]. Since these estimates hold for all λ, they hold in particular at a critical point for $P(\varphi)_2$, giving a logarithmic singularity. On the other hand, the long distance behavior at the critical point for $P(\varphi)_2$ models is not canonical, since $\langle\varphi(x)\varphi(y)\rangle \to$ const. as $|x - y| \to \infty$. Consequently, we do not expect that any $P(\varphi)_2$ model we have constructed is scale invariant. In fact, a scale invariant vacuum would ensure that scale transformations are unitarily implemented. This would ensure in turn that the long and short distance scaling properties were the same.

Let us assume that a critical point exists. Then we conclude that the theory at the critical point must contain a fundamental length. This length characterizes the distance at which the small distance asymptotic behavior is replaced by the long distance asymptotic behavior. Scale transformations change this length, so if a critical point exists, there are continuously many zero mass theories related to one another by scaling. One can attempt to force scale invariance by performing an infinite scale transformation. Do such limits exist? Some of the problems raised here are unresolved for the three dimensional Ising model, and a serious effort might start with this case.

The Renormalization Group. Above we parameterized zero mass $P(\varphi)_2$ theories by a fundamental length. An alternative description is based on the renormalization group, which itself has intrinsic interest. Can the Callen-Symanzik equations be used to investigate the long distance behavior of $P(\varphi)_2$ models?

2. FROM ESTIMATES TO PHYSICS

How do we obtain physical properties of particles from our expansions and bounds? In this lecture we show how properties of the one particle states follow from known cluster expansions. These basic estimates for quantum field models exhibit the decoupling $\exp(-d/\xi)$ of disjoint regions in Euclidean phase space. In two space-time dimensions $(d = 2)$, cluster expansions yield space-time decoupling, as in Part II. For $d = 3$, related bounds yield phase space decoupling and the positivity of φ_3^4.

We recall that the theory of a single type of particle with mass m has the energy-momentum spectrum

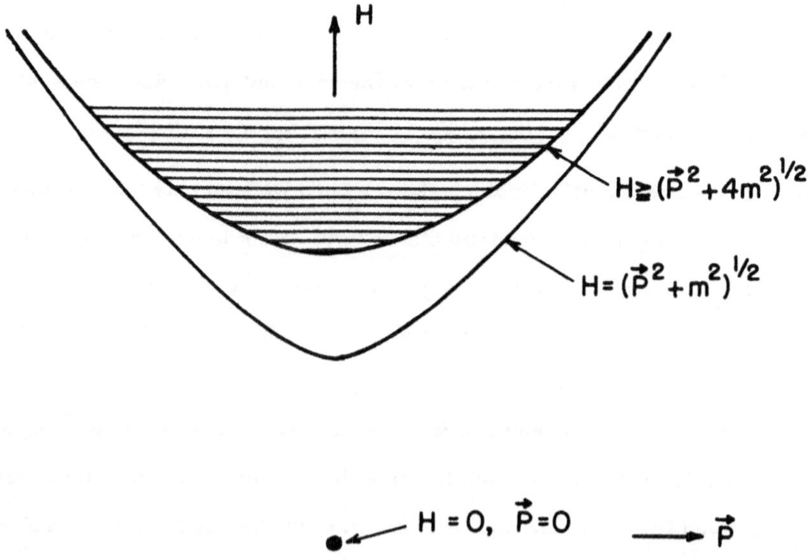

divided into three disjoint parts, the vacuum $\vec{P} = 0$, $H = 0$, the one particle hyperboloid $H^2 - \vec{P}^2 = m^2$ and the continuum $H^2 - \vec{P}^2 \geq (2m)^2$. The two particle states with momentum $\vec{p}_1$, $\vec{p}_2$ are conveniently parameterized by the relative momentum $p_R = \vec{p}_1 - \vec{p}_2$ and the total momentum $\vec{p}_T = \vec{p}_1 + \vec{p}_2$. The invariant mass for the two particle states is $2^{1/2}(\mu_1\mu_2 - \vec{p}_1 \cdot \vec{p}_2 + m^2)^{1/2}$, which for $\vec{p}_R = 0$ equals $2m$.

The mass operator $M = (H^2 - \vec{P}^2)^{1/2}$ has the corresponding spectrum

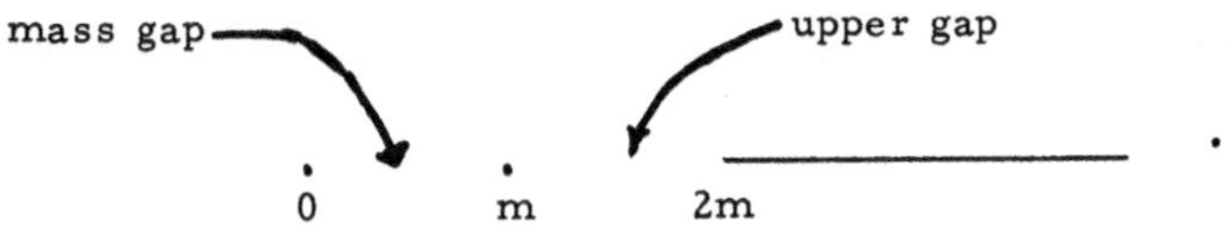

The eigenspace of 0 is the vacuum, and the eigenspace of m is defined to be the one particle space.

In order to establish spectral properties of H and M we use estimates proved by cluster expansions:

(1) Uniform vacuum cluster estimates yield convergence as the volume $\Lambda \to R^2$, and cluster estimates carry over to the infinite volume limit.

(2) The limiting Schwinger functions (for real coupling constants) satisfy the Osterwalder-Schrader axioms, and hence yield a Wightman theory. The cluster property of the vacuum (asymptotic factorization) yields uniqueness of the vacuum vector.

(3) The vacuum cluster expansion bounds the exponential decay to a factorizing vacuum and determines the mass gap. From the one particle cluster expansion, we obtain the upper mass gap and an isolated eigenvalue $M = m$.

In Section 2.1 we give some simple functional analysis. We apply these results in Section 2.2 to establish (1) - (3) above.

2.1. Functional Analysis

Let $0 \leq H = H^*$ and let E_a be the spectral projection for $[0, a]$. Let $\mathcal{D}$ be a dense subset of $\mathcal{H}$, and let $\mathcal{D}_0 \subset \mathcal{D}$ be given.

Proposition 2.1.1. Suppose that for each $\theta \in \mathcal{D}$, there exists $\chi \in \mathcal{D}_0$ and $\epsilon > 0$ such that

$$\langle \theta - \chi , e^{-tH}(\theta - \chi) \rangle \leq M_\theta e^{-(a+\epsilon)t} \quad . \tag{2.1.1}$$

Then $E_a \mathcal{D}_0$ is dense in $E_a \mathcal{H}$ and

$$\langle \theta - \chi , e^{-tH}(\theta - \chi) \rangle \leq \|\theta - \chi\|^2 e^{-(a+\epsilon)t} \quad .$$

Proof. For $\theta \in \mathcal{D}$,

$$\begin{aligned} \| E_a(\theta - \chi) \| &= \| E_a e^{tH} e^{-tH}(\theta - \chi) \| \\ &\leq e^{at} \langle \theta - \chi , e^{-2tH}(\theta - \chi) \rangle^{1/2} \\ &\leq M_\theta^{1/2} e^{-\epsilon t} \longrightarrow 0 \quad . \end{aligned}$$

Thus $E_a \mathcal{D}_0 = E_a \mathcal{D}$, which is dense in $E_a \mathcal{H}$. By applying the Schwarz inequality n times,

$$\begin{aligned} \langle \theta - \chi , e^{-tH}(\theta - \chi) \rangle &\leq \|\theta - \chi\| \, \|e^{-tH}(\theta - \chi)\| \\ &\leq \|\theta - \chi\|^{2-2^{-n}} . e^{-(a+\epsilon)t} M_\theta^{2^{-n}} \\ &\longrightarrow \|\theta - \chi\|^2 e^{-(a+\epsilon)t} \quad . \end{aligned}$$

We now let $\mathcal{H}$ be a Hilbert space carrying a unitary representation $U(a,\Lambda)$ of the inhomogeneous Lorentz group. Let $\mathcal{H}_0 \subset \mathcal{H}$ be a subspace of bounded energy and momentum, $P_0 \leq a$, $|P| \leq b$. Let $U(a)\mathcal{H}_0 \subset \mathcal{H}_0$ and $(\cup U(a,\Lambda)\mathcal{H}_0)^- = \mathcal{H}$.

Proposition 2.1.2. If $\mathcal{H}_0 \neq \{0\}$ contains a cyclic vector for the space translation subgroup $U(\vec{a})$, then the spectrum of $M \upharpoonright \mathcal{H}$ contains exactly one point and $U(a,\Lambda)$ is irreducible on $\mathcal{H}$.

Proof. The family $U(\vec{a})$ is maximal abelian on $\mathcal{H}_0$, so any commuting operator is a function of $\vec{P}$. In particular the energy momentum spectrum is a set of the form $\{H(\vec{P}), \vec{P}\}$, and by Lorentz invariance $H = (\vec{P}^2 + \bar{m}^2)^{1/2}$ for some $\bar{m}$. (Here we assume the nontriviality of $\mathcal{H}_0$.) Thus $M = \bar{m}$ on $\mathcal{H}_0$ and by Lorentz $M = \bar{m}$ on $\mathcal{H}$. Since reducibility would be accompanied by multiplicity in the mass spectrum, the representation $U(a,\Lambda)$ is irreducible.

2.2. Relevance to Physics

The Schwinger functions with a space time cutoff h are given by

$$S_h(x_1, \cdots, x_n) = \int \Phi(x_1) \cdots \Phi(x_n) dq_h \tag{2.2.1}$$

where dq_h is the measure

$$dq_h = \frac{e^{-V(h)} d\Phi}{\int e^{-V(h)} d\Phi} , \tag{2.2.2}$$

$d\Phi$ is the Gaussian measure with mean zero and covariance $(-\Delta + m_0^2)^{-1} = C$, and

$$V(h) = \lambda \int : P(\Phi(x)) : h(x) dx$$

is the $P(\varphi)_2$ Euclidean action. If $h(x)$ is the characteristic function for a set $\Lambda \subset R^2$ with area $|\Lambda|$, then $V(h)$ is the action for Λ . We denote the corresponding Schwinger functions S_Λ .

We state the vacuum cluster expansion, which bounds the rate of asymptotic factorization of the vacuum state. Let A be a function of Euclidean fields,

$$A = \prod_{i=1}^{n} \int : \Phi(x)^{n_i} : f_i(x) dx \tag{2.2.3}$$

where $f_i(x)$ is either an $L_2(R^2)$ function, or else $\delta_s(t) f_i(\vec{x})$, where $f_i(\vec{x})$ is $L_2(R)$. Let $\{\Delta\}$ be a cover of R^2 by unit lattice squares Δ , and define suppt. A as the smallest union of Δ 's containing suppt. $f_1 \cup \cdots \cup$ suppt. f_n . In the following, A , B have the form (2.2.3).

<u>Theorem</u> 2.2.1 (<u>Vacuum Cluster Expansion</u> [Gl Ja Sp 1, 2].) Let $\gamma = m_0 - \epsilon$, with $\epsilon > 0$. Consider $\lambda P(\varphi)_2$ models with $\lambda < \lambda(\epsilon, P, m_0)$. Let $d = \text{dist.}\{\text{suppt. } A, \text{suppt. } B\}$. Then there exists a constant $M_{A,B}$ such that

(2.2.4) $$\left|\int AB\,dq_h - \int A\,dq_h \int B\,dq_h\right| \le M_{A,B}\, e^{-\gamma d} \quad ,$$

uniformly in h .

[The constant $M_{A,B}$ can be bounded explicitly in terms of the f_i . We suppose each f_i is supported in a single Δ_i (an arbitrary A is a sum of such localized A's). We let $N(\Delta)$ be the sum of n_i's , for suppt. $f_i \in \Delta$. Let K_1 be given and let

$$\eta = \prod_{\Delta} (K_1 N(\Delta)!) \quad .$$

We define for a localized A ,

(2.2.5) $$\|A\| = \eta \prod_{i=1}^{n} \|f_i\| \quad .$$

Here $\|f\| = \|f\|_{L_2(R^2)}$ for $f \in L_2(R^2)$, or if $f(x) = \delta_s(t)f(\vec{x})$, $\|f\| = \|f(\vec{x})\|_{L_2(R)}$. Let us assume $n_i \le \bar{n}$. Then for K_1 sufficiently large,

(2.2.6) $$M_{A,B} \le \|A\| \ \|B\| \quad .$$

If it is not the case that $n_i \le \bar{n}$, we obtain (2.2.6) with η^3 replacing η in (2.2.5). Also

(2.2.7) $$\int A\,dq_h \le \|A\|$$

uniformly in h .]

Theorem 2.2.2. The Schwinger functions $S_\Lambda(x_1, \cdots, x_n)$ converge in $\mathcal{S}'(R^{2n})$ as $\Lambda \to R^2$, to limits $S(x_1, \cdots, x_n)$ obeying the Osterwalder-Schrader axioms.

Proof. As explained in the lectures of Osterwalder and Nelson, it is sufficient to prove convergence as $h \to 1$, and a simple φ bound that follows from the vacuum cluster expansion. Here we establish convergence. Let $0 \leq h_0 \leq h_1 \leq 1$ be two space-time cutoffs, with $h_1 - h_0$ supported on a bounded set Γ. Let $A = \Phi(f_1) \cdots \Phi(f_n)$, let suppt. $A \subset$ suppt. h_0 and let $d = \text{dist}(\Gamma, \text{suppt. } A)$. Let $\bar{\Gamma}$ be the set of lattice squares intersecting Γ.

Define the function

$$F(\alpha) = \int A dq_\alpha$$

where $dq_\alpha = dq_{h_\alpha}$ and $h_\alpha = \alpha h_1 + (1-\alpha) h_0$. Then by differentiating (2.2.2) we obtain

$$\left| \int A dq_1 - \int A dq_0 \right| = \left| \int_0^1 F'(\alpha) d\alpha \right|$$

$$= \int_0^1 d\alpha \sum_{\Delta \subset \bar{\Gamma}} \left[\int A V(\Delta) dq_\alpha - \int A dq_\alpha \int V(\Delta) dq_\alpha \right] .$$

By Theorem 2.2.1 and (2.2.6), the sum above is

$$\sup_{\Delta} M_{A, V(\Delta)} e^{-\gamma' d} \leq O(1) e^{-\gamma' d} ,$$

for $\gamma' = m_0 - 2\epsilon$. We let h_0, h_1 be characteristic functions of sets $\Lambda_0, \Lambda_1 \subset R^2$. Then $d \to \infty$ as $\Lambda_0, \Lambda_1 \to R^2$ and we obtain the desired convergence.

The λ dependence of the cluster expansion shows immediately that $S(f_1, \cdots, f_n)$ are continuous functions of λ. In fact, in Part II we use the cluster expansion to establish analyticity in λ in the half circle $0 < |\lambda| < \lambda_0$, $\text{Re}\,\lambda > 0$.

<u>Theorem</u> 2.2.3. (<u>Mass Gap</u>). For $\lambda P(\varphi)_2$ models with small coupling λ, the vacuum Ω spans the subspace of energy less than $\gamma = m_0(1-\epsilon)$.

<u>Proof.</u> Let $A = \Phi(f_1) \cdots \Phi(f_n)$, suppt. A compact, and let θ_A be the vector in the relativistic Hilbert space $\mathcal{H}$ associated with the Euclidean function A. The plan is to apply Proposition 2.1.1 with $\mathcal{D}$ the dense subset of $\mathcal{H}$ spanned by such

θ_A , and with $\mathcal{H}_0$ the subspace of $\mathcal{H}$ spanned by Ω .

Since the cluster estimates are uniform in h , they carry over to the infinite volume limit $h = 1$. We choose A_d to be the translate of A in the (Euclidean) time direction. Thus by the vacuum cluster expansion, Theorem 2.2.1, with $\gamma = m_0 - \epsilon$,

$$\langle \theta_A , e^{-dH} \theta_A \rangle = \int A^- A_d dq$$

$$= \int A^- dq \int A dq + O(M_A e^{-\gamma d})$$

$$= |\langle \theta_A , \Omega \rangle|^2 + O(M_A e^{-\gamma d}) \quad .$$

In other words, if $\theta_A^\perp = \theta_A - \langle \Omega , \theta_A \rangle \Omega$ is the component of θ_A orthogonal to Ω ,

$$\langle \theta_A^\perp , e^{-dH} \theta_A^\perp \rangle \leq O(M_A e^{-\gamma d}) \quad .$$

The theorem now follows as planned.

We have established stronger cluster properties, which provide an analysis of arbitrary intervals of the energy spectrum [Gl Ja Sp 1]. These expansions are defined inductively, rather than in closed form or in the form of Kirkwood-Salsburg equations. We now state the $n = 1$ expansion, or one particle cluster expansion. Let $\gamma = 2(m_0 - \epsilon)$ with $\epsilon > 0$.

Theorem 2.2.4. (One Particle Cluster Expansion). Given $\epsilon > 0$ consider $\lambda P(\varphi)_2$ models with $\lambda < \lambda(\epsilon , m_0 , P)$. Then given θ_A as above, there exists an $L_2(R)$ function h such that for $\chi = \langle \Omega , \theta_A \rangle \Omega + \varphi(h)\Omega$, we have

$$\langle \theta_A - \chi , e^{-tH}(\theta_A - \chi) \rangle \leq M_A e^{-\gamma t} \quad .$$

We apply Proposition 2.1.1 once again. We choose $\mathcal{H}$ as in the previous example, and $\mathcal{H}_0$ the span of $\{\Omega, \varphi(h)\Omega\}$, where $h \in L_2(R)$.

Corollary 2.2.5. The vectors $E_{2m_0-\epsilon}\mathcal{D}_0$ span states of energy $< 2m_0-\epsilon$.

Theorem 2.2.6. (Upper Mass Gap). For $\lambda P(\varphi)_2$ models with small coupling, the mass operator M has eigenvalues 0, m and no other spectrum in $[0, 2m_0-\epsilon]$.

Proof. Let $E = E_{2m_0-\epsilon}(I-E_0)$, let $\mathcal{H}_0 = E\mathcal{H}$ and let $\mathcal{H}$ equal the union of the Lorentz translates of $\mathcal{H}_0$. Below we obtain a cyclic vector χ for the space translation subgroup on $\mathcal{H}_0$. By Proposition 2.1.2, the spectrum of M on $\mathcal{H}$ contains exactly one point (unless $\mathcal{H}_0 = \{0\}$). We show $\mathcal{H}_0 \neq 0$: The two point function converges in $\mathcal{S}'$ to the free two point function as $\lambda \to 0$, using the λ dependence of the cluster bounds. Since the free theory has one particle states with mass m_0, the interacting theory must have spectrum in a neighborhood of m_0, for λ sufficiently small. Thus $\mathcal{H}_0 \neq 0$. M has the eigenvalues 0 and $m \in [m_0 - \epsilon, m_0 + \epsilon]$, and no other spectrum in $[0, 2m_0 - \epsilon]$.

To complete the proof we construct χ. Let $h_1 \in \mathcal{S}(R)$, $\tilde{h}_1 > 0$. We show that $E\varphi(h_1)\Omega$ is cyclic on $\mathcal{H}_0$. Let $h_a(x) = h(x-a)$. Then

$$U(\vec{a})\varphi(h)\Omega = \varphi(h_a)\Omega$$

and

$$\begin{aligned}\varphi(h_1 * h_2)\Omega &= \varphi\Big(\int h_1(\cdot - a)\, h_2(a)\, da\Big)\,\Omega \\ &= \int da\, h_2(a)\varphi(h_{1a})\Omega \\ &= \int da\, h_2(a)\, U(\vec{a})\varphi(h_1)\Omega \quad .\end{aligned}$$

Since E and $U(\vec{a})$ commute,

$$E\varphi(h_1 * h_2)\Omega = \int da\, h_2(a)\, e^{-iPa}\, E\varphi(h_1)\Omega$$

lies in the span of translates of $E\varphi(h_1)\Omega$. Since $(h_1 * h_2)^{\sim} = \tilde{h}_1 \tilde{h}_2$ are dense in C_0^∞ as h_2 ranges over C_0^∞, $\chi = E\varphi(h_1)\Omega$ is cyclic for $U(\vec{a})$ on $\mathcal{H}_0$. Here we have also used Corollary 2.2.5 to identigy $\mathcal{H}_0$ with the span of $E\varphi(f)\Omega$.

3. BOUND STATES AND RESONANCES

3.1 Introduction

An important problem in physics is how particles form composites, namely bound states and resonances. In atomic physics, familiar consequences of Coulomb forces and the Schrödinger Hamiltonian are atoms and molecules: their existence and their scattering. The spectrum of atomic and molecular Hamiltonians has been the subject of extensive mathematical analysis.

The realm of nuclear and elementary particle structure includes qualitatively similar ideas, but without detailed justification. Thus a crucial physical question is whether a particular quantum field model does or does not have bound states. For instance: Do mesons bind nucleons to form stable nuclei? Are nucleons bound states of quarks? Are the ρ and the η mesons really π meson resonances?

Little is known about such important questions in quantum field theory. In fact, no quantum field models are known to have bound states, and heuristic calculations based on perturbation theory and the Bethe-Salpeter equation are inconclusive.

In this lecture we give a physical picture of when to expect or not to expect bound states in $P(\varphi)_2$ models with weak coupling or a strong external field. We prove the absence of two particle bound states in weakly coupled, pure φ^4 models. We outline an argument to prove the presence of bound states in the presence of a strong external field, and certain other models.

Bound states are eigenvalues of the mass operator M, introduced in Chapter 2. Two particle bound states lie below the two particle continuum; otherwise no energetic reason would prevent their decay into free particles. (The decay of bound states in the mass continuum may, however, be forbidden by additional selection rules included in the interaction.) On the other hand, there is no physical interpretation of continuous mass spectrum in the spectral interval $[0, 2m)$. Hence none is believed to exist, and two particle bound states may occur in the "bound state interval" $(m, 2m)$ of the mass spectrum, as illustrated in Figure 7.

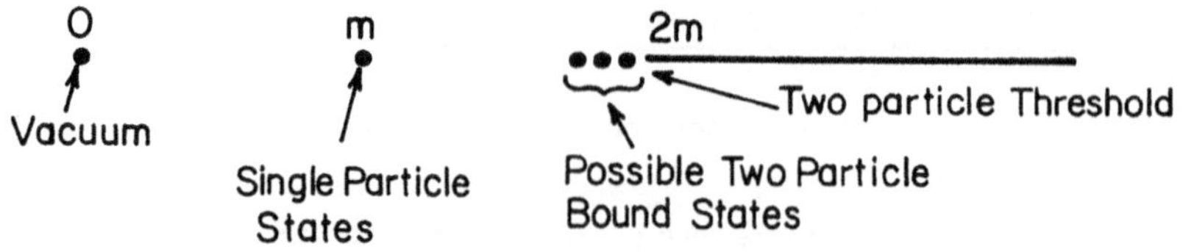

Figure 7. Spectrum of the mass operator M

In an even theory, e.g., φ^4, we can decompose the Hilbert space according to whether states are even or odd under the symmetry $\varphi \to -\varphi$. States with an even number of particles lie in the even subspace. Restricted to the odd subspace, M has the spectrum of Figure 8.

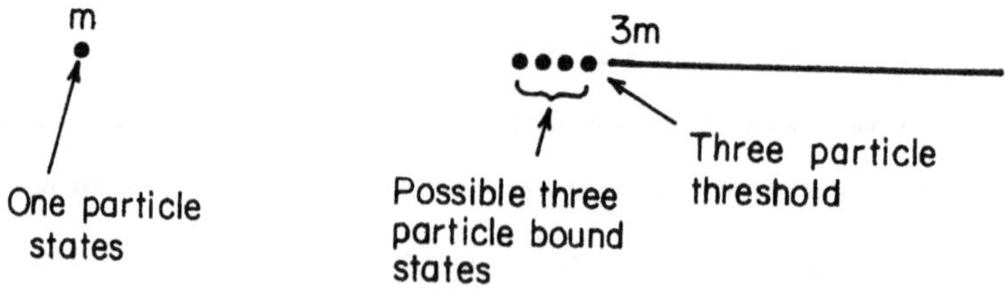

Figure 8. Mass Spectrum on the Odd Subspace of an Even Theory.

The resolvent $(M - z)^{-1} = R(z)$ of the mass operator is an analytic function of z, for $\operatorname{Im} z \neq 0$. It has a pole at each eigenvalue of M (particles and bound states) and presumably a cut starts at each n-particle threshold. The question of resonances concerns the analytic properties of $R(z)$ (or suitable matrix elements) after continuation across a threshold cut. A complex pole, close to the cut, is called a resonance. Such a pole appears in the scattering of particles as a peak in the cross section. Another interpretation of a resonance is an unstable particle. The real part of the position of the pole determines the mass of the resonance, while the distance to the real axis determines the lifetime. It is a challenging question to make a detailed investigation of resonances, and to determine: Are there coupling constants for which $P(\varphi)_2$ models have resonances?

The presence or absence of composite particle states depends on whether the interparticle forces are attractive or repulsive. We pose the related questions: Does the mutual interaction of two particles raise or lower their energy, compared with the state in which they are asymptotically far apart? If the energy is raised, binding does not occur. If the energy is lowered below the continuum, we expect a bound state. In Section 3.2 we motivate our point of view on this question by perturbation theory. In Section 3.3, we use cluster estimates and correlation inequalities to study the same question. In Section 3.4, we show how binding occurs.

Our picture of a two particle bound state is best understood in terms of the relative momentum $\vec{p}_R$. We describe three kinds of forces: attractive, repulsive and dispersive. The attractive and repulsive forces are self explanatory. The dispersive effect arises from the curvature of the mass hyperboloid. A state of two free particles, with $\vec{p}_T = 0$, has a total energy $(4m^2 + \vec{p}_R^{\,2})^{1/2}$, and in general, for small momentum, a two particle state has energy $2m + O(\vec{p}_R^{\,2} + \vec{p}_T^{\,2})$. This raising of the energy away from zero momentum is what we call the dispersive force. For bound states to occur, the attractive force must dominate the repulsive and dispersive forces.

We introduce a parameter δ to measure the spread of the bound state wave packet. For a momentum space distribution concentrated in $|\vec{p}_R| \leq \delta$, we have a configuration space spreading of order δ^{-1}. For weak coupling, we expect increased spreading in configuration space, as a bound state grows in size and disappears into the continuum. Thus we expect $\delta \to 0$ as $\lambda \to 0$. The binding forces have characteristic dependences on δ and λ: The dispersive effect is $O(\delta^2)$. In $P(\varphi)_2$ models, we find in perturbation theory that attractive and repulsive effects are $O(\delta)$, times the appropriate dimensionless coupling constants λ_j/m_0^2. We discuss the balance of these forces in Section 3.4.

3.2 Formal Perturbation Theory

For a $\lambda\varphi^4$ interaction, the first order shift in the two particle energy is given by the Feynman diagram

 ,

which is positive for $\lambda > 0$. In second order, we find the shift has two sorts of contributions, a second order mass shift with the disconnected Feynman diagrams

and a second order attractive (negative) contribution of the form

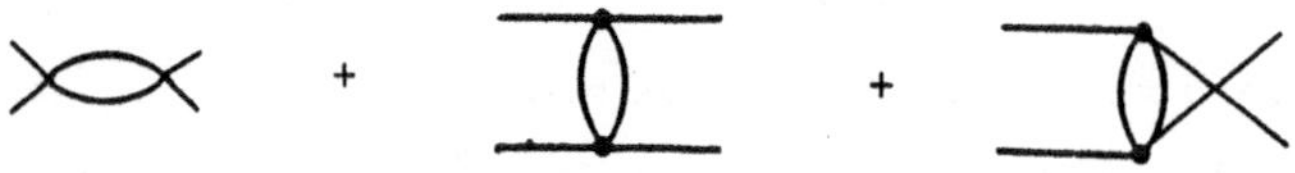

The first order repulsive shift dominates for small λ. Thus we do not expect two particle bound states to occur in weakly coupled φ_2^4 models, and we establish this in Section 3.3.

We remark that the mass shift diagrams above represent the second order mass renormalization of single particle states, i.e., the shift from m_0 to m_2. Of course, to second order, we measure our n-particle forces (energy shifts) with respect to nm_2. We do not include vacuum energy shifts, since they are eliminated by considering perturbations of the exact (coupling λ) ground state.

If we consider three particle interactions, in lowest order, diagrams of the form

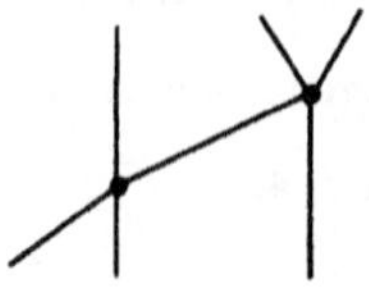

give an attractive three body force. However the diagram

gives a repulsive effect in the two particle subsystems. Since the two body force is first order, and the three body force is second order, we expect the repulsive force to dominate at small coupling. A three particle unstable state (resonance) is possible.

With a φ^3 interaction, the lowest order two body force is attractive

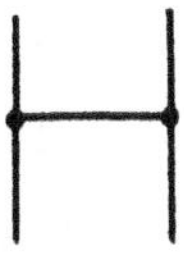

Similarly, n body forces in lowest order are attractive. For instance, in third order we have diagrams of the form

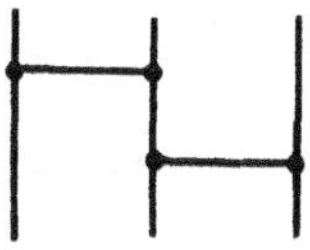

These attractive forces complement the attractive forces in two body subsystems, i. e., in the three body case,

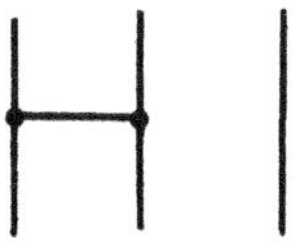

Thus we expect two particle bound states, and bound states of three or more particles if a selection rule prevents their decay. Otherwise, the attractive many body forces should yield many body resonances.

Of course, a pure φ^3 theory does not exist, because the energy is unbounded from below. However, if the φ^3 term in an interaction has a coefficient much larger than the other coupling constants, we expect that the φ^3 effects will dominate. Thus the above qualitative discussion applies to the $\lambda_1\varphi^3 + \lambda_2 P(\varphi)_2$ model, where $\lambda_1 >> \lambda_2$. In this case we expect bound states, and in particular, two particle bound states.

Closely related is the case of a $P(\varphi)_2$ model in an external field, i.e., the $P(\varphi)_2 - \mu\varphi$ model. By the transformation $\varphi \to \varphi + \text{const.}$ (implemented locally by $\exp(i\int \pi))$ we can eliminate the external field. For instance, the $\lambda\varphi^4 - \mu\varphi$ model is transformed into a $\lambda\varphi^4 + a\varphi^3 + b\varphi^2$ model, where $4\lambda a^3 + am^2 = \mu$. The mass term b also grows with μ, but by scaling it can be reduced to unity. Thus we conjecture: Bound states exist in the φ^4 model with a strong external field, $\mu >> \lambda$.

A similar analysis applies to an arbitrary $\lambda P(\varphi)_2 - \mu\varphi$ model. Transforming away the external field, we add to P a lower degree polynomial. For μ large, the dominant coefficients have degree 2 and 3. The degree 2 term gives a mass shift, while the degree 3 term yields an attractive potential in lowest order. Thus we conjecture: Bound states exist in $\lambda P(\varphi)_2$ models in external fields with $\mu \gg \lambda$.

Question: Do bound states occur in Y_2 models? We conjecture that this is the case.

3.3 On the Absence of Bound States

We consider the weakly coupled $\lambda\varphi_2^4$ model, and we prove that two particle bound states do not occur.

Theorem 3.3.1. Let λ/m_0^2 be sufficiently small in the $\lambda\varphi_2^4$ model. Then the mass operator $M = (H^2 - P^2)^{1/2}$ has no spectrum in the two particle bound state interval $(m, 2m)$.

From the uniqueness of the vacuum, we infer that the symmetry $\varphi \to -\varphi$ can be unitarily implemented, and that the Hilbert space $\mathcal{H}$ decomposes into even and odd subspaces $\mathcal{H}_e$, $\mathcal{H}_o$ each invariant under $U(a,\Lambda)$ and φ. Our theorem depends on three facts: Cluster expansions [Gl Ja Sp 1] reduce the problem to the consideration of the two point function for $\mathcal{H}_o$, and the four point function for $\mathcal{H}_e$. Second, an inequality that Lebowitz [Leb 2] proved for Ising models excludes the possibility that mass spectrum in the interval $(0, 2m)$ occurs in the four point function. Finally, cluster bounds exclude mass spectrum in the interval $(m, 2m)$ in the two point function.

The condition of weak coupling in Theorem 3.3.1 concerns the rate γ, of exponential decay $e^{-\gamma d}$, in the error term of the two particle cluster expansion. We show in [Gl Ja Sp 1] that $\gamma \to 3m_0$ and $m \to m_0$ as $\lambda/m_0^2 \to 0$. In Theorem 3.3.1 we require that λ/m_0^2 be sufficiently small to ensure $\gamma \geq 2m$.

More generally, we obtain for even $P(\varphi)_2$ models a larger mass gap on the odd subspace, as suggested in Figure 8 above.

Theorem 3.3.2. Consider an even $\lambda P(\varphi)_2$ model. Given $\epsilon > 0$, let λ/m_0^2 be sufficiently small to ensure $\gamma \geq 3m_0 - \epsilon$, for the rate γ of exponential decay for the error in the two particle cluster expansion. Then $M \restriction \mathcal{H}_o$ has no spectrum in the interval $(m, 3m_0 - \epsilon)$.

Let dq be the Euclidean measure for the $\lambda\varphi_2^4$ model, and for a function A of the Euclidean field Φ let

$$\langle A \rangle \equiv \int A\, dq \quad .$$

Proposition 3.3.3. For the $\lambda\varphi_2^4$ model,

$$(3.3.1) \quad \begin{aligned} &\langle \Phi(x_1)\Phi(x_2)\Phi(x_3)\Phi(x_4)\rangle - \langle \Phi(x_1)\Phi(x_2)\rangle\langle \Phi(x_3)\Phi(x_4)\rangle \\ &\leq \langle \Phi(x_1)\Phi(x_3)\rangle\langle \Phi(x_2)\Phi(x_4)\rangle + \langle \Phi(x_1)\Phi(x_4)\rangle\langle \Phi(x_2)\Phi(x_3)\rangle \ . \end{aligned}$$

Remark. Since $\langle \Phi(x)\rangle = 0$, this inequality states that the connected (truncated) four point function is negative. This bound is special to φ^4 models. In fact the philosophy of Section 3.2 suggests the presence of two particle bound states in $\varphi^6 - \varphi^4$ models.

The key inequality due to Lebowitz concerns independent spin variables $\sigma_i = \pm 1$ for a ferromagnetic Ising model. The energy of a spin configuration $\underset{\sim}{\sigma} = \{\sigma_i\}$, $1 \leq i \leq n$, is

$$H(\underset{\sim}{\sigma}) = -\sum_{i<j} J_{ij}\sigma_i\sigma_j \ ,$$

where $J_{ij} \geq 0$. For a function $f(\underset{\sim}{\sigma})$, let

$$\langle f \rangle = Z^{-1} \sum_{\underset{\sim}{\sigma}} f(\underset{\sim}{\sigma})\, e^{-H(\underset{\sim}{\sigma})} \ ,$$

where

$$Z = \sum_{\underset{\sim}{\sigma}} e^{-H(\underset{\sim}{\sigma})} \ .$$

Lebowitz proves [Leb 2]

$$\langle \sigma_i\sigma_j\sigma_k\sigma_\ell\rangle - \langle\sigma_i\sigma_j\rangle\langle\sigma_k\sigma_\ell\rangle \leq \langle\sigma_i\sigma_k\rangle\langle\sigma_j\sigma_\ell\rangle + \langle\sigma_i\sigma_\ell\rangle\langle\sigma_j\sigma_k\rangle \ .$$

The inequality (3.3.4) follows immediately, since Griffiths and Simon [Gr Si] have proved that the Euclidean φ_2^4 model is a limit of Ising models of the above form, where $\Phi(x)$ can be expressed as a limit of a sum of spin variable σ_i.

We recall that the relativistic time zero field $\varphi(f)$ equals the time zero Euclidean field $\Phi(f, t=0)$. We let $f_i \in \mathcal{S}(R)$ and define

$$(3.3.2) \quad \theta(f_1, f_2) = \varphi(f_1)\varphi(f_2)\Omega - \langle \Omega, \varphi(f_1)\varphi(f_2)\Omega \rangle \Omega \quad ,$$

where Ω is the vacuum vector.

<u>Corollary</u> 3.3.4. The vectors $\theta(f_1, f_2)$ have energy $\geq 2m$.

<u>Proof.</u> It is no loss of generality to choose f_i real and positive. By the Feynman-Kac formula,

$$\langle \theta, e^{-tH}\theta \rangle = \langle \varphi(f_1)\varphi(f_2)\Omega, e^{-tH}\varphi(f_1)\varphi(f_2)\Omega \rangle - |\langle \Omega, \varphi(f_1)\varphi(f_2)\Omega \rangle|^2$$

$$= \langle \Phi(g_1)\Phi(g_2)\Phi(g_3)\Phi(g_4) \rangle - \langle \Phi(g_1)\Phi(g_2) \rangle \langle \Phi(g_3)\Phi(g_4) \rangle \quad ,$$

where $x = (x_1, x_0)$ and

$$g_1(x) = f_1(x_1)\delta(x_0) \quad , \qquad g_2(x) = f_2(x_1)\delta(x_0) \quad ,$$

$$g_3(x) = f_1(x_1)\delta_t(x_0) \quad , \qquad g_4(x) = f_2(x_1)\delta_t(x_0) \quad .$$

By (3.3.1) and the Feynman-Kac formula,

$$\langle \theta, e^{-tH}\theta \rangle \leq \prod_{i=1}^{2} \langle \varphi(f_i)\Omega, e^{-tH}\varphi(f_i)\Omega \rangle$$

$$+ |\langle \varphi(f_1)\Omega, e^{-tH}\varphi(f_2)\Omega \rangle|^2 \quad .$$

Since $\langle \Omega, \varphi\Omega \rangle = 0$, the spectrum condition yields $|\langle \varphi(f_i)\Omega, e^{-tH}\varphi(f_j)\Omega \rangle| \leq O(1)e^{-tm}$ and therefore

$$\langle \theta, e^{-tH}\theta \rangle \leq O(1)e^{-2tm} \quad ,$$

completing the proof. We remark that only vacuum cluster expansions are necessary to this point.

Next we state a result [Gl Ja Sp 1] which follows from the two particle cluster expansion. We let $\epsilon > 0$, and we let E be the spectral projection for the energy interval $[0, 3m_0 - \epsilon]$ in an even $\lambda P(\varphi)_2$ model. We assume λ/m_0^2 sufficiently small to ensure a decay rate $\gamma = 3m_0 - \epsilon$ in the two particle cluster expansion.

Proposition 3.3.5. With the above assumptions, linear combinations of the vector Ω and $e^{tH}E\theta(f_1, f_2)$ are dense in $E\mathcal{H}_e$. Also the vectors $E\varphi(f)\Omega$ are dense in $E\mathcal{H}_o$.

We remark that in [Gl Ja Sp 1] we prove a weaker result for $E\mathcal{H}_o$, namely that vectors $e^{tH}E\varphi(f)\Omega$ span $E\mathcal{H}_o$. A simple modification of Theorem 4.2, [Gl Ja Sp 1] can be used to bring first degree polynomials in the n-particle cluster expansion to time zero. This yields Proposition 3.3.5, for $n = 2$.

Proof of the Theorems. Suppose that $M \upharpoonright \mathcal{H}_e$ has mass spectrum in the interval $(m, 2m)$. By Lorentz invariance, there is a nonzero vector $\psi \in \mathcal{H}_e$ corresponding to that spectral interval and with energy $< 2m$. By Proposition 3.3.5, ψ is a limit of sums of vectors $e^{tH}E\theta(f_1, f_2)$. By Corollary 3.3.4, $\psi = 0$, proving Theorem 3.3.1 on $\mathcal{H}_e$.

Finally we show $M \upharpoonright E\mathcal{H}_o$ has only one point in its spectrum, namely m. By Proposition 3.3.5, the vectors $\mathcal{K}_0 = \{E\varphi(f)\Omega\}$ span $E\mathcal{H}_o$. We let $\mathcal{K}$ be the closure of the union of Lorentz translates of $\mathcal{K}_0$. Our assertion then follows by Proposition 2.1.2. Theorems 3.3.2 and 3.3.1 then follow by Lorentz invariance.

3.4 On the Presence of Bound States

The ideas of Section 3.2 suggest the presence of bound states in certain $P(\varphi)_2$ models. We give two methods to establish the existence of mass spectrum in the two particle bound state interval $(m, 2m)$. As we mentioned above, there is no physical interpretation of continuous spectrum in this interval, so the existence of spectrum presumably ensures the existence of eigenvalues, i.e., bound states.

Variational Method. The first method is to choose an approximate bound state wave function θ, with the properties: (i) $\|\theta\| \geq 1$; (ii) θ is orthogonal to the vacuum and one particle states; and (iii) $\langle \theta, M\theta\rangle < 2m$. Since $M \leq H$, we may replace the bound on $\langle \theta, M\theta\rangle$ by

$$\langle \theta, H\theta\rangle < 2m \quad . \tag{3.4.1}$$

In a theory with weak coupling, the cluster expansion shows that the low momentum part of the mass interval $(m, 2m)$ is spanned by vectors

$$\theta = \alpha\Omega + a^*(f)\Omega + e^{tH}\, a^*(f_1)\, a^*(f_2)\Omega \quad , \tag{3.4.2}$$

see [Gl Ja Sp 1] and Section 3.3. Here a^* is a time zero creation operator. If, in addition, $P(\varphi)$ is even, we may choose $f = 0$ and $\alpha = -\langle \Omega, a^*(f_1)a^*(f_2)\Omega\rangle$. Alternatively, we can replace a^* by the time zero field φ in (3.4.2).

With this variational method, we eliminate H from $\langle \theta, H\theta\rangle$ by using $H\Omega = 0$ and the canonical commutation relations. For instance,

$$[H, a^*(f)] = a^*(\mu f) + [H_I, a^*(f)] \quad , \qquad \text{where} \quad \mu = (\vec{p}^{\,2} + m_0^2)^{1/2} \quad .$$

If $\chi = a^*(f)\Omega$, then

$$\begin{aligned} \langle \chi, H\chi\rangle &= \langle f, \mu f\rangle_{L_2} + \langle \Omega, a^*(\mu f)\, a(\bar{f})\Omega\rangle \\ &\quad + \langle a^*(f)\Omega, [H_I, a^*(f)]\Omega\rangle \end{aligned} \tag{3.4.3}$$

We estimate vacuum expectation values of Wick ordered monomials

$$W = \int a^*(x_1) \cdots a(x_n) w(x)\, dx \quad ,$$

by the cluster expansion [Gl Ja Sp 1]. In fact, before estimation, we expand $\langle \Omega, W\Omega \rangle$ using integration by parts, to isolate low order dependence in the coupling λ, see Chapter 4. For instance, in second order, we obtain a second order mass-shift correction to $\langle f, \mu f \rangle_{L_2}$.

In this manner, we need not calculate the physical mass m exactly, but we can obtain explicitly the relevant low order corrections to m_0. (Here we assume that m is asymptotic to m_0.) Furthermore, let us assume that f is scaled to give momentum localization $O(\delta)$, namely $f(\vec{p}) = \delta^{-1/2} h(\vec{p}/\delta)$. (In Section 3.1 we explained that $\delta \to 0$ as $\lambda \to 0$.) Then

$$\langle f, \mu f \rangle = m_0 \|f\|^2 + O(\delta^2) \quad ,$$

which exhibits the momentum dispersion about $\vec{p} = 0$ of the single particle state. Similarly, the second order mass correction will equal $m_2 \lambda^2 \|f\|^2 + O(\lambda^2 \delta^2)$.

We sketch our proof for the $\lambda(\varphi^6 - \varphi^4)$ interaction. We take

$$\theta = a^*(f)^2 \Omega - \langle \Omega, a^*(f)^2 \Omega \rangle \Omega \quad , \qquad \text{with } \|f\|_{L_2} = 2^{-1/4}$$

which satisfies (i), (ii) above. We study

$$\langle \theta, H\theta \rangle = \langle \theta, \{a^*(\mu f) a^*(f) + a^*(f) a^*(\mu f)\}\Omega \rangle + \left\langle \theta, \left[\lambda \int \varphi^6 - \lambda \int \varphi^4, a^*(f)^2\right]\Omega \right\rangle \quad ,$$

and integrate by parts. We isolate, in closed form, all terms of degree 0, 1 or 2 in λ. The mass terms have the form

$$2\{m_0 + \lambda^2 m_2 + O(\delta^2 + \lambda^3 + \delta^2 \lambda^2)\} \quad .$$

The attractive contribution from diagrams of the form

lowers the energy by $-O(\delta\lambda)$. Other contributions are $O(\lambda^2\delta)$ or higher order. We choose $\delta = \lambda^{1+\epsilon}$. Then for small λ, the decrease in energy $-O(\delta\lambda) = -O(\lambda^{2+\epsilon})$ dominates the dispersive effect $O(\delta^2) = O(\lambda^{2+2\epsilon})$ and the repulsive effects $O(\lambda^2\delta + \lambda^3) \le O(\lambda^3)$. The operator parts of these estimates result from a variant of the cluster expansion. This completes our sketch of the proof that bound state spectrum exists in the weakly coupled $\lambda(\varphi^6 - \varphi^4)$ model.

Similar arguments should hold for $\lambda\varphi^6$. In this case, however, the attraction is $O(\lambda^2\delta)$. We must therefore isolate the fourth order mass shift and we set $\delta = \lambda^{2+\epsilon}$. For the interaction $\lambda\varphi^3 + \lambda^6\varphi^4$, we must orthogonalize θ to the one particle states (at least to third order in λ). We would then isolate the fourth order mass renormalization and take $\delta = \lambda^{2+\epsilon}$. We thank B. Simon for observing that an even theory is technically simpler.

Cluster Method. In an even $P(\varphi)_2$ model, for θ of the form (3.3.2),

$$\langle \theta, e^{-tH}\theta\rangle = \langle \Phi(g_1)\Phi(g_2)\Phi(g_3)\Phi(g_4)\rangle_C + \langle \Phi(g_1)\Phi(g_3)\rangle\langle \Phi(g_2)\Phi(g_4)\rangle + \langle \Phi(g_1)\Phi(g_4)\rangle\langle \Phi(g_2)\Phi(g_3)\rangle \; .$$

where $\langle\cdot\rangle_C$ denotes the connected (truncated) part. Thus $\langle\theta, e^{-tH}\theta\rangle$ exhibits the two particle decay $O(e^{-2mt})$, unless

$$\langle \Phi(g_1)\cdots\Phi(g_4)\rangle_C \ge O(e^{-2(m-\epsilon)t}) \; . \tag{3.4.4}$$

Using the Bethe-Salpeter equation, we can isolate in $\lambda(\varphi^6 - \varphi^4)$ models a slowly decaying part of $\langle\Phi(g_1)\cdots\Phi(g_4)\rangle_C$, given by (positive) φ^4 contributions. We propose using cluster expansions to estimate the errors. The inequality (3.4.4) would establish the existence of mass spectrum on $\mathcal{H}_e$ in the interval $(0, 2m-\epsilon]$. This proposed calculation appears interesting. However, unlike the variational proof above, we presently have no error estimates using this method. Conversely, we remark that the existence of two-particle bound state spectrum in a weakly coupled even $P(\varphi)_2$ model (as established by the variational method) ensures (3.4.4).

4. PHASE SPACE LOCALIZATION AND RENORMALIZATION

4.1 Results for φ_3^4

In a series of related papers, we have given convergent expansions [Gl Ja Sp 1, 2] and convergent upper bounds [Gl Ja IV, 8] for quantum field models. These expansions and bounds deal with the problem of removing cutoffs $\varkappa$, Λ, namely in taking infinite volume limits in phase space. Most of this conference has dealt with the $\Lambda \to R^2$ limit. However the $\varkappa \to \infty$ limit in Y_2 and in higher dimensional models presents the most challenging problems, for both physics and for mathematics; we hope these ultraviolet problems will be the focus of increasing attention in constructive field theory. In this section we describe the results for φ_3^4: Let $d\Phi_C$ be the Gaussian measure with covariance C, and let $d\Phi$ denote the choice $C = (-\Delta + m_0^2)^{-1}$.

Let $V(\Lambda,\varkappa)$ denote the Euclidean action, the sum of the φ^4 interaction V_I and the counterterms V_C. Then

$$V_I = \lambda \int_{\Lambda \subset R^3} :\Phi_\varkappa^4: \, dx$$

and V_C are the Green's function counterterms given in second and third order perturbation theory. The partition function for the action $V = V_I + V_C$, namely

$$Z(\Lambda,\varkappa) = \int e^{-V(\Lambda,\varkappa)} \, d\Phi \quad ,$$

contains the ultraviolet divergent counterterms.

Theorem 4.1.1 [Gl Ja 8]. For $0 \le \lambda$

(4.1.1) $$Z(\Lambda,\varkappa) \le e^{\alpha|\Lambda|} \quad ,$$

uniformly in $\varkappa$. For λ bounded, (4.1.1) is uniform in λ also.

We now let $H(\mathcal{V})$ denote the renormalized φ_3^4 Hamiltonian, defined formally by

$$H(\mathcal{V}) = H_0 + \lambda \int_{\mathcal{V}\subset R^2} :\varphi^4: \, d\vec{x} - \frac{1}{2}\delta m_2^2 \int :\varphi^2: \, d\vec{x} - E_2 - E_3 \ .$$

Here δm_2^2, E_2 and E_3 are the Hamiltonian counterterms in second and third order perturbation theory. (These counterterms differ by a constant and a transient from the Green's function counterterms, see [Gl Ja 8].)

Corollary 4.1.2. The Hamiltonian $H(\mathcal{V})$ is bounded from below by a constant proportional to the volume $|\mathcal{V}|$,

(4.1.2) $$0 \le H(\mathcal{V}) + O(|\mathcal{V}|) \ .$$

The corollary follows from the theorem and the fact that $\langle \Omega(\mathcal{V},\kappa), \Omega_0\rangle \neq 0$. In fact

$$\langle \Omega_0, e^{-tH(\mathcal{V},\kappa)}\Omega_0\rangle = e^{-tE(\mathcal{V},\kappa) - A(\mathcal{V},\kappa) + T(\mathcal{V},\kappa,t)} \ ,$$

where $E(\mathcal{V},\kappa)$ is the partially renormalized vacuum energy, vanishing in second and third order, and convergent as $\kappa\to\infty$ for fixed volume. The constant A in $e^{-A(\mathcal{V},\kappa)} = |\langle \Omega_0, \Omega(\mathcal{V},\kappa)\rangle|^2$ is the logarithmically divergent wave function renormalization constant, independent of t. Also $T(\mathcal{V},\kappa,t)$ is a transient that is bounded as $\kappa\to\infty$ and $o(1)$ as $t\to\infty$.

As $|\mathcal{V}|\to\infty$, the constants $E(\mathcal{V},\kappa)$, $A(\mathcal{V},\kappa)$ and $T(\mathcal{V},\kappa,t)$ diverge. The second order, i.e., the ultraviolet divergent, part of $A(\mathcal{V},\kappa)$ has been cancelled in $Z(\Lambda,\kappa)$.

These results have been extended by Joel Feldman [Fel 2], who proved

Theorem 4.1.3. The finite volume partition function $Z(\Lambda,\kappa)$ and

(4.1.3) $$Z(\Lambda,\kappa)S(\Lambda,\kappa;f_1,\dots,f_n) = \int \Phi(f_1)\dots\Phi(f_n)\, e^{-V(\Lambda,\kappa)}\, d\Phi$$

converge as $\kappa\to\infty$. The limits are continuous in λ and satisfy

(4.1.4) $$|Z(\Lambda)S(\Lambda;f_1,\cdots,f_n)| \le n!\Big(\prod_i \|f_i\| \, e^{O(|\Lambda|)}\Big)$$

for a Schwartz space norm $\|\cdot\|$.

From continuity in λ and $Z(\Lambda) = 1$ for $\lambda = 0$, we conclude that for Λ fixed and λ sufficiently small,

$$Z(\Lambda) > 1/2 \quad .$$

Thus for Λ fixed and λ small, the approximate Schwinger functions $S(\Lambda, f_1, \ldots, f_n)$ do not vanish identically and

$$|S(\Lambda;\, f_1, \ldots, f_n)| \le n! \prod_i \|f_i\| . \tag{4.1.5}$$

<u>Corollary</u> 4.1.4 [Fe 2]. For small λ, volume Λ Schwinger functions are the moments of a unique measure on $\mathcal{S}'(R^3)$, namely

$$dq = \lim_{\varkappa \to \infty} Z(\Lambda, \varkappa)^{-1} e^{-V(\Lambda, \varkappa)} d\Phi$$

$$= \lim_{\varkappa \to \infty} dq_{\Lambda, \varkappa} \quad .$$

The corollary is based on a study of the perturbation of Z in an external Euclidean field, namely on the study of the generating functional

$$Z(h) = \int e^{\Phi(h)} dq_\Lambda$$

for the (disconnected) Schwinger functions. This functional was studied in $P(\varphi)_2$ by Fröhlich [Fr 2]; see also [Gl Ja 13].

Of course, we are interested in the $\Lambda \to R^3$ limit of these Schwinger functions $S(\Lambda; \cdot)$ and of measures dq_Λ, in order to obtain the full relativistic theory. We conjecture that the Kirkwood-Salsburg equations of Part II can be generalized to φ_3^4 and yield the limit. In fact the local estimates of Theorem 4.1.3 and Corollary 4.1.4 are exactly the type of local estimates which the cluster expansion for $P(\varphi)_2$ uses as input. We conjecture that $Z(\Lambda) \ge \exp(-O|\Lambda|)$ for small λ. We expect that such estimates lead to the Wightman axioms for φ_3^4.

4.2. Elementary Expansion Steps

The proof of the estimates for $P(\varphi)_2$, as well as those for φ_3^4, results from the use of four elementary identities and bounds concerning the non-Gaussian measure

$$e^{-V(\Lambda,\varkappa)}\, d\Phi_C \quad . \tag{4.2.1}$$

The four steps are

I. Change of covariance C.

II. Change of exponent V.

III. Wick ordering bound.

IV. Integration by parts.

The four steps are combined to yield expansions or bounds. The difficult part of the construction is generally the question of how to combine these steps to isolate the desired property of the model, at the same time to ensure convergence. We use three expansion techniques:

a) Explicit expansions. We prescribe definite elementary steps to yield an expansion, as the expansion for ZS in Part II.

b) Neumann series. The Kirkwood-Salsburg equations of Part II, obtained by explicit expansion of Z, yield a Neumann series $(I - \mathcal{K})^{-1} = \Sigma_n \mathcal{K}^n$ for their solution.

c) Inductively defined expansions. We prescribe for each possible term (integral) in our expansion, rules to expand it into a sum of similar terms.

There is considerable freedom in the definition of our expansions and bounds. The inductively defined expansions leave the widest lattitude of choice, since they are not tied to recovering an expansion expressible in closed form, or to obtaining the inverse of an operator. In addition, the inductive expansions and bounds yield the most detailed information about our models, including the positivity of φ_3^4 [Gl Ja 8] and the φ-bounds for all couplings [Gl Ja IV]. These expansions and bounds are not tied to the use of particular boundary conditions on the covariance

C, but have more general validity.

We now give the elementary steps; the first two steps are merely the fundamental theorem of calculus:

I. Change of Covariance. Let $C_\alpha = \alpha C_1 + (1-\alpha)C_0$ be a family of interpolating covariances, and let

$$d\Phi_{C_1} = d\Phi_{C_0} + \int_0^1 d\alpha \frac{d}{d\alpha} d\Phi_{C_\alpha}$$

$$= d\Phi_{C_0} + \frac{1}{2}(C_1 - C_0)\cdot \Delta_\Phi \int_0^1 d\alpha \, d\Phi_{C_\alpha} \quad .$$

This formula has been used to deal with the infinite volume limit, see Part II. It is established by integration by parts on function space [Di Gl]; see also the proof of IV below. We do not use Step I in this chapter.

II. Change of V. Let $V_\alpha = \alpha \in [0,1]$ be a differentiable family of interpolating Euclidean actions. Then

$$e^{-V_1} = e^{-V_0} + \int_0^1 \frac{d}{d\alpha} e^{-V_\alpha} d\alpha$$

(4.2.2)

$$= e^{-V_0} - \int_0^1 \frac{dV_\alpha}{d\alpha} e^{-V_\alpha} d\alpha \quad .$$

We use this identity to lower an upper momentum cutoff in the action V, in the positivity proofs for $P(\varphi)_2$ [See Gl Ja 7, IV] and φ_3^4 [Gl Ja 8], and we call this formula the perturbation or Duhamel identity.

Iterating (4.2.2) leads to the unrenormalized perturbation series. With an ultraviolet cutoff, this series diverges because of the $O(n!^2)$ diagrams arising in n^{th} order. For example, with one degree of freedom,

$$\int e^{-q^2-\lambda q^4} dq \neq \sum_{n=0}^{\infty} \frac{(-\lambda)^n}{n!} \int e^{-q^2} q^{4n} dq \quad ,$$

since the series on the right side diverges. It is therefore necessary to truncate perturbation theory, for which we use step III below.

III. Wick Bound. For $:\Phi_\kappa^4:$, we have

$$e^{-V(\Lambda,\kappa)} \le \begin{cases} e^{O(\log^2 \kappa)|\Lambda|} & d = 2 \\ e^{O(\kappa^2)|\Lambda|} & d = 3 \end{cases} \tag{4.2.3}$$

This bound follows by integrating

$$:\Phi_\kappa^4: \;=\; (\Phi_\kappa - 3c_\kappa)^2 - 6c_\kappa^2 \;\ge\; -6c_\kappa^2$$

over the space time volume Λ. Here

$$c_\kappa = \int \Phi_\kappa(x)^2 \, d\Phi = C_\kappa(x,x) \le \begin{cases} O(\log \kappa) & d = 2 \\ O(\kappa) & d = 3 \end{cases}$$

This Wick bound is used to raise the lower momentum cutoff ρ in the exponent $V(\Lambda,\kappa,\rho)$. Our expansions terminate if $\kappa = \rho$.

IV. Integration by Parts.

$$\int \Phi(x)\, F(\Phi)\, d\Phi_C = \int dy \int C(x,y) \frac{\partial F}{\partial \Phi(y)}\, d\Phi_C \; . \tag{4.2.4}$$

We use this integration by parts formula to exhibit the cancellation of the divergent renormalization counterterms in $V_1 - V_0$ of (4.2.2). In [Gl Ja IV, 8] we use other forms of (4.2.4), called there the pull through and contraction formulas.

It is easy to establish (4.2.4) by studying finite dimensional approximations to the function space integral. We choose a Gaussian measure $d\Phi_N$ converging to $d\Phi_C$,

$$d\Phi_N = N \exp\left(-\tfrac{1}{2} \sum_{i,j} q_i C_{ij}^{-1} q_j\right)\left(\prod_k dq_k\right)$$

$$\equiv \nu_N dq_N \; ,$$

where C_{ij} is the covariance matrix and N is a normalization constant. Ordinary integration by parts then yields

$$\int \sum_j C_{ij}^{-1} q_j F(q)\, d\Phi_N = -\int F(q) \frac{\partial \nu_N}{\partial q_i}\, dq_N = \int \frac{\partial F(q)}{\partial q_i}\, d\Phi_N .$$

Inverting C,

$$\int q_j F(q)\, d\Phi_N = \int \sum_i C_{ji} \frac{\partial F}{\partial q_i}\, d\Phi_N ,$$

which converges to (4.2.4) as $d\Phi_N \to d\Phi_C$.

For Wick ordered monomials, we obtain similarly

$$\int :\Phi(x_1)\dots\Phi(x_n): F(\Phi)\, d\Phi_C = \int :\Phi(x_2)\dots\Phi(x_n): \int dx_1' \, C(x_1, x_1') \frac{\partial F}{\partial\Phi(x_1')}\, d\Phi_C .$$

As an example, we integrate by parts one $\Phi(x)$ factor in a simple expression,

$$\int :\Phi^4(x): e^{-\int :\Phi^4: dz}\, d\Phi_C = -4\int dx dy \int :\Phi^3(x): C(x-y) :\Phi^3(y): e^{-\int :\Phi^4: dz}\, d\Phi_C .$$

Further integration by parts yields

$$\int :\Phi^4(x): e^{-\int :\Phi^4: dz}\, d\Phi_C = 4! \int dx\, dy\, C(x-y)^4 \int e^{-\int :\Phi^4: dz}\, d\Phi_C \tag{4.2.5}$$

$$+ \text{ other terms .}$$

4.3. Synthesis of the Elementary Steps

We have two basic aims in combining the elementary expansion steps. First, we desire convergent expansions in a given space-time or phase space volume. Second, we desire polynomial decoupling of different localization regions. In Part II we present the vacuum cluster expansion for $P(\varphi)_2$ models in full detail. In that case, momentum localization is unnecessary (no ultraviolet divergences occur) and our localization regions are unions of unit lattice squares in space time. With no cutoff, distant regions decouple exponentially, and the decay rate determines the physical mass. In this section we present the basic ideas of phase space localization which we used to deal with the ultraviolet divergent φ_3^4 model, and which yield the results summarized in Section 4.1. For smooth cutoffs in momentum space, we obtain polynomial decoupling.

For simplicity we discuss the partition function Z. We fix the volume Λ and investigate how Z depends on the ultraviolet cutoff κ. In order to truncate the perturbation expansions, we introduce a lower cutoff ρ into the action V. To ensure that $V(\kappa, \rho)$ is bounded from below, we introduce the momentum cutoffs in a symmetric fashion: each momentum component in $V(\kappa, \rho)$ lies in the interval $[\rho, \kappa]$.

We perform our expansions on integrals of the form

$$\int R(\Phi) e^{-V(\kappa, \rho)} \, d\Phi \quad , \tag{4.3.1}$$

where R is a polynomial function of Φ. Each expansion step replaces (4.3.1) by a sum of similar terms. We use a high momentum (perturbation) expansion step to lower κ, and we use a low momentum (truncation) expansion step to raise ρ. At the start of the expansion, $\rho = 0$ and $\kappa = \kappa_0$. The expansion terminates when $\rho = \kappa$, i.e., $V(\kappa, \rho) = 0$, and (4.3.1) is reduced to a sum of Gaussian integrals $\int R \, d\Phi$. We estimate this sum uniformly in κ_0.

The rules for alternating the expansion steps are somewhat complicated. The main idea is to obtain a small contribution from each high energy vertex in R, by performing explicit renormalization cancellations. We avoid the $(n!)^r$ number of terms which would arise from iterating (4.2.2), by truncating the perturbation expansion.

The high momentum expansion. We use Step II to lower κ, taking $V_1 = V(\bar{\kappa}, \rho)$, $V_0 = V(\underline{\kappa}, \rho)$. The first term in (4.2.2) has the desired form. The second term has the same upper cutoff and a new vertex $\delta V = dV_\alpha/d\alpha$ in R. Since δV has a lower momentum cutoff at $\underline{\kappa}$, we desire that δV contributes a convergence factor $\underline{\kappa}^{-\epsilon}$ to our final estimates. We obtain the proof of this fact only after performing the renormalization cancellation of the divergent counterterms δV_C in δV. Using Step IV, we integrate by parts δV_I, namely the ϕ^4 part of δV. We also integrate any new V_I produced in R as a result of differentiating the exponent. We thus obtain in closed form (and we cancel) the ultraviolet divergent part of δV. For instance, in (4.2.5), we displayed the second order vacuum energy contribution. The third order vacuum and the mass counterterms occur among the "other terms" in (4.2.5). The vacuum energy contributions cancel exactly with the corresponding counterterm in δV_C. The mass renormalization diagram, after cancellation, leaves a remainder $O(\underline{\kappa}^{-\epsilon})$. The remaining "other terms" from this procedure are convergent and so contribute $\underline{\kappa}^{-\epsilon}$ to the final estimate.

In this manner, Steps II and IV combine to yield one order (i.e., one δV) in a renormalized perturbation expansion. Because of the large number of terms, it is necessary to truncate this expansion after introducing $\bar{\kappa}^{\delta}$ vertices δV.

The low momentum expansion. We truncate the perturbation series by raising the lower cutoff in the exponent. We use Step III to raise ρ from $\underline{\rho}$ to $\bar{\rho}$. The Wick bound is an L_∞ estimate on path space, so we expect to apply it in space-time cubes Δ on which (4.2.3) remains bounded, i.e., cubes for which

$$\bar{\rho}^2 |\Delta| \le O(1) \quad .$$

This restriction means that the localization length $L = |\Delta|^{1/3}$ satisfies

(4. 3. 2) $$L \le O(\bar{\rho}^{-2/3}) \quad ,$$

and defines our phase space localization. On the other hand, the uncertainty principle requires that

(4. 3. 3) $$O(\underline{\rho}^{-1}) \le L$$

for the localization to be proper, i. e., for the spreading $O(\underline{\rho}^{-1})$ of the wave packet (due to momentum localization) to be less than L. We note that (4. 3. 2) - (4. 3. 3) are compatible. This compatibility is actually another aspect of super-renormalizability. For the φ_4^4 model, (4. 3. 2) would be replaced by $L \le O(\bar{\rho}^{-1})$, for which our estimates are borderline.

Our analysis has shown that we must treat separately cubes Δ belonging to a space-time cover $\mathcal{D}$. (Also some Δ's tend to zero as $\varkappa_0 \to \infty$.) Thus we actually deal with upper and lower cutoff functions $\varkappa(\Delta)$, $\rho(\Delta)$ which are functions of Δ.

Furthermore, we remark that the Wick bound (4. 2. 3) deals only with the pure low momentum part of $\delta V = V(\varkappa, \underline{\rho}) - V(\varkappa, \bar{\rho})$. This low momentum part equals $V(\bar{\rho}, \underline{\rho})$, i. e., all momenta are less than $\bar{\rho}$. The cross terms in δV must also be removed in order to raise the lower cutoff in the exponent. We dominate the cross terms by the low momentum terms and by $V(\varkappa, \bar{\rho})$, the new exponent. This procedure has some complications, but poses no essential difficulty, see [Gl Ja 8]. In φ_4^4, however, it is such cross terms which yield the charge renormalization divergences, our biggest challenge.

<u>Independence of Phase Cells.</u> Finally we remark that the relevant distance parameter that we must use with phase space localization is the scaled distance d,

$$d = \text{Euclidean distance} \times \text{lower momentum cutoff}.$$

For smooth momentum cutoffs, scaling standard estimates gives $O(d^{-n})$ decay of correlations between different phase cells with proper localization. Any such decay for $n > 3$ is sufficient to control distance factors in sums over phase cells (whose diameter goes to zero as $\varkappa_0 \to \infty$). We remark that in the limiting theory without ultraviolet cutoff, we expect to recover exponential decoupling and a mass gap.

REFERENCES

[AHK 1] S. Albeverio and R. Hoegh-Krohn, Uniqueness of the physical vacuum and the Wightman functions in the infinite volume limit for some non-polynomial interactions, Commun. Math. Phys. 30, 171-200 (1973).

[AHK 2] ______, The scattering matrix for some non-polynomial interactions, I and II, Helv. Phys. Acta, to appear.

[AHK 3] ______, The Wightman axioms and the mass gap for strong interactions of exponential type in two dimensional space-time, Oslo reprint.

[AHK 4] ______, Asymptotic series for the scattering operator and asymptotic unitarity of the space cut-off interactions, Oslo preprint.

[Ba] R. Baumel, Private communication, see also [GuRoSi 3].

[BoGr] A. Bortz and R. Griffiths, Phase transitions in anisotropic classical Heisenberg ferromagnets, Commum. Math. Phys. 26, 102-108 (1972).

[Br 1] O. Bratelli, Conservation of estimates in quantum field theory, Commun. Math. Phys., to appear.

[Br 2] ______, Local norm convergence of states on the zero time boson fields, Courant preprint.

[Ca] J. Cannon, Continuous sample paths in quantum field theory, Rockefeller preprint.

[CaJa] J. Cannon and A. Jaffe, Lorentz covariance of the $(\varphi^4)_2$ quantum field theory, Commun. Math. Phys. 17, 261-321 (1970).

[Co 1] S. Coleman, Dilitations, Lectures given at the 1971 International Summer School of Physics "Ettore Majorana," to appear.

[Co 2] ______, There are no Goldstone bosons in two dimensions, Commun. Math. Phys. 31, 259-264 (1973).

[CoWe] S. Coleman and E. Weinberg, Radiative corrections as the origin of spontaneous symmetry breaking, Phys. Rev. D 7, 1888-1910, (1973).

[ConTh] F. Constantinescu and W. Thalheimer, Euclidean Green's functions for Jaffe fields, Frankfurt preprint.

[De Ma] C. De Dominicis and P. Martin, Stationary entropy principle and renormalization in normal and superfluid systems, I, II. Jour. Math. Phys. 5, 14-30, 31-59 (1964).

[Di 1] J. Dimock, Estimates, renormalized currents and field equations for the Yukawa_2 field theory, Ann. Phys. 72, 177-242 (1972).

[Di 2] ______, Spectrum of local Hamiltonians in the Yukawa_2 field theory, J. Math. Phys. 13, 477-481 (1972).

[Di 3] ______, Perturbation series asymptotic to Schwinger functions in $P(\varphi)_2$, Courant preprint.

[DiGl] J. Dimock and J. Glimm, Measures on the Schwartz distribution space and application to $P(\varphi)_2$ field theories, Courant preprint.

[Dob 1] R. L. Dobrushin, Gibbsian random fields for lattice systems with pairwise interactions, Funct. Anal. and Appl., 2, 292-301 (1968).

[Dob 2] ______, The problem of uniqueness of a Gibbsian random field and the problem of phase transitions, Funct. Anal. and Appl., 2, 302-312 (1968).

[Dob 3] ______, The description of the random field by its conditional distributions and its regularity conditions, Theor. Prob. and Appl., 13, 201-229 (1968).

[DobMin] R. Dobrushin and R. Minlos, Construction of a one dimensional quantum field via a continuous Markov field, Moscow preprint.

[DHR] S. Doplicher, R. Haag and J. Roberts, Fields, observables and gauge transformations II, Commun. Math. Phys., 15, 173-200 (1969).

[Ec] J.-P. Eckmann, Representations of the CCR in the $(\varphi^4)_3$ model: independence of space cutoff, Commun. Math. Phys. 25, 1-61 (1972).

[EcOs] J.-P. Eckmann and K. Osterwalder, On the uniqueness of the Hamiltonian and of the representation of the CCR for the quartic boson interaction in three dimensions, Helv. Phys. Acta 44, 884-909 (1971).

[FaPo] L. Faddeev and V. Popov, Feynman diagrams for the Yang-Mills field, Phys. Letters 25B, 29-30 (1967).

[Fed] P. Federbrush, Positivity for some generalized Yukawa models in one space dimension, Michigan reprint.

[Fe 1] J. Feldman, A relativistic Feynman-Kac formula, Nuclear Physics B 52, 608-614 (1973).

[Fe 2] ______, The $\lambda\varphi_3^4$ field theory in a finite volume, Harvard preprint.

[FiWi] M. Fisher and K. Wilson, Critical exponents in 3.99 dimensions, Phys Rev. Lett. 28, 240-243 (1972).

[Fr 1a] J. Fröhlich, On the infrared problem in a model of scalar electrons and massless, scalar bosons, Annales de l'Institut Henri Poincare, to appear.

[Fr 1b] ______, Existence of dressed one electron states in a class of persistent models, Fortschritte der Physik, to appear.

[Fr 2] ______, Schwinger functions and their generating functionals, Harvard preprint.

[FoKaGi] C. Fortuin, P. Kastelyn and J. Ginibre, Correlation inequalities on some partially ordered sets, Commun. Math. Phys. 22, 89-103 (1971).

[Gi] J. Ginibre, On some recent work of Dobrushin, in: Systemes a un nombre infini de degrés de liberté, ed. by L. Michel and D. Ruelle, CNRS, Paris, 1970.

[Gl 1] J. Glimm, Boson fields with nonlinear self-interaction in two dimensions, Commun. Math. Phys. 8, 12-25 (1968).

[Gl 2] ______, Yukawa coupling of quantum fields in two dimensions, I, Commun. Math. Phys. 5, 343-386 (1967).

[Gl 3] ______, Yukawa coupling of quantum fields in two dimensions, II, Commun. Math. Phys. 6, 120-127 (1967).

[Gl 4] ______, Boson fields with the $:\varphi^4:$ interaction in three dimensions, Commun. Math. Phys. 10, 1-47 (1968).

[GlJa I] J. Glimm and A. Jaffe, A $\lambda(\varphi^4)_2$ quantum field theory without cutoffs I, Phys. Rev., 176, 1945-1951 (1968).

[GlJa II] ______, The $\lambda(\varphi^4)_2$ quantum field theory without cutoffs II, The field operators and the approximate vacuum, Ann. Math. 91, 362-401 (1970).

[GlJa III] ______, The $\lambda(\varphi^4)_2$ quantum field theory without cutoffs III, The physical vacuum, Acta Math. 125, 203-261 (1970).

[GlJa IV] ______, The $\lambda(\varphi^4)_2$ quantum field theory without cutoffs IV, Perturbations of the Hamiltonian, J. Math. Phys. 13, 1558-1584 (1972).

[GlJa 5] ______, The energy momentum spectrum and vacuum expectation values in quantum field theory II, Commun. Math. Phys. 22, 1-22 (1971).

[GlJa 6] ______, What is renormalization? in Symposium on Partial Differential Equations, Berkeley 1971, American Mathematical Society, Providence, 1973.

[GlJa 7] ______, Positivity and self-adjointness of the $P(\varphi)_2$ Hamiltonian, Commun. Math. Phys. 22, 253-258 (1971).

[GlJa 8] ______, Positivity of the φ_3^4 Hamiltonian, Fortschritte der Physik 21, 327-376 (1973).

[GlJa 9] ______, Self-adjointness of the Yukawa_2 Hamiltonian, Ann. of Phys. 60, 321-383 (1970).

[GlJa 10] ______, The Yukawa_2 quantum field theory without cutoffs, J. Funct. Analysis 7, 323-357 (1971).

[GlJa 11] ______, Quantum field models, in: Statistical mechanics and quantum field theory, ed. by C. de Witt and R. Stora, Gordon and Breach, New York, 1971.

[GlJa 12] ______, Boson quantum field models, in: Mathematics of contemporary physics, ed. by R. Streater, Academic Press, New York, 1972.

[GlJa 13] ______, Entropy principle for vertex functions in quantum field models.

[GlJaSp 1] J. Glimm, A. Jaffe and T. Spencer, The Wightman axioms and particle structure in the $P(\varphi)_2$ quantum field model, Ann. Math., to appear.

[GlJaSp 2] ______, The particle structure of the weakly coupled $P(\varphi)_2$ models and other applications of high temperature expansions, Part II: The cluster expansion, these Erice lectures.

[Go] J. Goldstone, Field theories with "superconductor" solutions, Nuovo Cimento 19, 154-164 (1961).

[GoSaWe] J. Goldstone, A. Salam and S. Weinberg, Broken symmetries, Phys. Rev. 127, 965-970 (1962).

[Gr 1] R. Griffiths, Phase transitions, in: Statistical mechanics and quantum field theory, Les Houches 1970, ed. by C. De Witt and R. Stora, Gordon and Breach, New York, 1971.

[Gr 2] ______, Rigorous results for Ising ferromagnets for arbitrary spin, J. Math. Phys., 10, 1559-1565 (1969).

[GrHuSh] R. Griffiths, C. Hurst and S. Sherman, Concavity of magnetization of an Ising ferromagnet in a positive external field, J. Math. Phys. 11, 790-795 (1970).

[GrSi] R. Griffiths and B. Simon, The $(\varphi^4)_2$ field theory as a classical Ising model, Commun. Math. Phys. to appear.

[Gro 1] L. Gross, The relativistic polar without cutoffs, Commun. Math. Phys. 31, 25-74 (1973).

[Gro 2] ______, Existence and uniqueness of physical ground states, J. Funct. Anal. 10 52-109 (1972).

[Gro 3] ______, Logarithmic Sobolev inequalities, Cornell preprint.

[Gro 4] ______, Analytic vectors for representations of the canonical commutation relations and non-degeneracy of ground states, Cornell preprint.

[Gu 1] F. Guerra, Uniqueness of the vacuum energy density and Van Hove phenomenon in the infinite volume limit for two dimensional self-coupled Bose fields, Phys. Rev. Letts. 28, 1213 (1972).

[GuRu] F. Guerra and P. Ruggiero, A new interpretation of the Euclidean-Markov field in the framework of physical Minkowski space-time, Salerno preprint.

[GuRoSi 1] F. Guerra, L. Rosen and B. Simon, Nelson's symmetry and the infinite volume behaviour of the vacuum in $P(\varphi)_2$, Commun. Math. Phys. 27, 10-22 (1972).

[GuRoSi 2] ______, The vacuum energy for $P(\varphi)_2$: infinite volume limit and coupling constant dependance, Commun. Math. Phys. 29, 233-247 (1973).

[GuRoSi 3] ______, The $P(\varphi)_2$ Euclidean quantum field theory as classical statistical mechanics, Ann. Math., to appear.

[He 1] K. Hepp, Théorie de la renormalisation, Springer-Verlag, Heidelberg 1969 .

[He 2] ______, Renormalization theory, in Statistical mechanics and quantum field theory, ed. by C. De Witt and R. Stora, Gordon and Breach, New York, 1971.

[He 3] ______, Erice lectures, 1973.

[HKSi] R. Hoegh-Krohn and B. Simon, Hypercontractive semi-groups and two dimensional self-coupled Bose fields, J. Funct. Anal. 9, 121-180 (1972).

[Ja 1] A. Jaffe, The dynamics of a cutoff $\lambda\varphi^4$ field theory, Princeton University Thesis.

[Ja 2] ______, Existence theorems for a cutoff $\lambda\varphi^4$ field theory, in: Mathematical theory of elementary particles, ed. by R. Goodman and I.Segal, M.I.T. Press, 1966.

[JaMcB] A. Jaffe and O. McBryan, What constructive quantum field theory has to say about currents, in: Proceedings of a conference on currents, Princeton,1971.

[JoL] G. Jona-Lasinio, Relativistic field theories with symmetry-breaking solutions, Nuovo Cimento Letters 34, 1790-1795 (1964).

[KaRoSw] D. Kastler, D. Robinson and A. Swieca, Conserved currents and associated symmetries; Goldstone's theorem. Commun. Math. Phys. 2, 108-120 (1966).

[KoWi] J. Kogut and K. Wilson, The renormalization group and the ϵ expansion, Phys. Reports, to appear.

[Leb 1] J. Lebowitz, Bounds on the correlations and analyticity properties of ferromagnetic Ising spin systems, Commun. Math. Phys. 28, 313-321 (1972).

[Leb 2] ______, GHS and other inequalities, Yeshiva preprint.

[Leb Pe] J. Lebowitz and O. Penrose, Decay of correlations, Yeshiva preprint.

[ML] A. Martin-Löf, Mixing properties, differentiability of the free energy and the central limit theorem for a pure phase in the Ising model at low temperature, Commun. Math. Phys. 32, 75-92 (1973).

[McB 1] O. McBryan, The vector currents in the Yukawa_2 quantum field theory with SU_3 symmetry, Harvard thesis.

[McB 2] ______, Generators for the Lorentz group in the $P(\varphi)_2$ theory, Toronto preprint.

[MinSin 1] R.A. Minlos and Y.G. Sinai, The phenomenon of "phase separation" at low temperatures in some lattice models of a gas I, Math Sbornik, Tom 73 (115), 335-395 (1967).

[MinSin 2] ______, The phenomenon of "phase separation" at low temperatures in some lattice models of a gas II, Trans. Moscow Math. Soc., 19, 121-196 (1968).

[Nel 1] E. Nelson, A quartic interaction in two dimensions, in: Mathematical theory of elementary particles, ed. by R. Goodman and I. Segal, M.I.T. Press, 1966.

[Nel 2] ______, Quantum fields and Markoff fields, in: Proceedings of Summer Institute of Partial Differential Equations, Berkeley 1971, Amer. Math. Soc., Providence, 1973.

[Nel 3] ______, Construction of quantum fields from Markoff fields, J. Funct. Anal. 12, 97-112 (1973).

[Nel 4] ______, The free Markoff field, J. Funct. Anal. 12, 211-227 (1973).

[Nel 5] ______, These Erice notes.

[New 1] C. Newman, The construction of stationary two-dimensional Markoff fields with applications to quantum field theory, J. Funct. Anal., to appear.

[New 2] ______, Zeroes of the partition function for generalized Ising systems, Courant preprint.

[OsSch 1] K. Osterwalder and R. Schrader, On the uniqueness of the energy density in the infinite volume limit for quantum field models, Helv. Phys. Acta, 45, 746-754 (1972).

[OsSch 2] ______, Feynman-Kac formula for Euclidean Fermi and Bose fields, Phys. Rev. Letts. 29, 1423-1425 (1972).

[OsSch 3] ______, Axioms for Euclidean Green's functions, Commun. Math. Phys. 31, 83-112 (1973).

[OsSch 4] ______, Euclidean Fermi fields and a Feynman-Kac formula for boson-fermion models, Helv. Phys. Acta, to appear.

[Os] K. Osterwalder, Duality for free Bose fields, Commun. Math. Phys. 29, 1-14 (1973).

[Oz] H. Ozkaynak, Euclidean fields for particles of arbitrary spin, Harvard preprint.

[Pa] Y. Park. Local Lorentz transformations of the $P(\varphi)_2$ model, Indiana preprint.

[Pe] R. Peierls, On Ising's model of ferromagnetism, Proc. Camb. Phil. Soc. 32, 477-481 (1936).

[Ro 1] L. Rosen, A $\lambda\varphi^{2n}$ field theory without cutoffs, Commun. Math. Phys. 16, 157-183 (1970).

[Ro 2] ______, The $(\varphi^{2n})_2$ quantum field theory: higher order estimates, Comm. Pure App. Math. 24, 417-457 (1971).

[Ro 3] ______, The $(\varphi^{2n})_2$ quantum field theory: Lorentz covariance, J. Math. Anal. and Appl. 38, 276-311 (1972).

[Ro 4] ______, Renormalization of the Hilbert space in the mass shift model, J. Math. Phys. 13, 918-927 (1972).

[Ru] D. Ruelle, Statistical mechanics, Benjamin, New York, 1969.

[SaSt] A. Salam and J. Strathdee, A renormalizable gauge model of lepton interactions, Nuovo Cimento, 11A, 397-435 (1972).

[Sch 1] R. Schrader, Yukawa quantum field theory in two space-time dimensions without cutoff, Annals of Physics 70, 412-457 (1972).

[Sch 2] ______, A remark on Yukawa plus boson self-interaction in two space-time dimensions, Commun. Math. Phys. 21, 164-170 (1971).

[Schw] S. Schweber, An introduction to relativistic quantum field theory, Row, Peterson and Co., New York, 1961.

[Se 1] I. Segal, Notes toward the construction of nonlinear relativistic quantum fields I; the Hamiltonian in two space-time dimensions as the generator of a C*-automorphism group, P.N.A.S., 57, 1178-1183 (1967).

[Se 2] ______, Construction of nonlinear local quantum processes I, Ann. Math. 92, 462-481 (1970).

[Si I] B. Simon, Correlation inequalities and the mass gap in $P(\varphi)_2$ I, Domination by the two point function, Commun. Math. Phys. 31, 127-136 (1973).

[Si II] ______, Correlation inequalities and the mass gap in $P(\varphi)_2$ II, Uniqueness of the vacuum for a class of strongly coupled theories, Toulon preprint.

[Sl] A. Sloan, The relativistic polaron without cutoffs in two space dimensions, Cornell thesis.

[Sp 1] T. Spencer, Perturbation of the $P(\varphi)_2$ quantum field Hamiltonian, J. Math. Phys. 14, 823-828 (1973).

[Sp 2] ______, The mass gap for the $P(\varphi)_2$ quantum field model with a strong external field.

[St] R. Streater, Connection between the spectrum condition and the Lorentz invariance of $P(\varphi)_2$, Commun. Math. Phys. 26, 109-120 (1972).

[St Wi] R. Streater and A. Wightman, PCT, spin and statistics and all that, Benjamin, New York, 1964.

[Sw] J. Swieca, Range of forces and broken symmetries in many-body systems, Commun. Math. Phys. 4, 1-7 (1967).

[Sy 1] K. Symanzik, On the many-particle structure of Green's functions in quantum field theory I, J. Math. Phys. 1, 249-273 (1960).

[Sy 2] ______, A modified model of Euclidean quantum field theory, N.Y.U. preprint, 1964.

[Sy 3] ______, Euclidean quantum field theory, in: Local quantum theory, proceedings of the International School of Physics "Enrico Fermi" Course 45, ed. by R. Jost, Academic Press, New York, 1969.

[Sy 4] ______, Renormalizable models with simple symmetry breaking I, Symmetry breaking by a source term, Commun. Math. Phys. 16, 48-80 (1970).

[Sy 5] ______, Small distance behaviour in field theory and power counting, Commun. Math. Phys. 18, 227-246 (1970).

[Sy 6] ______, Small distance behaviour, analysis and Wilson expansions, Commun. Math. Phys. 23, 49-86 (1971).

[Sy 7] ______, Small distance behaviour in field theory, in: Strong interaction physics, Springer tracts in modern physics, Vol. 57, Springer-Verlag, New York, 1971.

[Wi] A. Wightman, Lecture at the 1972 Coral Gables Conference, Gordon and Breach Sci. Pub., to appear.

* Supported in part by the National Science Foundation, Grant NSF-GP-24003.

† Supported in part by the National Science Foundation, Grant NSF-GP-40354X.

‡ Supported in part by the Alfred P. Sloan Foundation.

VI The Particle Structure of the Weakly Coupled $P(\varphi)_2$ Model and Other Applications of High Temperature Expansions: Part II. The Cluster Expansion

THE PARTICLE STRUCTURE OF THE WEAKLY COUPLED $P(\phi)_2$ MODEL AND OTHER APPLICATIONS OF HIGH TEMPERATURE EXPANSIONS

Part II: The cluster expansion

James Glimm*
Courant Institute, NYU
New York, N.Y.

Arthur Jaffe†
Harvard University
Cambridge, Mass.

Thomas Spencer‡
Courant Institute, NYU
New York, N.Y.

Contents

*Supported in part by the National Science Foundation, Grant NSF-GP-24003.

†Supported in part by the National Science Foundation, Grant NSF-GP- 40354X

‡Supported in part by the Alfred P. Sloan Foundation.

§1. INTRODUCTION

From the point of view of physics, there are essentially two problems in the study of quantum field models. The first problem is to find the phenomena of interest to physics (e.g. particles, bound states, resonances, broken symmetries, and asymptotic power series expansions) in the two and three dimensional quantum field models.[1] This problem includes obtaining the mathematical structure underlying these phenomena, and determining qualitatively which interactions and/or parameter values give rise to these phenomena and which do not. We believe this problem is feasible at the present time; indeed the initial progress is the subject of the present lectures. The second problem is the construction of nontrivial fields in four space time dimensions. Here it seems that some new ideas are needed to supplement those which have already been used in the construction of quantum fields.

These two problems are also interesting from the point of view of mathematics. In addition we mention a third problem, which is to place the solutions already constructed in a more general conceptual framework. In particular, elliptic equations with an infinite number of independent variables, Markov fields, non Gaussian integrals over functions of several variables, and C*-algebras arise naturally in the construction of quantum fields. The quantum field results may suggest directions for further development of these theories.

In Part II of these lectures, we give in detail a high temperature (cluster) expansion, in a form which yields an exponential cluster property for the Schwinger functions, as well as analyticity in the coupling constant λ, for λ in a bounded sector

$$0 < |\lambda| < \lambda_0 , \qquad -\pi/2 < \arg \lambda < \pi/2 ,$$

for the $P(\phi)_2$ quantum field theory. Other applications of this type of expansion to the study of quantum field models were developed in part I of these lectures; see also [4,5,6,14].

The expansion itself is related to the virial and cluster expansions [13, Chapter 4] of statistical mechanics. The analogy between statistical mechanics and $P(\Phi)$ quantum field theories has been known for some time. For a discussion of the relation of statistical mechanics to quantum field theory, see the lectures of Guerra, Rose

[1]For three space time dimensions, it is still necessary to complete the construction of the quantum fields. However enough progress has been made to eliminate any serious doubt as to their eventual existence.

and Simon, as well as Part I of these lectures.

These lectures develop a cluster expansion in the case of small coupling. We believe expansion techniques will apply for any parameter values which are sufficiently far from the critical point. Our belief is based partly on low temperature expansions developed by Peierls and by Minlos and Sinai [10] in statistical mechanics as well as work of Dobrushin and Minlos mentioned in part I. Furthermore Spencer [14] has extended the expansion presented here to the case of large external field.

Within their region of convergence expansion methods tend to give the most detailed information available. In fact these methods yield the main information known about the energy spectrum, above the energy level E = 0 of the vacuum. Correlation inequalities, the Lee-Yang theorem and other analyticity methods complement these techniques and often yield results outside the region of convergence. See [7,10,14] and other lectures of this series.

For the cluster expansion, the relevant formula from statistical mechanics is the following expansion for the density of the Gibbs ensemble:

$$(1.1)\quad \prod_{i<j} e^{-\beta V(x_i-x_j)} = \prod_{i<j}\left[1+\left(e^{-\beta V(x_i-x_j)}-1\right)\right]$$

$$= \sum_{\Gamma}\ \prod_{(i,j)\in\Gamma}\left(e^{-\beta V(x_i-x_j)}-1\right) .$$

Here Γ is a set of distinct unordered pairs (i,j), i.e. Mayer graphs, and the sum extends over all such graphs. This formula expresses the interaction between the distinct i and j ($i\neq j$) particles, $e^{-\beta V(x_i-x_j)}$, as a sum of a zero interaction term, 1, for which there is no coupling between the particles and a perturbation $e^{-\beta V(x_i-x_j)}-1$ which is small at high temperatures $kT=\beta^{-1}$. Heuristically, the role of the Gibbs density in the $P(\phi)_2$ field theory is replaced by the measure

$$(1.2)\quad e^{-\int[\mathcal{H}_0(x)+\lambda P(\Phi(x))]dx}\prod_{x\in R^2} d\Phi(x) .$$

Here

(1.3) $$\mathcal{H}_o(x) = \frac{1}{2}\left(\nabla\Phi(x)^2 + m_o^2\Phi(x)^2\right)$$

and the formal expression

(1.4) $$e^{-\int \mathcal{H}_o(x)dx} \prod_{x\in R^2} d\Phi(x)$$

denotes the Gaussian measure on $\mathcal{S}'(R^2)$ with mean zero and covariance

(1.5) $$C_\emptyset = (-\Delta + m_o^2)^{-1} .$$

We also use the notation $d\Phi_C$ for the Gaussian measure with covariance C, so that (1.4) $= d\Phi_{C_\emptyset}$. We recognize (1.2) as the canonical Gibbs density for transverse vibrations of an elastic membrane subject to the nonlinear restoring force

$$F = -m_o^2\Phi(x) - \lambda P'\left(\Phi(x)\right) ,$$

after integration over the momentum variables $\Pi(x)$.

In (1.2), the coupling between distinct points comes entirely from the $\nabla\Phi$ term in $\mathcal{H}_o(x)$ and so $e^{-\int\nabla\Phi^2}$ in (1.2) plays the role of $e^{-\beta V}$ in (1.1). Our cluster expansion is constructed in the spirit of (1.1). In the completely decoupled theory, the Laplacian in

$$\int \mathcal{H}_o(x)dx = \frac{1}{2}\int \Phi(x)\left(-\Delta + m_o^2\right)\Phi(x)dx$$

is replaced by zero. The resulting ultralocal theory [9,12] seems to be very singular relative to the theory defined by (1.2). We reduce and control the singularity of the difference between the coupled and decoupled theories in two separate steps. The first step introduces a lattice structure into R^2 and into the expansion generalizing (1.1). We do not introduce the lattice into the measure (1.2) and so the resulting expansion is an expansion for the continuum infinite volume $P(\Phi)_2$ theory, rather than for a lattice approximation to this theory. Let Γ denote a set of lattice lines, joining nearest neighbor lattice points in Z^2, let Δ_Γ be the Laplace operator with Dirichlet[2] boundary conditions on Γ, and let

(1.6) $$C_\Gamma = (-\Delta_\Gamma + m_o^2)^{-1} .$$

[2] From the point of view of this discussion, Neumann data might seem more appropriate, but we use Dirichlet data because it is technically easier.

Then $d\Phi_{C_\Gamma}$ plays the role of the decoupled measure, with decoupling along the curve Γ.[3] In summary, the lattice structure gives discrete variables in the sum and product

$$\sum_{\Gamma} \prod_{(i,j)\in\Gamma}$$

in (1.1), even when this formula is applied to the continuum $P(\Phi)_2$ model.

The second step regularizes the differences corresponding to

$$e^{-\beta V(x_i - x_j)} - 1$$

in (1.1). The difference between two Gaussian measures can be expressed as

$$d\Phi_{C_1} - d\Phi_{C_2} = \int_0^1 \frac{d}{ds}\, d\Phi_{C(s)}$$

where

$$C(s) = sC_1 + (1-s)C_2 .$$

This formula applies equally well to the non Gaussian measures

$$e^{-\lambda \int P(\Phi(x))dx}\, d\Phi_C$$

of interest to us. There is a simple formula for the evaluation of $\frac{d}{ds}\, d\Phi_{C(s)}$, namely [1]

$$\frac{d}{ds} \int F d\Phi_{C(s)} = \frac{1}{2} \int \left(C'(s)\cdot\Delta_\Phi\right) F\, d\Phi_{C(s)} \tag{1.7}$$

where

$$C'(s)\cdot\Delta_\Phi F = \frac{d}{ds} \int C(s,x,y) \frac{\delta^2}{\delta\Phi(x)\delta\Phi(y)} F\, dxdy \tag{1.8}$$

$$= \int [C_1(x,y) - C_2(x,y)] \frac{\delta^2}{\delta\Phi(x)\delta\Phi(y)} F\, dxdy$$

and $C(s,x,y)$ is the kernel of $C(s)$.

The proof of this formula contains an integration by parts in

[3] In comparison with (1.1), Γ should be replaced by $(Z^2)^* \sim \Gamma$, where $(Z^2)^*$ is the set of all nearest neighbor bonds in Z^2.

the function space integral. In the case of interest to us, $C_1 = C_{\Gamma_1}$ and $C_2 = C_{\Gamma_2}$, and the integration by parts regularizes[4] $\frac{d}{ds}\, d\Phi_{C(s)}$.

[4]Formally, $\frac{d}{ds}\, d\Phi_{C(s)} = J(\Phi) d\Phi_{C(s)}$, where the Jacobean J is

$$J(\Phi) = \frac{1}{2} \int :\Phi(x)[C(s)^{-1}C'(s)C(s)^{-1}](x,y)\Phi(y):dx\, dy$$

$$= -\frac{1}{2} \int :\Phi(x)[\tfrac{d}{ds}\, C^{-1}(s)](x,y)\Phi(y):dx\, dy \;.$$

Since C_{Γ_1} and C_{Γ_2} are mutually singular, for $\Gamma_1 \neq \Gamma_2$, some singularity in $J(\Phi)$ should be anticipated.

§2. THE MAIN RESULTS

For small values of λ/m_0^2, we prove an exponential cluster property on the Schwinger functions. We prove the cluster property in a finite volume with bounds independent of the volume. From these results, it follows easily that the Schwinger functions converge in the infinite volume limit, and that the infinite volume Schwinger functions satisfy an exponential cluster property. (They are also independent of the boundary conditions.) See [6]. Using the Osterwalder-Schrader reconstruction theorem, the infinite volume $P(\phi)_2$ field theory is constructed from the Schwinger functions, and in this theory the Wightman axioms are satisfied and the physical mass is strictly positive. We also show that the Schwinger functions are analytic in λ, for λ in the bounded sector

$$(2.1) \qquad 0 < |\lambda| < \varepsilon \,, \qquad -\pi/2 < \arg \lambda < \pi/2 \,.$$

Let $\mathcal{C}$ be the set of convex combinations of Dirichlet covariance operators (1.6). The important properties of the measures $d\Phi_C$, $C \in \mathcal{C}$, are summarized in §9 (as preparation for the estimate of §10). In order to make elementary definitions, we mention merely that for Λ a bounded measurable set and for $C \in \mathcal{C}$,

$$(2.2) \qquad V(\Lambda) = \int_\Lambda :P(\Phi(x)):dx \in L_p(\mathcal{S}', d\Phi_C)$$

and

$$(2.3) \qquad e^{-\lambda V(\Lambda)} \in L_p(\mathcal{S}', d\Phi_C)$$

for all $p \in [1,\infty)$ and $\operatorname{Re} \lambda \geq 0$. (See Prop. 9.3, Th. 9.5.) In (2.2) the polynomial P is arbitrary, while in (2.3), P is semibounded. The Wick ordering in (2.2) is taken relative to the fixed measure $d\Phi_{C_\emptyset}$ of (1.4), even when $C \neq C_\emptyset$. To simplify the discussion, we take Λ to be a union of lattice squares, and in the limits $\Lambda \to \infty$, we further restrict Λ to be a square centered at the origin.[5] Let

$$(2.4) \qquad Z(\Lambda) = Z(\Lambda, C) = \int e^{-\lambda V(\Lambda)} d\Phi_C \,.$$

In Theorem 6.1 below we show that $Z(\Lambda) \neq 0$. Let

$$(2.5) \qquad dq_\Lambda = dq_{\Lambda,C} = Z(\Lambda,C)^{-1} e^{-\lambda V(\Lambda)} d\Phi_C \,.$$

[5]More general cutoffs, as required for the proof of Euclidean covariance of the limit $\Lambda \to \infty$, can be introduced by inserting a nonnegative L_∞ function h in the integral (2.2). Such an h does not affect the subsequent discussion.

The finite volume Schwinger functions are by definition the moments of the measure dq_Λ:

$$(2.6) \qquad S_\Lambda(x_1,\dots,x_n) = \int \Phi(x_1)\dots\Phi(x_n)\, dq_\Lambda \ .$$

S_Λ is defined as a tempered distribution in $\mathcal{S}'(R^{2n})$. To see this, take a test function of the form $w(x) = w_1(x_1)\dots w_n(x_n)$ and integrate against (2.6). The n+1 factors

$$\Phi(w\ ) = \int \Phi(x_i)w_i(x_i)\, dx_i \qquad \text{and} \qquad e^{-\lambda V(\Lambda)}$$

in the integrand are bounded by Hölder's inequality, (2.3), and the following consequence of Prop. 9.3:

$$\|\Phi(w_i)\|_{L_p(\mathcal{S}')} \le M_p \|w_i\|_{L_2(R^2)} \ .$$

In addition to the monomials in (2.4), it is useful to integrate products of Wick ordered polynomials, namely

$$(2.7) \qquad A = \int :\Phi(x_1)^{n_1}:\dots:\Phi(x_j)^{n_j}:w(x_1,\dots,x_j)dx \ .$$

We assume $w \in \mathcal{S}(R^{2j})$, although weaker bounds would suffice. We define suppt A to be the intersection of all closed subsets $C \subset R^2$ with

$$(2.8) \qquad \text{suppt } w \subset C \times \dots \times C \qquad (j \text{ factors})$$

Theorem 2.1. Let λ belong to the closure of the half circle (2.1) and let ε/m_0^2 be sufficiently small. Let A and B be functions on $\mathcal{S}'$ of the form (2.7). Let d be the width of a strip in R^2 separating suppt A and suppt B. There is a constant $M = M_{A,B}$ and a positive constant m independent of A and B such that

$$\left| \int AB\, dq_{\Lambda,C_\emptyset} - \int A\, dq_{\Lambda,C_\emptyset} \int B\, dq_{\Lambda,C_\emptyset} \right| \le M_{A,B}\, e^{-md} \ ,$$

uniformly in Λ, as $\Lambda \to \infty$. Furthermore, M is independent of translations in either A or B.

Theorem 2.2. Let λ belong to the closure of the half circle (2.1), and let ε/m_0^2 be sufficiently small. For A of the form (2.7), $\left| \int A\, dq_{\Lambda,C} \right|$ is bounded uniformly in λ and in Λ, as $\Lambda \to \infty$.

From analyticity of the finite volume Schwinger functions, from

Vitali's theorem and from the convergence, as $\Lambda \to \infty$, for λ real and small, we have

Corollary 2.3. The Schwinger functions are analytic in λ in (2.1) for ε/m_o^2 small.

§3. THE CLUSTER EXPANSION

The proof of the main theorems are based on a cluster expansion, which we now derive. Let $\mathcal{B}$ be a set of line segments in R^2. We are interested in two examples: Either $\mathcal{B} = (Z^2)^*$, the set of all lattice lines (bonds) joining nearest neighbor lattice sites in Z^2 or $\mathcal{B} = (Z^2)^* \sim \Gamma$ with Γ a finite subset of $(Z^2)^*$. We identify a subset $\Gamma \subset \mathcal{B}$ with the subset

$$\Gamma = \bigsqcup_{b \in \Gamma} b \subset R^2 .$$

The subsets $\Gamma \subset \mathcal{B}$ label terms in our expansion. The term labeled by Γ is decoupled across

$$\Gamma^c = \mathcal{B} \sim \Gamma \tag{3.1}$$

and formed by choosing a Gaussian measure with Dirichlet covariance on Γ^c. Thus the lines b in Γ^c are called Dirichlet lines. For the lines $b \in \Gamma$ there are differences between coupled and decoupled measures as in (1.1). These differences are expressed in terms of derivatives by the fundamental theorem of calculus. Thus the $b \in \Gamma$ are called derivative lines.

The covariance operators we consider are convex combinations of the operators C_Γ. For each $b \in \mathcal{B}$, we introduce a parameter $s_b \in [0,1]$ to measure the strength of the coupling across b. The parameter value $s_b = 0$ corresponds to zero Dirichlet data on b, hence to zero coupling across b, while $s_b = 1$ corresponds to full coupling across b. With

$$s = (s_b)_{b \in \mathcal{B}} \quad , \tag{3.2}$$

we define the multiparameter family of covariance operators

$$C(s) = \sum_{\Gamma \subset \mathcal{B}} \prod_{b \in \Gamma} s_b \prod_{b \in \Gamma^c} (1 - s_b) C_{\Gamma^c} . \tag{3.3}$$

Since the coefficients in (3.3) are the terms in the expansion of

$$1 = \prod_{b \in \mathcal{B}} 1 = \prod_{b \in \mathcal{B}} [s_b + (1-s_b)] ,$$

(3.3) is a convex sum as asserted. The free covariance is now denoted

$$C_\emptyset = C(1,1,\ldots) = (-\Delta + m_o^2)^{-1}$$

and the completely decoupled covariance is

$$C_{\mathcal{B}} = C(0,0,\ldots) = (-\Delta_{\mathcal{B}} + m_o^2)^{-1} .$$

The Schwinger functions and the partition function are, in a natural way, functions of s, and we use the notation

$$(3.4)\qquad \begin{cases} Z(s)S_s(x) = Z(\Lambda,s)S_{\Lambda,s}(x) = \int \Pi\Phi(x_i)e^{-\lambda V(\Lambda)}d\Phi_s \\ Z(s) = Z(\Lambda,s) = \int e^{-\lambda V(\Lambda)}d\Phi_s \end{cases}$$

where

$$d\Phi_s = d\Phi_{C(s)}$$

The goal of the cluster expansion is to express coupled quantities ($s_b \equiv 1$) in terms of decoupled (and hence finite volume) quantities. The decoupled quantities have $s_b = 0$ for many $b \in \mathcal{B}$, and in order to formulate them, it is convenient to define

$$s(\Gamma) = \left(s(\Gamma)_b\right)_{b\in\mathcal{B}}$$

by the formula

$$s(\Gamma)_b = \begin{cases} s_b, & b \in \Gamma \\ 0, & b \notin \Gamma . \end{cases}$$

For Γ finite, $s(\Gamma)$ specifies Dirichlet boundary conditions at large distances (on Γ^c), while s can be thought of as giving general boundary conditions on Γ^c, and agreeing with $s(\Gamma)$ on Γ. Thus the following definition is a way of saying that F is independent of the boundary conditions at ∞.

<u>Definition 3.1</u>. A function F(s) is called regular at infinity if for each s,

$$(3.6)\qquad F(s) = \lim_{\{\Gamma\nearrow\mathcal{B}:\ \Gamma \text{ finite}\}} F\left(s(\Gamma)\right) .$$

<u>Proposition 3.1</u>. The functions (3.4) are regular at infinity. Here S converges in $\mathcal{S}'$, and $\mathcal{B} \subseteq (Z^2)^*$.

Because Λ is fixed and bounded, the limit (3.6) is elementary. We omit the proof, since it is a consequence of techniques to be developed in Chapters 7,9.

The first step in the cluster expansion is to apply the fundamental theorem of calculus to the finite number of nonzero parameters in

$F(s(\Gamma))$. Define

$$\partial^{\Gamma} = \prod_{b\in\Gamma} \frac{d}{ds_b} \tag{3.7}$$

and for two coordinate values s and σ define order coordinatewise, so that

$$\sigma \leq s \Leftrightarrow \sigma_b \leq s_b , \qquad \forall b \in \mathcal{B}.$$

<u>Proposition 3.2</u>. Let $F(s)$ be smooth and regular at infinity. Then

$$F(s) = \sum_{\{\Gamma\subset\mathcal{B}:\Gamma \text{ is finite}\}} \int_{0\leq\sigma\leq s(\Gamma)} \partial^{\Gamma} F(\sigma(\Gamma))\, d\sigma . \tag{3.8}$$

<u>Proof</u>. Let $G(s)$ denote the right side of (3.8). We assert that $F(s(B)) = G(s(B))$, for B any finite subset of $\mathcal{B}$. Now $G(s(B))$ is just the sum in (3.8), restricted to sets $\Gamma \subset B$. Since $F(s)$ is regular at infinity, convergence of the sum in (3.8) follows from the assertion. Also (3.8) follows by limits in the assertion, as $B \nearrow \mathcal{B}$.

For a function $f(s_b)$ of a single variable, let

$$(\delta^b f)(s_b) = f(s_b) - f(0) = \int_0^{s_b} \partial^b f(\sigma_b)\, d\sigma_b$$

$$(E_0^b f)(s_b) = f(0) .$$

Then $I = E_0^b + \delta^b$ (fundamental theorem of calculus) and so

$$I = \prod_{b\in B} (E_0^b + \delta^b) = \sum_{\Gamma\subset B} E_0^{B\sim\Gamma} \delta^{\Gamma} \tag{3.9}$$

where

$$\delta^{\Gamma} = \prod_{b\in\Gamma} \delta^b , \qquad E_0^{B\sim\Gamma} = \prod_{b\in B\sim\Gamma} E_0^b .$$

It is easy to see that (3.9) yields the desired identity $F(s(B)) = G(s(B))$.

The next step in the cluster expansion is a factorization and partial resummation of (3.8). Write

$$R^2 \sim \Gamma^c = X_1 \cup X_2 \cup \dots \cup X_r \tag{3.10}$$

so that each X_i is a union of connected components such that $X_i \cap X_j = \emptyset$ for $i \neq j$.

Definition 3.2. A function $F(\Lambda,s)$ decouples at $s=0$ provided

$$(3.11) \qquad F(\Lambda,s(\Gamma)) = \prod_{i=1}^{r} F(\Lambda \cap X_i,\ s(\Gamma \cap X_i))$$

whenever (3.10) holds.

Proposition 3.3. Given (3.10), the measures $d\Phi_{s(\Gamma)}$ and $e^{-\lambda V(\Lambda)}d\Phi_{s(\Gamma)}$ each factor into a product of r measures, and the ith factor measure is defined on $\mathcal{S}'(X_i)$. To avoid regularity questions, we assume $\mathcal{B} \subseteq (Z^2)^*$.

Sketch of Proof. Since $C(s(\Gamma))$ has zero Dirichlet data on Γ^c, $C(s(\Gamma))$ is the direct sum of the operators

$$C(s(\Gamma)) \restriction L_2(X_i) \quad .$$

The factorization of $d\Phi_{s(\Gamma)}$ follows from this fact, and can be seen from the standard formula for evaluation of a Gaussian integral of a polynomial (Theorem 9.1). Because $:P(\Phi(x)):$ is a local function, $\exp(-\lambda V(\Lambda))$ also factors, and so does $\exp(-\lambda V)d\Phi_{s(\Gamma)}$.

Corollary 3.4. The functions ZS and Z of (3.4) decouple at $s=0$.

The cluster expansion represents F as a sum of products of contributions from connected graphs. We start with some set X_o of interest (e.g. $X_o = \{x_1,\ldots,x_n\}$ in (3.4)). The graphs which do not meet X_o are resummed, giving a single factor of F in an exterior region.

We now carry out this resummation in more detail. Substitute (3.11) in the expansion (3.8). This yields

$$(3.12) \qquad F(\Lambda,s) = \sum_{\Gamma} \prod_{i=1}^{r} \int_0^{s(\Gamma_i)} \partial^{\Gamma_i} F(\Lambda \cap X_i, \sigma(\Gamma_i))\, d\sigma$$

where $\Gamma_i = \Gamma \cap X_i$. If the X_i are connected, this is the sum over products of connected graphs. Now we choose X_1 in (3.10) to be the union of all components meeting X_o and let X_2 be the union of the remaining components. The resummation consists of holding X_1 and Γ_1 fixed, while summing over all choices of Γ_2. Explicitly,

$$F(\Lambda,s) = \sum_{X_1,\Gamma_1} \int_0^{s(\Gamma_1)} \partial^{\Gamma_1} F(\Lambda \cap X_1, \sigma(\Gamma_1))\, d\sigma$$

$$\times \sum_{\Gamma_2} \int_0^{s(\Gamma_2)} \partial^{\Gamma_2} F(\Lambda \cap X_2, \sigma(\Gamma_2))\, d\sigma \ .$$

The Γ_2 sum runs over all finite sets Γ_2 of bonds in $\mathcal{B} \sim X_1^-$. For this reason, the Γ_2 sum can be evaluated by (3.8) as $F(\Lambda \cap X_2, s(\mathcal{B} \sim X_1^-))$. Setting $X = X_1^-$ and writing Γ for Γ_1, the expansion has the form

$$(3.13) \quad \begin{cases} F(\Lambda,s) = \sum_X K(X_o,X) F(\Lambda \sim X, s(\mathcal{B} \sim X)) \\ K(X_o,X) = \sum_\Gamma \int_0^{s(\Gamma)} \partial^\Gamma F(\Lambda \cap X, \sigma(\Gamma))\, d\sigma \ . \end{cases}$$

In these sums, X ranges over finite unions of closed lattice squares while Γ ranges over finite subsets of $\mathcal{B}$ such that

(i) Each component of $X \sim \Gamma^c$ meets X_o

(ii) $\Gamma \subset \text{Int } X$.

If no such Γ exists, for a given X, then $K(X_o,X) \equiv 0$.

<u>Theorem 3.5</u>. Let X_o be bounded and let F be smooth, regular at infinity and decouple at s=0. Then the cluster expansion (3.13) holds.

<u>Example 1</u>. Let $\mathcal{B} = (Z^2)^*$ be the set of all lattice lines and let $X_o = \{x_1,\ldots,x_n\}$. We substitute ZS = F. See (3.4).
Observe that s=1

$$(3.14) \quad F(\Lambda \sim X, s(\mathcal{B} \sim X)) = \int e^{-\lambda V(\Lambda \sim X)} d\Phi_{s(\mathcal{B} \sim X)}$$

$$= Z(\Lambda \sim X, s(\mathcal{B} \sim X))$$

$$= Z(\Lambda \sim X, C_{\partial X}) \equiv Z_{\partial X}(\Lambda \sim X)$$

since a change in the data within X does not affect the integral. Thus if we divide by Z, (3.13) yields

$$(3.15) \quad S_\Lambda(x) = \sum_{X,\Gamma} \int \partial^\Gamma \int \Pi_i \, \Phi(x_i) \, e^{-\lambda V(\Lambda\cap X)} \, d\Phi_{s(\Gamma)} \, ds(\Gamma)$$

$$\times \, Z_{\partial X}(\Lambda \sim X)/Z(\Lambda) \; .$$

If we write $X \sim \Gamma^c$ as a union of connected components, the integral in (3.15) can be factored, as in (3.12).

Example 2. Let $\Gamma_1 \subset (Z^2)^*$ and $\Gamma_2 = \Gamma_1 \sim b_1$ be finite sets of lattice lattice bonds, where b_1 is the first element of Γ_1 in some lexicographic order of $(Z^2)^*$. We use the cluster expansion (in all bonds $b \neq b_1$) to study the difference $Z_{\Gamma_1} - Z_{\Gamma_2}$ where

$$Z_\Gamma = \int e^{-\lambda V(\Lambda)} \, d\Phi_{C_\Gamma}$$

and so we take $\mathcal{B} = (Z^2)^* \sim b_1$, $X_0 = b_1$. Z is now a function of the pair (s, s_{b_1}). We define

$$F(\Lambda,s) = \begin{cases} Z(\Lambda,s(\Gamma^c),0) - Z(\Lambda,s(\Gamma^c),1) \; , & b_1 \subset \Lambda \\ Z(\Lambda,s(\Gamma^c),0) \; , & b_1 \cap \Lambda = \emptyset \end{cases}$$

Ths definition is arranged so that

$$F(s=1) = Z_{\Gamma_1} - Z_{\Gamma_2} \; , \quad \text{for} \quad b \subset \Lambda \; .$$

Also F is smooth, regular at infinity and decouples at s = 0. (The last property depends on the exclusion of b_1 from $\mathcal{B}$.) We note that F is independent of the variables s_b for $b \in \Gamma_2$, and so $\partial^{b_j} F = 0$, for $b \in \Gamma_2$. Thus in all nonzero terms, Γ_2 consists of Dirichlet lines in the cluster expansion for F, and in (3.13), we may impose the further restriction on $\sum_\Gamma$:

(iii) $\Gamma \cap \Gamma_2 = \emptyset$.

From (3.13), we have

$$F(\Lambda, s=1) = Z_{\Gamma_1}(\Lambda) - Z_{\Gamma_2}(\Lambda)$$

$$= \sum_X K(b_1,\Gamma_1,X)F(\Lambda\sim X,\ s(\mathcal{B}\sim X))$$

$$= \sum_X K(b_1,\Gamma_1,X)Z_{\Gamma_1 \cup X^*}(\Lambda\sim X)$$

Here X^* is the set of lattice lines in X and the last line is justified by the fact that the second factor above is independent of s_b, $b \in \text{Int } X$. We multiply and divide by

$$Z_{\partial\Delta}(\Delta)^{|\Lambda \cap X|}$$

where Δ is a single lattice square. Since

$$Z_{\partial\Delta}(\Delta)^{|\Lambda \cap X|}\ Z_{\Gamma_1 \cup X^*}(\Lambda\sim X) = Z_{\Gamma_1 \cup X^*}(\Lambda) ,$$

we obtain (with a new K)

$$(3.16) \quad Z_{\Gamma_1} = Z_{\Gamma_1\sim b_1} + \sum_X K(b_1,\Gamma_1,X)Z_{\Gamma_1 \cup X^*}(\Lambda)$$

and now

$$(3.17) \quad K(b_1,\Gamma_1,X) = Z_{\partial\Delta}(\Delta)^{-|\Lambda \cap X|}\int \sum_\Gamma \partial^{\Gamma \cup b_1} Z(\Lambda \cap X, s(\Gamma \cup b_1))\,ds(\Gamma \cup b_1).$$

These equations have a structure intermediate between the Kirkwood-Salsburg and the Mayer-Montroll equations. They will be studied in §6 in order to obtain bounds on Z_Γ/Z, i.e. the second factor in (3.15).

§4. CLUSTERING AND ANALYTICITY: PROOF OF THE MAIN RESULTS

We prove the cluster property and analyticity -- the main results of these lectures -- in this section, using as hypothesis convergence of the cluster expansion, (3.15). Let $T(x,\Lambda,X,\Gamma)$ denote the X-Γ term in (3.15), so that

$$S_\Lambda(x) = \sum_{X,\Gamma} T(x,\Lambda,X,\Gamma) . \qquad (4.1)$$

For each test function $w \in \mathcal{S}(R^{2n})$, the series

$$\int S_\Lambda(x)\ w(x)\quad x = \sum_{X,\Gamma} \langle w,T\rangle \qquad (4.2)$$

converges absolutely. The rate of convergence is governed by $|X|$, the area of X. With $K > 0$, we prove that

$$\sum_{\{X,\Gamma:|X|\geq D\}} |\langle w,T\rangle| \leq |w|\ e^{-K(D-n)} . \qquad (4.3)$$

where $|w|$ is some (n-dependent) $\mathcal{S}$-norm on w.

Theorem 4.1. Let $K > 0$ be given. Let m_0 be large and ε be small (depending on K) and let λ belong to the closure of (2.1). There is an $\mathcal{S}$-norm w such that (4.3) holds uniformly in λ, m_0, Λ and D, and $|w|$ is invariant under translations in any of its variables.

The proof allows integrands A of the form (2.7) to be substituted in (3.15). Thus Theorem 2.2 and Corollary 2.3 follow from the case $K = 1$, $D = 1$ of (4.3).

Theorem 2.1 also follows from the convergence of the cluster expansion. The proof for the two point function with an even interaction is conceptually easier, but still illustrates the main idea, so we consider that case first.

Proof of Theorem 2.1, for n = 2, P even, assuming Theorem 4.1. Because P, the interaction polynomial in (1.2), is even, $\Phi(x) \to -\Phi(x)$ is a symmetry of the theory. For Gaussian integrals defined by the factorizing measure $\exp(-\lambda V)\ d\Phi_{C(s(\Gamma))}$ (cf. Proposition 3.3), more is true:

$$\Phi(x) \to \sigma(x)\ \Phi(x) \qquad (4.4)$$

is a symmetry, where

$$\sigma(x) = \pm 1; \quad \sigma(x) = \text{const. on each } X_i .$$

Because of the symmetry $\Phi \to -\Phi$, $S_\Lambda(x_1,\ldots,x_n) = 0$ for n odd. Similarly,

$$\int \Pi\ \Phi(x_i)\ e^{-\lambda V(\Lambda\cap X)}\ d\Phi_{C(s(\Gamma))} = 0$$

unless each of the connected components $X_1,\ldots,X_r$ contains an even number of x_i's. Since $\partial^\Gamma 0 = 0$, the same restriction $T(x,\Lambda,X,\Gamma) = 0$ applies unless each X_j has an even number of x_i's. Since each X_j has at least one x_i by (i) of §3, each X_j must have at least two x_i's. Thus for n = 2, there is only one X_j. In other words $X \sim \Gamma^c$ is connected. Let d be defined by Theorem 2.1 and let $w = w_1 \otimes w_2$. Because X is connected and suppt $w_1 \cap X \neq \emptyset$, we have $d \leq |X| + 1$. Thus by (4.2-3),

$$\left|\int S_\Lambda(x_1,x_2)w_1(x_1)w_2(x_2)dx\right| \leq |w|e^{-K(D-2)} \leq M_w e^{-Kd}$$

Since the one-point functions $S_\Lambda(x_1)$ vanish for P even, this bound completes the proof of Theorem 2.1.

<u>Proof of Theorem 2.1</u> (general case) assuming Th. 4.1. The idea, as before, is to reduce the cluster expansion to terms involving only one connected component $X_1 = X$. The terms with two or more connected components are to vanish, because of some symmetry. Since the general case (P not even) does not have such a symmetry, we follow Ginibre [2] and introduce a new theory containing an artificial symmetry.

To construct the new theory, let $d\Phi^*_{C^*}$ denote an isomorphic copy of the measure $d\Phi_C$, $(C^* \cong C)$ defined on an isomorphic copy $\mathcal{S}'^*$ of $\mathcal{S}'$. The new theory has free measure $d\Phi_C \times d\Phi^*_{C^*}$, covariance

$$C \otimes I + I \otimes C^* = C^\sim$$

and normalized physical measure

$$Z^{\sim -1} e^{-V(\Lambda)} e^{-V(\Lambda)^*} d\Phi_C \times d\Phi^*_{C^*} = dq^\sim$$

and a field

$$\Phi^\sim = \Phi \otimes I + I \otimes \Phi^* .$$

This theory is invariant (even) under the symmetry $\Phi \leftrightarrow \Phi^*$, which interchanges the two factors.

We now apply the cluster expansion to the expression

$$Z^\sim \int (A-A^*)(B-B^*)dq^\sim$$

The covariance operators which arise in this expansion have the form

$$C(s)^\sim = C(s) \otimes I + I \otimes C(s)^*$$

so that the symmetry $\Phi \leftrightarrow \Phi^*$ is preserved in the Gaussian measure of each term of the expansion. (However the expansion is modified by choosing $\mathcal{B} = (Z\ )^* \sim \Gamma$, where Γ is the set of lattice lines in two connected sets, one containing suppt A = suppt A-A*, and the other containing suppt B. Thus for each component X_i, $X_i \supset$ suppt A or $X_i \cap$ suppt A $= \emptyset$, and the same for suppt B. With this restriction, there are at most two components. Since the n in (4.3) is a bound on the number of components, as one can check, n = 2 in (4.3).) Now consider a term in (3.15) with components $X_1, X_2, \ldots$ satisfying

$$\text{suppt } A \subset X_i \ , \qquad \text{suppt } B \subset X_j \ , \qquad i \neq j \ . \qquad (4.5)$$

For this term, the symmetry $\Phi \leftrightarrow \Phi^*$ can be applied separately in each X_i. However A-A* is odd for the X_i symmetry, so terms satisfying (4.5) must vanish. The nonvanishing terms, which violate (4.5) contain a component $X_i = X$ with $d \leq 0|X|$, and so for $d \geq 4$,

$$\left|\int (A-A^*)(B-B^*)dq^{\sim}\right| \leq M_{AB}e^{-md} \ .$$

Expanding the integrand on the left yields four terms, and in each term the dq integral factors. We evaluate the result as

$$2\left|\int ABdq_{\Lambda,C} - \int Adq_{\Lambda,C}\int Bdq_{\Lambda,C}\right| \ ,$$

and Theorem 2.1 follows.

§5. CONVERGENCE: THE MAIN IDEAS

There are two main ideas in part II of these lectures. The first is the formula (3.17) giving the cluster expansion for the Schwinger functions. The second is the estimates which lead to the convergence of this expansion, uniformly as $\Lambda \to \infty$. In this section we state these estimates as a series of three propositions, and using them, give the proof of Theorem 4.1. The simpler estimates are then proved, while the harder ones are postponed for later sections. The most difficult of these estimates is contained in Prop. 5.3. Since it is in some sense the core of the paper, we discuss the ideas involved in its proof at the end of this section.

The proof of convergence depends on estimates of the following types:

(a) Combinatoric estimates to count the number of terms in the expansion, and especially to count or bound the number of terms of some special type.

(b) Single particle estimates, to bound kernels of covariance operators, and their derivatives, $\partial^{\Gamma}C(s)$.

(c) Estimates on function space integrals. Typically estimates of type (a) and (b) will be components of an estimate of type (c).

Returning to specifics, (4.3) or (3.15) is a sum in which each term is a product of two factors: a ratio of partition functions times a function space integral. The first proposition is purely combinatoric - type (a)- and it counts the number of terms with $|X|$ = area X held fixed. The second and third propositions bound the two factors making up a term in (3.15). These latter two propositions are hybrids, since their proof uses estimates of types (a), (b) and (c).

Proposition 5.1. The number of terms in (3.15-3.16) with a fixed value of $|X|$ is bounded by $e^{K_1|X|} \equiv O(1)e^{19|X|}$.

Proposition 5.2. There is a constant K_2, independent of λ in the closure of (2.1), and of Λ and m_0, such that for ε small and m_0 large,

$$\left|Z_{\partial X}(\Lambda \sim X)/Z(\Lambda)\right| \leq e^{K_2|X|} .$$

Proposition 5.3. There is a constant K_3 and a norm $|w|$ on test functions such that for any $K > 0$, for any Λ and for λ in the closure of (2.1) with m_0 large (the m_0 bound depends on K)

$$\left| < \int \partial^{\Gamma} \int \prod_{i=1}^{n} \Phi(x_i)\, e^{-\lambda V(\Lambda)} d\Phi_{C(s(\Gamma))}\, ds(\Gamma), w> \right| \leq e^{-K|\Gamma| + K_3|\Lambda|}\, |w| .$$

(The m_0 bound depends on K, and $|w|$ is invariant under translation in any of the variables of w.)

Remark. Wick polynomials, as in (2.7), are also allowed in the integrand above.

Proof of Theorem 4.1. We replace Λ by $\Lambda \cap X$ in Prop. 5.3. For the X in (3.15), we have $X = \bigcup_{i=1}^{r} X_i^-$ with $r \leq n$ and X_i connected. Moreover

$$\Gamma \subset \bigcup_{i=1}^{r} \text{Int } X_i^-$$

and so "many" of the lattice lines in X_i^- belong to Γ. In fact since $X_i^- \sim \Gamma^c = X_i$ is connected,

(5.1) $|X_i| - 1 \leq 2|\Gamma \cap \text{Int } X_i^-|$ and $|X| - n \leq 2|\Gamma|$.

Thus we replace the upper bound in Prop. 5.3 by

$$e^{-K(|X|-n)} \, |w| ,$$

with a new choice of K and $|w|$. Theorem 4.1 follows directly from this bound combined with Prop. 5.1 and 5.2.

Proof of Proposition 5.1. We consider (3.15). First we bound the number of ways of choosing the component X_i containing a particular x_j, $1 \leq j \leq n$. We identify each lattice square Δ with a lattice point located at its center, and we identify each lattice line $b \in \mathcal{B}$ lying between two squares Δ and Δ' with the line joining the center of Δ to the center of Δ'. Hence we must count the number of ways of drawing a connected graph with nearest neighbor bonds. We assert that each such graph may be constructed by starting at x_j and drawing an oriented path formed of unit line segments which traverses each segment at most twice. In fact if we regard the lattice sites as islands and the line segments as bridges, this simply follows from the solution to the Konigsberg bridge problem[6]. The number of connected paths of length ℓ formed from lattice line segments and starting at x_j is at most 4^{ℓ}. Since $\ell \leq 8|X_i|$, the number of choices for X_i is at most $O(1)2^{16|X_i|}$. Since the number of choices for Γ is at most $4^{|X|}$, the number of choices for pairs X,Γ is at most

$$O(1)2^{|X|}4^{|X|}\prod_i 2^{16|X_i|} = O(1)2^{19|X|} ,$$

since $2^{|X|}$ bounds the number of choices of $|X_i|$ with $|X|$ given. This completes the proof. The case (3.16) is similar.

Proposition 5.2 is proved in §6 using equations related to the

[6] We thank O. Bratteli for this observation. For a solution of the Konigsberg bridge problem see, D.W. Blackett, Elementary Topology, Academic Press, New York, 1967, page 159.

Kirkwood-Salsburg equations. We only remark here that the change in partition function involves both a change in the interaction region and a change in covariance ($\Lambda\sim X \to \Lambda$ and $C_{\partial X} \to C$).

Discussion of Prop. 5.3. Each derivative d/ds_b in ∂^Γ differentiates either the measure or the integrand. Derivatives of the measure are evaluated by (1.7), and produce s-dependent kernels $C'(s)$ in the integrand. Repeated differentiation yields

(a) A sum of terms coming from the iterated functional derivatives $\partial^2/\partial\phi^2$ in (1.7).

(b) A sum over terms arising from possible repeated derivatives $\partial^{\Gamma_1}C(s)$ of each $C'(s)$ introduced by (1.7).

Each derivative d/ds_b also produces convergence in one of two ways. Differentiation of the measure introduces a kernel C', and

(c) C' and C are small in the sense that

$$(c_1)\ \|C'(x,y)\|_{L_p} \le O(m_0^{-\varepsilon})$$

$$(c_2)\ 0 \le C'(x,y) = \partial^b C(x,y) \le e^{-m_0(\mathrm{dist}(x,b)+\mathrm{dist}(y,b))}$$

$$(c_3)\ 0 \le C(x,y) \le e^{-m_0|x-y|}, \quad |x-y| \ge 1 .$$

Repeated differentiation of $C(s)$ produces the further convergence

(d) $\partial^\Gamma C$ is small in the sense that

$$(d_1)\ \|\partial^\Gamma C(x,y)\|_{L_p} \le O(m_0^{-\varepsilon|\Gamma|})$$

$$(d_2)\ 0 \le \partial^\Gamma C(x,y) \le e^{-m_0 d}$$

where d is the length of the shortest path connecting x and y and passing through each $b\in\Gamma$. In fact from the Weiner integral representation for $(-\Delta+m_0^2)^{-1}$, $\partial^\Gamma C(s)$ is the integral over Wiener paths from x to y passing through each $b\in\Gamma$, see §7-8.

We use (c_2) to control (a). In fact by (c_2), only x and y near b contribute to the Laplacian $C'(s)\cdot\Delta_\phi$ of (1.7). By the nature of the expansion, there is at most one Laplacian Δ_ϕ per bond $b\in\mathcal{B}$, hence a bounded power, Δ_ϕ^n, locally. Evaluation of Δ_ϕ^n leads to $K(n) < \infty$ terms, and this evaluation, repeated over disjoint local regions leads to $e^{O(\mathrm{Vol})}$ terms. We use (d_2) to control (b). The basic convergence, $m_0^{-\varepsilon|\Gamma|}$, comes from (d_1).

After performing these steps, there remains a function space integral of the form $\int Re^{-\lambda V}d\Phi$. Then

$$\left| \int R e^{-\lambda V(\Lambda)} \, d\Phi \right| \leq \left(\int R^2 \, d\Phi \right)^{1/2} \left(\int e^{-2\mathrm{Re}\lambda V(\Lambda)} \, d\Phi \right)^{1/2}$$

and each factor on the right is $e^{O(\mathrm{Vol})}$. Now (c_3) produces the decoupling of distant local regions in R^2.

With some practice, estimates such as Proposition 5.3 can be done by inspection. In Sections 7-9 we develop some general machinery, the goal of which is to make these estimates more routine.

§6. AN EQUATION OF KIRKWOOD-SALSBURG TYPE

Equations (3.16-3.17) can be rewritten as a Banach space equation

$$\rho = Z(\Lambda)1 + \mathcal{K}\rho \tag{6.1}$$

with a unique solution

$$\rho = (1-\mathcal{K})^{-1} Z(\Lambda)1$$

satisfying the bound

$$\|\rho\| \le \|(I-\mathcal{K})^{-1}\| \, |Z(\Lambda)| \le 4|Z(\Lambda)| \tag{6.2}$$

This bound is essentially Proposition 5.2, as we shall see.

Let $\mathcal{X}$ be the Banach space of functions f defined on finite subsets $\Gamma \subset (Z^2)^*$. For $f \in \mathcal{X}$ let $(f_n)_{n\geq 0}$ denote the restriction of f to n element subsets. The norm on $\mathcal{X}$ is defined to be

$$\|f\| = \sup_{\substack{n,\Gamma \\ |\Gamma|=n}} 2^{-n} |f_n(\Gamma)| .$$

We define $\rho_\Lambda \equiv (\rho_{\Lambda,n})_{n\geq 0}$ to be the function

$$\Gamma \to Z_\Gamma(\Lambda) .$$

Theorem 6.1. Let $|\lambda| \le \varepsilon$, $\mathrm{Re}\,\lambda \ge 0$ where $\varepsilon > 0$ is small and let m_0 be sufficiently large. Then $Z(\Lambda) \neq 0$, and ρ_Λ defined above is the unique solution in $\mathcal{X}$ of equation (6.1). Moreover ρ_Λ satisfies the bounds (6.2).

By the dominated convergence theorem, $Z_{\partial\Delta}(\Delta) \to 1$ as $\varepsilon \to 0$. Thus for small ε,

$$\tfrac{1}{2} \le |Z_{\partial\Delta}(\Delta)| \le 2 .$$

This restriction fixes the ε in Theorem 6.1, and in fact is the only restriction on ε in these lectures.

Proof of Proposition 5.2, assuming Theorem 6.1.

$$\begin{aligned}|Z_{\partial X}(\Lambda \sim X)/Z(\Lambda)| &= |Z_{X^*}(\Lambda)/Z(\Lambda)| \; |Z_{\partial\Delta}(\Delta)|^{-|\Lambda\cap X|} \\ &\le 2^{|X^*|} \|(1-\mathcal{K})^{-1}\| \; |Z_{\partial\Delta}(\Delta)|^{-|\Lambda\cap X|} \\ &\le e^{K_2|X|} .\end{aligned}$$

Proof of Theorem 6.1. Let $1 = (1,0,0,\ldots) \in \mathcal{X}$, and define $\mathcal{K}$ by the equations

$$(\mathcal{K}f)_n(\Gamma) = f_{n-1}(\Gamma \sim b_1) + \sum_{m=1}^{\infty} \sum_{X,|X|=n} K(b_1,\Gamma,X)\, f_{|X^* \cup \Gamma|}(X^* \cup \Gamma) \tag{6.4}$$

for $n \geq 2$ where the sum over X (as in (3.13) and (3.17)) ranges over connected unions of lattice squares such that $b_1 \in X$. For $n = 1$, we omit the first term on the right of (6.4) and for $n = 0$, $(\mathcal{K}f)_0 \equiv 0$.

We now show that $\mathcal{K}$ is a contraction, $\|\mathcal{K}\| \leq 3/4$, in order to obtain (6.2). It suffices to show that

$$\frac{1}{2} + \sup_{\Gamma \subset \mathcal{B}} 2^{-|\Gamma|} \sum_{m=1}^{\infty} \sum_{X,|X|=m} |K(b_1,\Gamma,x)|\, 2^{|\Gamma \cup X|} \leq \frac{3}{4} \tag{6.5}$$

The proof of (6.5) is similar to the proof of Theorem 4.1 given in §5. In particular it depends on Propositions 5.1 and 5.3. By Proposition 5.1 there are at most $e^{K_1|X|}$ terms for each fixed value of $|X|$. From (3.17) each term has at least one derivative ∂^{b_1}. Using this fact and (5.1) the bound from Propositions 5.1 and 5.3 is

$$|K(b_1,\Gamma,X)| \leq e^{-K(|X|+1)}\, e^{K_3|X|} .$$

For K sufficiently large (6.5) follows. We complete the proof by showing $Z(\Lambda) \neq 0$. By our choice of ε, $Z_{\Lambda^*} = |Z_{\partial\Lambda}(\Delta)|^{|\Lambda|} \neq 0$. Thus $\rho_\Lambda \neq 0$ and by (6.2), $Z(\Lambda) \neq 0$.

§7. COVARIANCE OPERATORS

The basic facts for the kernel $C_\emptyset(x,y)$ of $C_\emptyset = (-\Delta+m_0^2)^{-1}$ are

(a) $C_\emptyset(x,y)$ is a function of the difference $|x-y|$.

(b) $0 < C_\emptyset(x,y)$

(c) for $|x-y| \geq 1$, $C_\emptyset \leq e^{-m_0(1-\delta)|x-y|}$

(d) as $|x-y| \to 0$, $C_\emptyset \sim \log |x-y|$

(e) for $r > 2/(2-\delta)$, $\delta > 0$, $C_\emptyset \in \mathcal{B}^{loc}_{r,\delta}(R^4)$.

By definition, the space $\mathcal{B}^{loc}_{r,\delta}(R^4)$ is the space of distributions $\psi \in \mathcal{S}'$ which satisfy

$$\|\zeta\psi\|_{\mathcal{B}_{r,\delta}} = \|(1+k^2)^{\delta/2}\,(\zeta\psi)^{\sim}\|_{L_r} < \infty$$

whenever $\zeta \in C_0^\infty$. For r and r' dual indices and for $1 \leq r' \leq 2$, $\mathcal{B}^{loc}_{r,\delta}$ is a space of functions with fractional $L_{r'}$ derivatives, by the Hausdorff-Young inequality. See [8, Chapter 2] for a general theory of such spaces. (e) follows from the fact that as a function of x-y, $C_\emptyset$ has the Fourier transform $(k^2+m^2)^{-1}$.

Property (a), which is based on the translation invariance of the Laplacian, does not extend to the operators $C \in \mathcal{C}$. Recall that $\mathcal{C}$ is the set of all convex combinations of the Dirichlet covariance operators (1.6), or in other words, operators of the form C(s). Properties (b) and (c) are valid for all $C \in \mathcal{C}$, while (d) is valid as an upper bound for all $C \in \mathcal{C}$.

Proposition 7.1. The kernel $C(x,y)$ of $C \in \mathcal{C}$ satisfies

$$0 \leq C(x,y) \leq C_\emptyset(x,y) .$$

Proof. Let $dz^T_{x,y}$ be the conditional Wiener density on paths $z(\tau)$ starting at x at $\tau = 0$ and ending at y at $\tau = T$. Let J^T_b be the function

$$J^T_b(z) = \begin{cases} 0 & \text{if } z(\tau) \in b,\ 0 \leq \tau \leq T \\ 1 & \text{otherwise} \end{cases} \tag{7.1}$$

defined on Wiener paths. Then C_Γ and $C(s)$ have the representations

$$C_\Gamma(x,y) = \int_0^\infty e^{-m_0^2 T} \int \Pi_{b\in\Gamma}\, J^T_b(z)\, dz^T_{x,y}\, dT \tag{7.2}$$

$$C(s)(x,y) = \int_0^\infty e^{-m_0^2 T} \int \Pi_{b\in\mathcal{B}}(s_b + (1-s_b)J^T_b)\, dz^T_{x,y}\, dT . \tag{7.3}$$

See [0, §5.3, 6.1 and 6.3].

The inequality of Proposition 7.1 comes from substituting

$$0 \le (s_b + (1-s_b)J_b^T) \le 1$$

in the formula above for C(s).

The next estimate combines the exponential decay (c) with a local bound coming from (d). We label each lattice square $\Delta = \Delta_j \subset R^2$ by the lattice point $j \in Z^2$ in the lower right corner of Δ. Any lattice square (or any set $X \subset R^2$), will be identified with the multiplication on $L_2(R^2)$ by its characteristic function. Now let $j = (j_1, j_2) \in Z^4$ be a pair of lattice points. We define the localized covariance operator

$$C(j) = \Delta_{j_1} C \Delta_{j_2} \tag{7.4}$$

and the distance

$$d(j) = \text{Dist}(\Delta_{j_1}, \Delta_{j_2}) , \tag{7.5}$$

which measures the nonlocality of C(j).

<u>Proposition</u> 7.2. For $1 \le q < \infty$ and $\delta > 0$, there is a constant $K_4(q,\delta)$ independent of $m_0 \ge 1$ and $C \in \mathcal{C}$ such that

$$\| C(j) \|_{L_q(\Delta_{j_1} \times \Delta_{j_2})} \le K_4 m_0^{-2/q} \exp(-m_0(1-\delta)d(j)) .$$

<u>Proof</u>. Because of Proposition 7.1, we may assume $C = C_\emptyset$. For $d(j) > 0$, necessarily $d(j) \ge 1$, and the required bound follows from property (c). The factor $m_0^{-2/q}$ in the proposition comes from choosing δ in (c) smaller than the δ in the proposition. Next consider the case $d(j) = 0$ (i.e. equal or adjacent squares). We have

$$\begin{aligned} \| C(x,y) \|_{L_q(\Delta_{j_1} \times \Delta_{j_2})} &\le K_3 \| C(\cdot,0) \|_{L_q(R^2)} \\ &\le K_4(q) m_0^{-2/q} . \end{aligned}$$

The last inequality follows from the fact that $(-\Delta+1)^{-1}(x,0) \in L_q$ for $q \in [1,\infty)$ and from the identity

$$(-\Delta+m_0^2)^{-1}(x,0) = (-\Delta+1)^{-1}(m_0 x,0) .$$

<u>Remark</u>. The differentiated covariance $\partial^\Gamma C$ also satisfies the bound of Proposition 7.1, as we can see from (7.3) and the inequality

$$0 \le \frac{d}{ds_b}(s_b + (1-s_b)J_b) = 1-J_b \le 1 .$$

Hence $\partial^\Gamma C$ also satisfies the bound of Proposition 7.2.

Local regularity of the covariance $C \in \mathcal{C}$ is used to justify the definition and basic formal properties (e.g. integration by parts) of Gaussian integrals and it is used in the bound $|Z| \le e^{O|\Lambda|}$. We obtain fractional derivatives of $C \in \mathcal{C}$, with bounds slightly weaker than (e).

<u>Proposition</u> 7.3. $C \in \mathcal{B}^{loc}_{2,\delta}$ for $0 \le \delta < 1/2$, with bounds independent of C.

<u>Proof</u>. We start with the observation that

$$0 \le -\Delta \le -\Delta_\Gamma .$$

Consequently $C_\emptyset^{-1/2} C_\Gamma C_\emptyset^{-1/2}$ is a bounded operator, with norm at most one. By convex combinations, $\|B\| \le 1$ also, where

$$B = C_\emptyset^{-1/2}\, C\, C_\emptyset^{-1/2} .$$

Let

$$A = C_\emptyset^{-(1/4)+\varepsilon} \zeta C_\emptyset^{1/2},$$

for $\zeta \in C_0^\infty(R^2)$. We use the fact that $A\,A^*$ is Hilbert Schmidt to calculate the Hilbert Schmidt norm

$$(7.6) \quad \|(-\Delta+m_0^2)^{-\varepsilon+1/4}\, \zeta\, C\, \zeta(-\Delta+m_0^2)^{-\varepsilon+1/4}\|^2_{HS}$$

$$= \mathrm{Tr}\; A\,B\,A^*A\,B\,A^* \;=\; \mathrm{Tr}\; B\,A^*A\,B\,A^*A$$

$$\le \mathrm{Tr}\;(B\,A^*A)^*B\,A^*A = \mathrm{Tr}\; A^*A\,B^2A^*A$$

$$\le \|B\|^2\; \mathrm{Tr}\,(A^*A)^2 \;\le\; \|A^*A\|^2_{HS} .$$

This inequality proves the proposition (and slightly more).

The reader may skip the next result, since it is not needed in what follows.

<u>Proposition</u> 7.4. For any $q \in (4/3,\infty)$, there is a $\delta = \delta(q) > 0$ with $C \in \mathcal{B}^{loc}_{q,\delta}$ and the bounds are independent of $C \in \mathcal{C}$.

<u>Proof</u>. With $k = (k_1,k_2) \in R^4$, define

$$f(k) = 1+k^2 , \quad g(k) = (1+k_1^2)(1+k_2^2) ,$$

$$h(k) = (\zeta\, C\, \zeta)^\sim(k) .$$

Then

$$g^{-\varepsilon-1} \in L_1 , \quad f/g \in L_\infty .$$

By Proposition 7.1 and the fact that $\zeta C_\emptyset \zeta \in L_1$,

$$h \in L_\infty$$

and by (7.6),

$$g^{-\varepsilon+1/4} h \in L_2 \quad \text{and} \quad h \in L_2 .$$

We apply these facts to the integral

$$(7.7) \quad \int |f^{\delta/2} h|^q = \int |g^{-\varepsilon+1/4} h| \, |\tfrac{f}{g}|^{\delta q/2} |g|^{(\delta q/2)+\varepsilon-1/4} |h|^{q-1} .$$

First take $\delta > 0$ and suppose that $h \in L_{q_1}$, $q_1 > q > 4/3$. The factors in (7.7) belong to L_2, $L_{4+\varepsilon}$, and $L_{q_1/q-1} \cap L_\infty$. For ε small and q near q_1, this implies $h \in L_q$ also. Since $h \in L_q$ for q in an interval, $h \in L_q$ for $q \in (4/3,\infty]$. Now with $h \in L_q$, $q > 4/3$, (7.7) shows that $f^{\delta/2} h \in L_q$ also, for small δ. This completes the proof.

<u>Proposition</u> 7.5. For $1 \le q < \infty$,

$$c(x) = \lim_{y\to x} C(x,y) - C_\emptyset(x,y) \in L_q^{\text{loc}} .$$

There is a constant $K_5(q)$, independent of $m_0 \ge 1$ and $C \in \mathcal{C}$ such that for any lattice square Δ,

$$\| c \|_{L_q(\Delta)} \le K_5 \, m_0^{-1/q} .$$

<u>Proof</u>. By scaling, we may take $m_0 = 1$, as in Proposition 7.2. Let $\Gamma = \mathcal{B}$ be the set of all lattice lines. For $x \notin \Gamma$,

$$\begin{aligned} 0 &\le - c(x) \le C_\emptyset(x,x) - C_\Gamma(x,x) \\ &\le O(1 + |\log \text{dist}(x,\Gamma)|) . \end{aligned}$$

This inequality completes the proof. It is proved in [7] as follows. For $x \notin \Gamma$, $y \notin \Gamma$,

$$(-\Delta_y + m_0^2)[C_\emptyset(x,y) - C_\Gamma(x,y)] = 0 .$$

Hence by the maximum principle, and the fact that $C_\Gamma(x,y) = 0$ for $y \in \Gamma$,

$$0 \le C_\emptyset(x,y) - C_\Gamma(x,y) \le \sup_{y\in\Gamma} C_\emptyset(x,y) \le O(1+|\log \text{dist}(x,\Gamma)|).$$

§8. DERIVATIVES OF COVARIANCE OPERATORS

For the differentiated covariance operator $\partial^{\gamma}C$, there should be a strong decay, $\sim e^{-m_0 d}$, where d is the length of the shortest path in R^2 joining x to y and passing through each lattice line segment $b \in \gamma$. This can be seen by inspection from the Wiener integral representation

$$(8.1)\quad (\partial^{\gamma}C(s))(x,y) = \int_0^{\infty} e^{-m_0^2 T} \int \Pi_{b\in\gamma}(1-J_b^T) \times \Pi_{b\in \mathcal{B}\sim\gamma}[s_b + (1-s_b)J_b^T]\, dz_{x,y}^T\, dT\ ;$$

We need the improved bounds on $\partial^{\gamma}C$ for two reasons. The first is to localize x and y, with γ given. For this purpose

$$(8.2)\qquad d(j,\gamma) = \sup_{b\in\gamma} \{\mathrm{Dist}(\Delta_{j_1},b) + \mathrm{Dist}(\Delta_{j_2},b)\}$$

is sufficient, as a crude lower bound on d.

We now explain the second use of bounds on $\partial^{\gamma}C$. Let $\mathcal{P}(\Gamma)$ be the set of all partitions π of the set of lattice line segments Γ. In Proposition 5.3, we are called on to bound $\partial^{\Gamma}\int F\, d\Phi_s$, which by Leibnitz' rule and by (1.7) is just

$$(8.3)\qquad \partial^{\Gamma}\int F\, d\Phi_s = \sum_{\pi\in\mathcal{P}(\Gamma)} \int \left(\Pi_{\gamma\in\pi} \tfrac{1}{2}\, \partial^{\gamma}C\cdot\Delta_{\Phi}\right) F\, d\Phi_s\ .$$

The second use of the bounds on $\partial^{\gamma}C$ is to control $\sum_{\pi\in\mathcal{P}(\Gamma)}$. As in §7, we also find a factor $m_0^{-O|\gamma|}$, which yields the overall convergence of the expansion.

<u>Proposition</u> 8.1. Let $1 \leq q < \infty$ and let m_0 be sufficiently large. There are constants $K_6(q,\gamma)$ and $K_7(q)$, independent of m_0, such that

$$(8.4)\qquad \|\partial^{\gamma}C\|_{L_q(\Delta_{j_1}\times\Delta_{j_2})} \leq K_6(q,\gamma)\, m_0^{-|\gamma|/2q} \exp(-m_0\, d(j,\gamma)/2)$$

$$(8.5)\qquad \sum_{\pi\in\mathcal{P}(\Gamma)} \Pi_{\gamma\in\pi} K_6(q,\gamma) \leq e^{K_7(q)|\Gamma|}\ .$$

<u>Proof</u>. We use the Wiener integral representation (8.1) for $\partial^{\gamma}C$. The proof consists of estimates on the Wiener measure of paths $z(\tau)$ which cross the lattice lines $b \in \gamma$ in some definite order together with combinatoric arguments to count the number of ways the lines

$b \in \gamma$ can be so ordered.

Let $L(\gamma)$ be the set of all possible linear orderings of the lines $b \in \gamma$, and for $\ell \in L(\gamma)$, let $\mathcal{W}(\ell)$ be the set of Wiener paths which cross all lines $b \in \gamma$, and whose order of first crossing is ℓ. Then

$$(8.6) \qquad 0 \le \partial^\gamma C(s) \le \int_0^\infty e^{-m_0^2 T} \int \prod_{b\in\gamma} (1-J_b^T(z))\, dz_{x,y}^T \, dT = \partial^\gamma C_\emptyset ,$$

and

$$(8.7) \qquad \partial^\gamma C_\emptyset(x,y) = \sum_{\ell\in L(\gamma)} \int_0^\infty e^{-m_0^2 T} \int_{\mathcal{W}(\ell)} dz_{x,y}^T \, dT .$$

Let $b_1, b_2, \ldots$ be the elements of γ, as ordered by ℓ. Let b_2' be the first of the b's not touching $b_1 = b_1'$, let b_3' be the first of the b's after b_2' and not touching b_2', etc. Set

$$a_j = \mathrm{Dist}(b_{j+1}', b_j') , \qquad 1 \le j \le m$$

and define

$$|\ell| = \sum_{i=1}^m a_i .$$

If there is no such b_2', we set $|\ell| = 0$, by convention.

With these definitions, we bound the $\ell \in L(\gamma)$ term in (8.7), for $|\ell| \ge 1$, by

$$\int e^{-m_0^2 T} \int_{\sum t_i = T} \Pi_i (2\pi t_i)^{-1} \exp\left(-\tfrac{1}{2} \sum_{i=1}^m \frac{a_i^2}{t_i}\right) dt\, dT$$

$$\le K_8^m \sup_{T, \sum t_i = T} e^{-(m_0^2-2)T} e^{-\frac{1}{3}\sum_{i=1}^m a_i^2/t_i}$$

since

$$\int_{\sum t_i = T} dt \le e^T , \qquad \int e^{-T}\, dT \le 1 ,$$

and for all $a_i \ge 1$,

$$(2\pi t_i)^{-1} e^{-a_i^2/6t_i} \le K_8 < \infty ,$$

thereby defining K_8.

Using the method of Lagrange multipliers to evaluate the maxima, we bound the $\ell \in L(\gamma)$ term by

$$K_8^m \exp\left((m_0^2-2)^{1/2} |\ell|\right) ,$$

for $|\ell| \ge 1$. For $|\ell| = 0$, we use the remark following Prop. 7.2.

There is an entirely similar estimate, based on the distance $d(j,\gamma)$ of (8.2), and taking geometric means of these two bounds yields

$$\|\partial^\gamma C\|_{L_q(\Delta_1\times\Delta_2)} \le \sum_{\ell\in L(\gamma)} K_8^{|\gamma|} e^{-m_0|\ell|/(2+\delta)} e^{-m_0 d(j,\gamma)/(2+\delta)} \tag{8.8}$$

for m_0 large. If $|\ell| \ge 1$ for all $\ell \in L(\gamma)$, then we can include a factor $m_0^{-|\gamma|}$ on the right side of (8.8), by increasing δ. If $|\ell| < 1$ for some ℓ, then $|\ell| = 0$, and in this case $|\gamma| \le 4$. With $|\gamma| \le 4$ and $d(j,\gamma) \ge 1$, we can still include the factor $m_0^{-|\gamma|}$ in (8.8) by increasing δ. Finally for $|\gamma| \le 4$ and $d(j,\gamma) = 0$, the factor $m_0^{-|\gamma|/2q} \le m_0^{-2/q}$ in (8.4) comes from scaling, as in Proposition 7.2.

We define

$$K_6(q,\gamma) = K_4 \sum_{\ell\in L(\gamma)} K_8^{|\gamma|} e^{-m_0|\ell|/(2+\delta)} \tag{8.9}$$

With this definition, (8.4) follows; in the case $d(j,\gamma) = 0$ and $|\ell| = 0$ for some ℓ, $K_6 \ge K_4$, and we use the bound of Proposition 7.2 to establish (8.4).

We complete the proof by establishing (8.5) as a separate proposition.

<u>Proposition</u> 8.2. For m_0 sufficiently large,

$$\sum_{\pi\in\mathcal{P}(\Gamma)} \prod_{\gamma\in\pi} \sum_{\ell\in L(\gamma)} e^{-m_0|\ell|/3} \le e^{K_9|\Gamma|} . \tag{8.10}$$

<u>Proof</u>. Let $\mathcal{L}(\Gamma)$ be the set of linear orderings defined on subsets of Γ. Thus $L(\Gamma) \subsetneq \mathcal{L}(\Gamma)$. As before, we define $|\ell|$ for $\ell \in \mathcal{L}(\Gamma)$. We assert that the number of $\ell \in \mathcal{L}(\Gamma)$ with $|\ell| \le r$ is bounded by

$$|\Gamma|\, e^{K_{10}(r+1)} . \tag{8.11}$$

Using (8.11), we complete the proof. Let $A_\ell = \exp(-m_0|\ell|/3)$. Expanding $\sum\prod\sum A_\ell$ in (8.10), we get a sum of terms of the form

$$A_{\ell_1} A_{\ell_2} \cdots A_{\ell_j}$$

where the ℓ_j are distinct elements of $\mathcal{L}(\Gamma)$. Adding all terms of this form, we bound (8.10) by

$$\sum\prod\sum A_\ell \le \sum A_{\ell_1}\cdots A_{\ell_j} = \prod_{\ell\in\mathcal{L}(\Gamma)} (1+A_\ell)$$

$$\le \prod_{\ell\in\mathcal{L}(\Gamma)} \exp A_\ell = \exp \sum_{\ell\in\mathcal{L}(\Gamma)} A_\ell \le \exp\,(O(1)|\Gamma|) .$$

Here in the last expression, we used the bound (8.11) to estimate $\sum_{\ell\in\mathcal{L}(\Gamma)} A_\ell$ and we choose m_0 sufficiently large.

Next we establish (8.11). Suppose the integer part $[a_i]$ of the distances a_i are given. We choose $b_1 = b_1'$ in $|\Gamma|$ ways, and we choose the b's between b_1' and b_2' in O(1) ways, since they all must overlap b_1. Next b_2' is chosen in $O(1)\ [a_1]$ ways, namely from the lattice line segments b with

$$[a_1] \leq \text{Dist}(b,b_1') < [a_1]+1 \ .$$

Continuing in this fashion, we choose all the b's in

$$|\Gamma|\ \Pi_i\ O(1)\ [a_i] \leq |\Gamma|\ e^{O(1)\sum [a_i]} \leq |\Gamma| e^{O(1)r}$$

ways. Finally we count the number of choices of the $[a_i]$. This is the number of ways of choosing integers $r_i \geq 1$ with $\sum r_i \leq r$, namely 2^r. In fact suppose $\sum r_i = r$, and we distribute the r units in $\sum r_i$ as follows: The first 1 goes into a_1 (no choice). The second 1 goes to a_1 or a_2 (one binary choice). If the jth 1 goes to a_i, the j+1st goes to a_i or a_{i+1} (one binary choice). Thus there are r-1 binary choices, or 2^{r-1} ways to choose r_i with $\sum r_i = r \geq 1$. Summing $j = \sum r_i$ gives $\sum_{j=1}^{r} 2^{j-1} = 2^j - 1$. Finally, we get one more choice from $|\ell| = 0$ (no a_i's).

§9. GAUSSIAN INTEGRALS

The integral of a polynomial with respect to a Gaussian measure can be evaluated in closed form. The closed form expression is a sum, and each term in the sum is labelled by a graph. We will encounter complicated polynomials of high degree, and the resulting graphs will also be complicated. However we present some very simple estimates for such polynomials; the structure of these estimates can be seen easily from the associated graphs.

We define a localized monomial to be a polynomial of the form

$$R = \int \prod_{i=1}^{r} :\Phi(x_i)^{n_i}: \, w(x)\, dx \,, \tag{9.1}$$

where $w(x)$ is supported in a product $\Delta_{j_1} \times \ldots \times \Delta_{j_r}$ of lattice squares. We also require $w \in L_{1+\varepsilon}$ and it is convenient but not essential to assume a bound

$$n_i \le \bar{n} \,, \quad 1 \le i \le r \,. \tag{9.2}$$

the bound (9.2) does not restrict r, nor the total degree of R. Polynomials which arise naturally are not usually of this form because the kernels w are not localized. However any polynomial can be written as a sum of localized monomials.

Associated with R of (9.1) is a graph G(R) consisting of r vertices and at the ith vertex, we draw n_i legs. See Figure 1.

Fig. 1. $G\left(\int :\Phi(x_1)^4: \, :\Phi(x_2)^4: \, w(x)\, dx\right)$.

In order to evaluate $\int R \, d\Phi_C$, we integrate by parts:

$$\int \Phi(x) R \, d\Phi_C = \int C(x,y) \frac{\delta R}{\delta \Phi(y)} \, d\Phi_C \, dy \,.$$

This formula can be proved by passing to the Fock space[7] of the measure $d\Phi_C$, expanding Φ as a sum of a creation and an annihilation operator and using the canonical commutation relations. See also Theorem 9.1 below. We integrate by parts to reduce the degree of the monomial R we want to integrate. After $(\sum_{i=1}^{r} n_i)/2$ partial integrations, the monomial is replaced by a sum of constants, and since

[7] See for example Theorem 3.5 of J. Glimm and A. Jaffe, Boson quantum field models, in Mathematics of Contemporary Physics. Ed. by R. Streater, Academic Press, New York, 1972.

$\int d\Phi_C = 1$, the integral is evaluated explicitly.

In applying this procedure, we encounter Φ's in a Wick ordered factor $:\Phi(x_i)^{n_i}:$ in R. For such Φ's, we use the formula

$$(9.3) \quad \int :\Phi(x)^n: R \, d\Phi_C = (n-1)\, c(x) \int :\Phi(x)^{n-2}: R \, d\Phi_C + \int :\Phi(x)^{n-1}: C(x,y) \frac{\delta R}{\delta\Phi(y)} d\Phi_C \, dy ,$$

with

$$c(x) = C(x,x) - C_\emptyset(x,x)$$

defined by Proposition 7.5. The first term arises from the difference between the covariance $C_\emptyset$ in : : and the covariance in $\int \ldots d\Phi_C$. The second term is exactly as before.

The integration by parts formula (9.3) has a simple expression in terms of graphs. In case the $\Phi(x)$ is a factor in $:\Phi(x_i)^{n_i}:$, we label the terms on the right side of (9.3) by drawing a line connecting one leg of the x_i-vertex to a distinct leg at the same or a distinct vertex. The graph with a line from the x_i to the x_j vertex labels each of the n_j terms

$$(9.4) \quad \int C(x_i,x_j) :\Phi(x_i)^{n_i-1}: :\Phi(x_j)^{n_j-1}: \prod_{\ell\neq i,j} :\Phi(x_\ell)^{n_\ell}: w(x) \, dx$$

coming from a single integration by parts in (9.1). See Figure 2.

Fig. 2. Integration by Parts.

As an example, we evaluate the integral of Figure 2. After four integrations by parts, we have

$$\iint :\Phi(x_1)^4: :\Phi(x_2)^4: w \, dx \, d\Phi_C = 4! \int C(x_1,x_2)^4 w \, dx$$

$$+ 2! \binom{4}{2}^2 \int c(x_1)\, c(x_2)\, C(x_1,x_2)^2 w \, dx + \binom{4}{2}^2 \int c(x_1)^2 c(x_2)^2 w \, dx .$$

The absolute value of the first term is bounded by

$$4! \left| \int C(x_1,x_2)^4 \, w(x_1,x_2) \, dx \right| \le 4! \, \|C\|^4_{L_8(\Delta_{j_1}\times\Delta_{j_2})} \|w\|_{L_2}$$

$$\le 4! \, K_4 m_0^{-1} \, e^{-m_0(1-\delta)4d(j)} \, \|w\|_{L_2}$$

if suppt $w \subset \Delta_{j_1} \times \Delta_{j_2}$. In the last line, we used Proposition 7.2 to bound $\|C\|_{L_q(\Delta_{j_1}\times\Delta_{j_2})}$. See Figure 3.

Fig. 3. Evaluation of a Gaussian Integral.

To evaluate $\int R\, d\Phi_C$ in the general case, we form the set $\mathcal{V}(R)$ of all vacuum graphs. A vacuum graph is a graph obtained from $G(R)$ by joining pairwise distinct legs until all legs are so joined. If the total number of legs is odd, $\mathcal{V}(R) = \emptyset$ (and $\int R\, d\Phi_C = 0$). For each $G \in \mathcal{V}(R)$, define

$$(9.5) \quad I(G,w,C) = \int w(x)\, \Pi_{\ell'}\, c(x_{i(\ell')})\Pi_\ell\, C(x_{i_1(\ell)},x_{i_2(\ell)})\, dx\ .$$

The first product, $\Pi_{\ell'}$, runs over all lines ℓ' of G which connect two legs of a single vertex, labeled $i(\ell')$. The second product runs over all lines ℓ joining pairs $(i_1(\ell),i_2(\ell))$ of distinct vertices.

We now state the formula for evaluation of Gaussian integrals. The above discussion of integration by parts and its graphical interpretation gives the formal derivation.

<u>Theorem</u> 9.1. Let w be localized and let $w \in L_q$ for some $q > 1$. Then $R \in L_r(\mathcal{S}', d\Phi_C)$, $r \in [1,\infty)$, and

$$(9.6) \qquad \int R\, d\Phi_C = \sum_{G\in\mathcal{V}(R)} I(G,w,C)\ .$$

The proof has a combinatoric aspect (sketched above) and an analytic aspect. The latter is an approximation argument, which uses a modified (momentum cutoff) form of (9.6) to prove that the approximants converge. For this reason, we bound the integrals $I(G,w,C)$ before sketching the proof of Theorem 9.1.

We consider a function of the form

$$F(x_1,\ldots,x_n) = \Pi_\ell\, F_\ell(x_{i_1(\ell)},x_{i_2(\ell)})\, \Pi_i\, \Delta_i(x_i)$$

where $\Delta_i(x)$ is the characteristic function of a set of area 1, and suppose that for each index ℓ,

$$1 \le i_1(\ell) < i_2(\ell) \le n\ .$$

<u>Lemma</u> 9.2. With the above notation,

$$\|F\|_{L_q} \le \Pi_\ell\, \|F_\ell\|_{L_{q_\ell}(\Delta_{i_1(\ell)},\Delta_{i_2(\ell)})}$$

provided that for each i,

$$q^{-1} \geq \sum \left\{q_\ell^{-1};\ i_1(\ell) = i \text{ or } i_2(\ell) = i\right\} .$$

Proof: With the substitution $F \to F^q$, we are reduced to the case $q = 1$. We apply Hölder's inequality successively in each of the n variables (vertices) of F, and we use the language of graphs, in order to visualize this process. Each factor F_ℓ corresponds to a line joining the $i_1(\ell)$ vertex to the $i_2(\ell)$ vertex. Let

$$\mathcal{L}_i = \{\ell:\ i_1(\ell) < i = i_2(\ell)\}$$

$$\mathcal{L}_i' = \{\ell:\ i_1(\ell) = i < i_2(\ell)\} .$$

The hypothesis concerning q_ℓ is

$$\sum_{\ell\in\mathcal{L}_i\cup\mathcal{L}_i'} q_\ell^{-1} \geq 1$$

($\mathcal{L}_i = \emptyset = \mathcal{L}_i'$ is also allowed, trivially.) Thus

$$\int_{\Delta_i} \prod_{\ell\in\mathcal{L}_i} |F_\ell(x_{i_1(\ell)},x_i)| \prod_{\ell\in\mathcal{L}_i'} \|F_\ell(x_i,\cdot)\|_{L_{q_\ell}(\Delta_{i_2(\ell)})} dx_i$$

$$\leq \prod_{\ell\in\mathcal{L}_i} \|F_\ell(x_{i_1(\ell)},\cdot)\|_{L_{q_\ell}(\Delta_i)} \prod_{\ell\in\mathcal{L}_i'} \|F_\ell\|_{L_{q_\ell}(\Delta_{i_1(\ell)}\times\Delta_{i_2(\ell)})}$$

by Hölder's inequality. A finite induction on i (decreasing from $i = n$ to $i = 1$) now completes the proof.

Remark. An obvious modification allows some of the factors F_ℓ to depend on a single variable, i.e. $F_\ell = F_\ell(x_{i(\ell)})$.

Sketch of proof of Theorem 9.1. We introduce momentum cutoffs, $\Phi(x_i) \to \int \Phi(y)\, j_k(y-x_i)\, dy$, and omit the x-integration. This replaces R by a polynomial cylinder function $R_k(x)$, based on a finite dimensional subspace of $\mathcal{S}$. $R_k(x)$ is integrable by the definition of $d\Phi_C$, and for $R_k(x)$ the formulas (9.3) and (9.6) are valid by the above combinatorial arguments. Thus (9.6) holds for $R_k = \int R_k(x)\, w(x)\, dx$, and can be used to evaluate the right side of the inequality

$$\int |R_k-R_{k'}|\, d\Phi_C \leq \left(\int (R_k-R_{k'})^2\, d\Phi_C\right)^{1/2} .$$

Removing the momentum cutoff in one linear factor Φ at a time, the estimates of Propositions 7.2 and 7.3 are sufficient. For details see [1].

We remark that the stronger hypothesis of [1], $C \in \mathcal{B}^{loc}_{q,\delta(q)}$, $\forall q > 1$ (used in [1] to control C-Wick ordering) is not required for the present lectures, since only $C_\emptyset = (-\Delta+m_0^2)^{-1}$ Wick ordering is used. See also Proposition 7.5.

Proposition 9.3. Let w be localized. Then

$$(9.7) \quad \left| \int R \, d\Phi_C \right| \le \| w \|_{L_p} \sum_{G \in \mathcal{V}(R)} \Pi_{\ell'} \| c \|_{L_q(\Delta_{j(\ell')})} \times \Pi_\ell \| C \|_{L_q(\Delta_{j_1(\ell)}, \Delta_{j_2(\ell)})}$$

where $q \ge p'\bar{n}$, $\bar{n}$ is defined by (9.2) and $j(\ell) = j_{i(\ell)}$ is the localization of the ith vertex.

Proof. By Theorem 9.1, it is sufficient to bound each integral I(G,w,C). We use the Hölder inequality to separate w from the c and C factors. The integral of the c and C factors is bounded by Lemma 9.2 and the remark following it. This completes the proof.

There are $O(\sum n_i/2)!$ graphs in $\mathcal{V}(R)$. Because of the exponential decay in C as $|x-y| \to \infty$, in general most of these graphs are very small. Efficient estimates must take advantage of this fact. For each lattice square Δ, we define

$$N(\Delta) = \sum \{n_i : \Delta_{j_i} = \Delta\}$$

as the number of legs (linear factors $\Phi(x_i)$) of R localized in Δ.

Theorem 9.4. Let w be localized. There is a constant K_{11} such that

$$\left| \int R \, d\Phi_C \right| \le \| w \|_{L_p} [\Pi_\Delta N(\Delta)! \, (K_{11} \, m_0^{-1/2} \, q)^{N(\Delta)}]$$

for $q = p'\bar{n}$, $\bar{n}$ defined by (9.2).

Proof. We bound the L_q norm in (9.7) by Propositions 7.2 and 7.5. Having done this, it is sufficient to show that

$$\sum_{G \in \mathcal{V}(R)} \Pi_\ell \, e^{-m_0(1-\delta)d(j(\ell))} \le \Pi_\Delta(\text{const.}^{N(\Delta)} \, N(\Delta)!) \, .$$

Let ν denote a leg of R. Let Δ_ν be the square in which ν is localized and given $G \in \mathcal{V}(R)$, and let Δ'_ν be the square of the leg joined to ν by G. Also let $d(\nu) = \text{dist}(\Delta_\nu, \Delta'_\nu)$. Then the sum over $G \in \mathcal{V}(R)$ can be written as a sum over the choices Δ'_ν for each ν and a sum over the $N(\Delta'_\nu)$ possible contractions in Δ'_ν, for each ν.

The summand is independent of the choice of possible contractions within each square Δ ($\nu \leftrightarrow \Delta'_\nu$ held fixed). To estimate this sum we have only to count the number of terms. Since an arbitrary term can be obtained (nonuniquely, in general) from a single given term by permutation of the legs in each square Δ, there are at most $\Pi_\Delta N(\Delta)!$ terms in this sum. Hence it remains to show that

$$\sum_{\{\Delta'_\nu\}} \Pi_\nu \, e^{-m_0(1-\delta)d(\nu)/2} \leq \Pi_\nu(\text{const.}) ,$$

since each $d(j(\ell))$ occurs as a $d(\nu)$ for exactly two ν's. The summation index $\{\Delta'_\nu\}$ is simply a set of functions from legs to lattice squares. We increase the left side by enlarging the set of summation indices to include all such functions. Then we can interchange the sum and product, obtaining

$$\sum_{\{\Delta'_\nu\}} \Pi_\nu \, e^{-cd(\nu)} \leq \Pi_\nu \sum_{\{\Delta'_\nu :\ \nu \text{ fixed}\}} e^{-cd(\nu)} = \Pi_\nu \text{ const.}$$

We note that K_{11} is independent of s and m_0, for m_0 bounded away from zero.

Theorem 9.5. Let Λ be a union of lattice squares and let $\operatorname{Re}\lambda \geq 0$. Then $e^{-V(\Lambda)} \in L_p(\mathcal{S}', d\Phi_C)$ for all $p \in [1,\infty)$. There is a constant K_{12} independent of C such that

$$(9.8) \qquad |Z| \equiv \left| \int e^{-\lambda V} d\Phi_C \right| \leq e^{K_{12}|\Lambda|} .$$

With $\operatorname{Re}\lambda$ bounded and m_0 bounded away from zero, K_{10} can also be chosen independent of λ and m_0.

Remark. The simple proof of [1, §II.3, p. 22-27] is self contained. In fact, using a slight generalization Theorem 9.4 (incorporating Proposition 7.4), the proof of (9.8) is nearly identical to the (standard) proof that $e^{-V(\Lambda)} \in L_1(\mathcal{S}', d\Phi_C)$. This proof is close in structure to that of [3].

Corollary 9.6. For $p > 1$ and $q \geq p'\bar{n}$,

$$\left| \int R e^{-V(\Lambda)} d\Phi_C \right| \leq e^{K_{12}|\Lambda|} \|w\|_{L_p} [\Pi_\Delta N(\Delta)! \, (2K_{11} m_0^{-1/2q})^{N(\Delta)}].$$

Proof. By the Schwartz inequality,

$$\left| \int R\, e^{-V(\Lambda)} d\Phi_C \right| \leq \left(\int R^2 d\Phi_C \right)^{1/2} \left(\int e^{-2V(\Lambda)} d\Phi_C \right)^{1/2} .$$

The factors on the right are estimate by Theoresm 9.4 and 9.5.

Theorem 9.7. Let w be a localized kernel in L_p, $p > 1$, let Λ be a union of lattice squares and let $F = Re^{-V(\Lambda)}$ in (1.7). Then (1.7) is valid.

Sketch of Proof: In order to present the formal ideas, we suppose first that F is a polynomial. Then $\int F \, d\Phi_{C(s)}$ is given explicitly in terms of graphs by (9.5) - (9.6). Differentiating with respect to s_b in these formulas yields

$$(9.9) \quad \frac{d}{ds_b} \int F \, d\Phi_{C(s)} = \sum_{G \in \mathcal{V}(F)} \int \sum_{\ell} \left(\frac{d}{ds_b} C_\ell\right) \Pi_{\ell' \neq \ell} C_{\ell'} \, w \, dx$$

where C_ℓ denotes $C(x_{i_1(\ell)}, x_{i_2(\ell)})$.

The product $\Pi_{\ell' \neq \ell}$ in effect removes one line from the vacuum graph $G \in \mathcal{V}(F)$. Equivalently, one could remove from F the two legs joined by the line ℓ. However removing legs from F is the same as removing linear factors from F, or the same as differentiating F with respect to these linear factors. Thus we see that $\sum_\ell$, the sum over lines removed from $G \in \mathcal{V}(F)$, is equivalent to a sum over mixed second derivatives of F with respect to pairs of linear factors. Such a sum is just $\frac{1}{2} \Delta_\Phi F$, so we identify the right side of (9.9) as

$$\frac{1}{2} \int \left[\left(\frac{d}{ds_b} C\right) \cdot \Delta_\Phi\right] F \, d\Phi_{C(s)} \, .$$

The proof in the general case, $F = R\,e^{-V}$, is based on approximations, starting with F a polynomial. The control over these approximations is given by Corollary 9.6. For details see [1].

§10. CONVERGENCE: THE PROOF COMPLETED

Proof of Proposition 5.3. Without loss of generality, the kernel w is localized, and in this case we take $\|w\| = \|w\|_2$. The expression we want to estimate is

$$(10.1) \qquad < \int \partial^\Gamma \int \Pi_{i=1}^n \Phi(x_i)\, e^{-\lambda V(\Lambda)}\, d\Phi_{s(\Gamma)}\, ds(\Gamma),\, w> \quad .$$

Let $\mathcal{P}(\Gamma)$ be the set of all partitions π of Γ. By Leibnitz' rule and (1.7), (10.1) equals

$$(10.2) \quad <\int \sum_{\pi\in \mathcal{P}(\Gamma)} \int \left(\Pi_{\gamma\in\pi} \tfrac{1}{2}\, \partial^\gamma C\cdot\Delta_\Phi\right) \Pi\, \Phi(x_i) e^{-\lambda V(\Lambda)}\, d\Phi_{s(\Gamma)}\, ds(\Gamma), w>$$

where $C = C(s(\Gamma))$.

As in (7.3), we define

$$\partial^\gamma C(j_\gamma) = \Delta_{j_{1,\gamma}}\, \partial^\gamma C\, \Delta_{j_{2,\gamma}}$$

where $j = (j_{1,\gamma}, j_{2,\gamma}) \in Z^4$, so that the two derivatives in $\partial^\gamma C(j_\gamma)\cdot\Delta_\Phi$ are localized in Δ_{j_1} and Δ_{j_2} respectively, and

$$\partial^\gamma C = \sum_{j_\gamma} \partial^\gamma C(j_\gamma) \ .$$

We substitute this identity into (10.2) and expand.

The resulting sum is now indexed by localizations $\{j_\gamma\}$ and partitions $\pi \in \mathcal{P}(\Gamma)$. For a given term let $M = M(\pi,\{j_\gamma\})$ be the number of terms resulting from the differentiations Δ_Φ in (10.2). By Corollary 9.6 each of the resulting terms can be estimated by

$$\|w'\|_{L_p}\, e^{K_{12}|\Lambda|}\, \Pi_\Delta\, N(\Delta)!\, (2K_{11})^{N(\Delta)}$$

for $m_0 \geq 1$. Here w' is the w of (10.1) multiplied by the kernels $\partial^\gamma C$ arising in (10.2). From Proposition 8.1 and Lemma 9.2 we have (for $p < 2$, and q large),

$$\|w'\|_{L_p} < \|w\|_{L_2}\, \|\Pi_{\gamma\in\pi}\, \partial^\gamma C(j_\gamma)\|_{L_q}$$

$$\leq \|w\|_{L_2}\, m_0^{-|\Gamma|/2q}\, \Pi_{\gamma\in\pi}\, K_6(q,\gamma)\, e^{-m_0 d(j_\gamma,\gamma)/2} \quad .$$

Now using (8.5) to control the sum over $\pi \in \mathcal{P}(\Gamma)$ we can bound

(10.2) by

$$\|w\|_{L_2} e^{K_7|\Gamma|} m_0^{-|\Gamma|/2q} \sum_{\{j_\gamma\}} \max_{\pi\in\mathcal{P}(\Gamma)} M \prod_{\gamma\in\pi} e^{-m_0 d(j_\gamma,\gamma)/2} \prod_\Delta N(\Delta)!$$

The proof of Proposition 5.3 follows from two lemmas which control M and the sum over $\{j_\gamma\}$ respectively. Let $M(\Delta)$ be the number of elements in the set $\{j_{i,\gamma} : \Delta_{j_{i,\gamma}} = \Delta,\ i = 1 \text{ or } 2,\ \gamma \in \pi\}$.

<u>Lemma</u> 10.1. There exists a constant K_{13}, independent of m_0, such that

$$M \le e^{K_{13}|\Gamma|} \prod_\Delta (M(\Delta)!)^p$$

and

$$\prod_\Delta N(\Delta)! \le e^{K_{13}|\Gamma|} \prod_\Delta (M(\Delta)!)^p .$$

where p is the degree of the interaction polynomial P.

<u>Lemma</u> 10.2. Given $\pi \in \mathcal{P}(\Gamma)$ and $r > 0$, there exists a constant K_{14}, independent of m_0, such that

$$\sum_{\{j_\gamma\}} \prod_{\gamma\in\pi} e^{-m_0 d(j_\gamma,\gamma)/2} \prod_\Delta M(\Delta)!^r \le e^{K_{14}|\Gamma|} .$$

<u>Proof of Lemma</u> 10.1: Let $N_0(\Delta)$ be the number of x_i, $1\le i\le n$, which are localized in Δ. The number of terms resulting from $M(\Delta)$ differentiations in Δ is bounded by

$$(N_0(\Delta)+1)(N_0(\Delta)+p+1) \ldots (N_0(\Delta)+p(M(\Delta)-1)+1) .$$

Since $N_0(\Delta) \le \sum_\Delta N_0(\Delta) = n$, we have M, the total number of terms resulting from all $\partial/\partial\Phi(y)$ differentiations, bounded by

$$M \le \Pi_\Delta p^{pM(\Delta)} (N_0(\Delta)+1+pM(\Delta))^{N_0(\Delta)+1} M(\Delta)!^p ;$$

using the inequalities

$$(a+b)! \le (a+b)^a (b!) \quad \text{and} \quad (ab)! \le a^{ab}(b!)^a .$$

Furthermore with $N(\Delta)$, as defined in §9, the number of legs in Δ, after differentiation, we have

$$N(\Delta) \le N_0(\Delta) + (p-1)M(\Delta)$$

and so $\Pi N(\Delta)!$ is bounded as above.

Proof of Lemma 10.2: The sum $\sum_{\{j_\gamma\}}$ is controlled by the exponentially decreasing distance factor, so it is sufficient to show

$$\Pi_\Delta\ M(\Delta)!^r \le \Pi_\gamma\ e^{\text{const.}|\gamma|}\ e^{\text{const.}\sum_\gamma d(j,\gamma)}$$

with constants independent of m_0, λ, $\{j_\gamma\}$ and π.

Recall that $d(j,\gamma)$, defined by (8.2), contains the distance from $j_{1,\gamma}$ and $j_{2,\gamma}$ to some $b \in \gamma$. Thus for fixed Δ, there are at most $O(1)r^2$ values of γ within a fixed partition π such that

$$\Delta_{j_\nu,\gamma} = \Delta\ , \qquad \nu = 1 \text{ or } 2, \tag{10.4}$$

and $d(j,\gamma) \le r$. By definition there are $M(\Delta)$ γ's which satisfy (10.4). The most distant half $(= M(\Delta)/2)$ of these γ's must also satisfy

$$M(\Delta)^{1/2} \le \text{const.}\ d(j,\gamma) + \text{const.}$$

because the γ's are nonoverlapping. Hence

$$M(\Delta)^{3/2} \le \text{const.}\ \sum_\gamma \{d(j,\gamma)\colon \Delta_{j_\nu,\gamma} = \Delta\} + \text{const.}$$

and so the proof is completed by the inequality

$$\begin{aligned}\Pi_\Delta\ M(\Delta)!^r &\le \exp\left(r \sum_\Delta M(\Delta)\ \ell n\ M(\Delta)\right)\\ &\le \exp\left(O \sum_\Delta \{M(\Delta)^{1+\delta}\colon M(\Delta) > 0\}\right)\\ &\le \exp\left(O \sum_\gamma d(j,\gamma)\right) \exp\left(O|\Gamma|\right) .\end{aligned}$$

REFERENCES

0. Z. Ciesielski, Lectures on Brownian motion, heat conduction and potential theory, Aarhus Universitet, 1965.
1. J. Dimock and J. Glimm. Measures on the Schwartz distribution Space and Applications to $P(\phi)_2$ field theories.
2. J. Ginibre. General formulation of Griffiths inequalities. Comm. Math. Phys. 16 (1970) 310-328.
3. J. Glimm and A. Jaffe. The $\lambda(\phi)_2^4$ quantum field theory without cutoffs. III The physical vacuum. Acta Math. 125 (1970) 203-261.
4. J. Glimm and A. Jaffe. The $\lambda(\phi)_2^4$ quantum field theory without cutoffs, IV. Perturbations of the Hamiltonian. J. Math. Phys. 13 (1972) 1568-1584.
5. J. Glimm and A. Jaffe. Positivity of the ϕ_3^4 Hamiltonian. Fort. der Physik. To appear.
6. J. Glimm, A. Jaffe and T. Spencer. The Wightman axioms and particle structure in the $P(\phi)_2$ quantum field model. To appear.
7. F. Guerra. L. Rosen and B. Simon. The $P(\phi)_2$ quantum field theory as classical statistical mechanics.
8. L. Hörmander. Linear Partial Differential Operators. Springer-Verlag, Berlin, 1964.
9. J. Klauder. Ultralocal scalar field models. Comm. Math. Phys. 18 (1970) 307-318.
10. J. Lebowitz and O. Penrose. Decay of correlations. Preprint.
11. R. Minlos and Ja. Sinai. The phenomenon of phase separation at low temperatures in some lattice models of a gas II. Trans. Moscow Math. Soc. Vol. 19 (1968), 121-196.
12. C. Newman. Ultralocal quantum field theory in terms of currents. Comm. Math. Phys. 26 (1972) 169-204.
13. D. Ruelle. Statistical Mechanics. Benjamin, New York, 1969.
14. T. Spencer. The mass gap for the $P(\phi)_2$ quantum field model with a strong external field. Preprint.
15. K. Wilson and J. Kogut. The renormalization group and the ε-expansion. Phys. Reports, to appear.

VII Particles and Bound States and Progress Toward Unitarity and Scaling

PARTICLES AND BOUND STATES AND PROGRESS TOWARD UNITARITY AND SCALING

James Glimm[1]
Rockefeller University
New York, N.Y. 10021

Arthur Jaffe[2]
Harvard University
Cambridge, MA 02138

Abstract

We present a survey of recent developments in constructive quantum field theory.

Introduction. The program of constructive quantum field theory starts with an approximate field theory whose existence is known and constructs a Lorentz covariant limit as the approximations are removed [50]. Frequently it has been convenient to work in the path space world of imaginary time [14,15,42]. The Osterwalder-Schrader axioms [34,35] give sufficient conditions on the Euclidean Green's functions (i.e. the path space theory in the case of bosons) to allow analytic continuation back to real (Minkowski) time and a verification of all Wightman axioms.

This program has been carried out in a number of models in space-time dimension $d < 4$. Once a model has been constructed, the interesting questions involve its detailed properties and how these properties depend on the parameters. For space-time dimension $d = 2$, considerable insight has been obtained into several models including the Sine-Gordon equation [11]. For $d = 3$, recent work yields the first nontrivial model [9]. We describe some of these recent results below, but first we propose a program for $d = 4$.

Consider a lattice ϕ_4^4 model, with lattice spacing ε . Using correlation inequalities of Lebowitz [18,30,47] or the Lee Yang theorem [32,33], we can bound the n-point Schwinger functions

[1] Supported in part by the National Science Foundation under Grant MPS 74-13252.

[2] Supported in part by the National Science Foundation under Grant MPS 73-05037.

$$S^{(n)}(x_1,\ldots,x_n) = \int \phi(x_1)\ldots\phi(x_n)dq ,$$

defined as moments of the measure

$$dq = \frac{e^{-V(\phi)}\prod_x d\phi(x)}{\int e^{-V(\phi)}\prod_x d\phi(x)} ,$$

$$V(\phi) = \sum_x [\tfrac{1}{2} : (\nabla\phi)^2 + m_0^2\phi^2 : + \lambda : \phi^4(x) :] .$$

In fact $S^{(n)}$ is bounded by a sum of products of two point functions. Thus a sufficient condition for the existence of a ϕ^4 field theory (by the method of compactness and subsequences) is a bound, uniform in the lattice spacing ε, on the lattice approximation two point function. In order to prevent the identical vanishing of all n-point functions in the limit $\varepsilon \to 0$, we hold the physical mass fixed, and since $d = 4$, we perform a field strength renormalization.

Field strength renormalization assumes the existence of a one particle pole in the Fourier transform of the two point function, of strength $Z = Z(\varepsilon) \neq 0$. We define

$$S^{(n)}_{ren} = Z^{-n/2}S^{(n)} ,$$

and as above $S^{(n)}_{ren}$ is bounded in terms of $S^{(2)}_{ren} = Z^{-1}S^{(2)}$. Thus the essential missing steps are (i) existence of the one particle pole, so that Z is defined and not zero, for $\varepsilon > 0$, and (ii) control over $Z(\varepsilon)$ as $\varepsilon \to 0$. Since one expects $Z(\varepsilon) \to 0$ (infinite field strength renormalization), it is necessary to show that the rest of the mass spectrum in $S^{(2)}$ has a spectral weight converging to zero (in each bounded mass interval) with ε, so that $S^{(2)}_{ren}$ remains bounded.

The coupling constant renormalization, $\lambda = \lambda(\varepsilon) \to \infty$ as $\varepsilon \to 0$, should not be required for existence of the $\varepsilon \to 0$ limit, but if it is not performed, the limit should be a free field, for $d = 4$. See [24] for further discussion.

In the case $d = 2,3$, we can apply these same ideas to construct the "scaling limit" of (superrenormalizable) ϕ^4 models. In this case we do not require a cutoff for the approximate theories, but vary the bare parameters so $Z \to 0$ and $\lambda \to \infty$, while the physical mass remains fixed and the dimensionless charge approaches its critical value (characterized by the onset of symmetry breaking). We now consider the case $d = 2$ in more detail.

<u>The scaling limit in ϕ_2^4</u> [24]. Consider an interaction Lagrangian

$$L^{\text{Int}} = -\frac{1}{2}\sigma\int :\phi^2:dx - \lambda\int :\phi^4:dx .$$

Here σ is related to the bare mass m_0, but does not equal m_0^2, since the free Lagrangian will contain a mass parameter also; see the appendix of [20] for a discussion. It is expected that a critical value $\sigma_c = \sigma_c(\lambda)$ of σ exists so that as $\sigma \downarrow \sigma_c$, the physical mass m goes to zero. The scaling limit combines the limit $\sigma \downarrow \sigma_c$ with an infinite scale transformation,

$$\lambda \to s\lambda, \quad \sigma \to s\sigma, \quad m^2 \to sm^2,$$

so that m is held fixed. The correct choice of s is given by the conditions

$$\lambda/\sigma \to \lambda/\sigma_c(\lambda) = (\lambda/\sigma(\lambda))_c ,$$

and in this limit $\lambda \to \infty$, $\sigma_c(\lambda) \to \infty$. Note that $\lambda/\sigma_c(\lambda)$ is dimensionless, hence a pure number independent of λ. In this limit all unrenormalized correlation functions converge (after passage to a subsequence). In taking the scaling limit, we also perform a field strength renormalization, so the one particle pole has residue 1. We expect $Z(\sigma) \to 0$ in the scaling limit.

As before, control over the renormalized two point function is sufficient for the existence of the scaling limit, and again the missing steps are (i), (ii) above. Other sufficient conditions are (α) negativity of the six point vertex function $\Gamma^{(6)} \leq 0$, or (β) better than one particle decay of the inverse propagator $-\Gamma^{(2)}(x)$ (which for $x \neq 0$ equals the proper self energy part $\Pi(x)$) or (γ) absence of level crossings and control over CDD zeros of the momentum space two point function [24], and also [23]. Thus (α)-(γ) are central open issues.

A numerical analysis of the ϕ_1^4 model (anharmonic oscillator) is consistent with the validity of parts of (γ) [28]. An explicit calculation shows that $\Gamma^{(6)} < 0$ in the one-dimensional Ising model [39]. A related correlation inequality $S_T^{(6)} \geq 0$ (positivity of the connected part of $S^{(6)}$) has been announced by Cartier; recently Percus [38] and Sylvester [47] have given proofs. Computer studies of the Ising model by Sylvester indicate furthermore that $(-1)^{n+1}S_T^{(2n)} \geq 0$, [48]. The question of correlation inequalities will be discussed at greater length in the lecture of Simon.

In addition to its relation to the problem of constructing ϕ_4^4, the scaling limit is important in the renormalization group approach to the study of critical exponents; see for example [37]. In this connection we note that a number of critical exponents have been bounded from below by their canonical (mean field) values [17]. Furthermore we prove a priori bounds on renormalized coupling constants [21].

The problems (i) and (ii) above arise also in the scaling limit of the d-dimensional Ising model, $2 \leq d \leq 4$. For I_2, asymptotic calculations using Toeplitz determinants indicate that the required bounds are in fact satisfied [31, 49]. The model I_1 is already scale invariant. A formal interchange of limits identifies the scaling limit of I_d with the infinitely scaled ϕ_d^4 model. This statement provides a basis for the idea that spin 1/2 Ising model critical exponents are independent of the details of the lattice structure and are equal to those defined by a ϕ^4 field theory. It also suggests that $\phi^6, \phi^8, \ldots$ tri and multi-critical points are associated with higher spin Ising model tri and multi-critical points.

Particles and Unitarity [25,26]. Let $M = (H^2 - \vec{P}^2)^{\frac{1}{2}}$ be the mass operator. For weakly coupled $P(\phi)_2$ models, the spectrum of M is completely known in the interval $[0, 2m-\varepsilon]$. At zero, M has a simple eigenvalue with eigenvector Ω, the vacuum state. At m, M has an isolated eigenvalue, and the corresponding eigenspace, the space of one particle states, carries an irreducible representation of the Poincaré group. M has no other spectrum below $2m-\varepsilon$; here $\varepsilon \to 0$ and $m \to m_0$ (the bare mass) as the coupling tends to zero. The existence of an isometric S matrix and n-particle in and out states then follows from the Haag-Ruelle scattering theorem.

These results about particles (spectrum of M) are proved using a "cluster expansion," similar to the high temperature expansions in statistical mechanics [25,26]. These expansions, furthermore, give spectral information about the mass (generalized) eigenvectors on any bounded spectral interval, for λ sufficiently small. The main consequence of these expansions is the fact that the Euclidean correlation functions decay (become uncorrelated) as the points separate into clusters, and rate of decay is exponential in the separation distance. Let

$$X = x_1, \ldots, x_n, \qquad Y = y_1, \ldots, y_m$$

$$\Phi(X) = \prod_{i=1}^{n} \phi(x_i), \qquad \Phi(Y) = \prod_{j=1}^{m} \phi(y_j).$$

A typical estimate is

(1) $$\int \Phi(X)\Phi(Y)d\phi - \int \Phi(X)d\phi \int \Phi(Y)d\phi = O(e^{-md})$$

where $m > 0$ is independent of X,Y and

$$d = \text{dist}(X,Y)$$

is the separation distance. The subtraction in (1) can be recognized as a Euclidean vacuum subtraction. Let $\Omega_E \equiv 1$ be the Euclidean vacuum state and let $< , >$ be the Euclidean inner product defined by the measure $d\phi$. Let P_0 be the orthogonal projection onto Ω_E. Then (1) can be written

(2) $$<\Phi(X),(I-P_0)\Phi(Y)> = O(e^{-md}) .$$

In fact, the exponential decay rate in (1),(2) is related to the spectrum of M. Define m_1 as the infimum over the exponential decay rates for different choices of X,Y. Then m_1 is the one particle mass (mass gap) in the spectrum of M.

We generalize (2) by replacing the projection $I-P_0$ with the projection onto the orthogonal complement of the subspace spanned by polynomials of degree $< n$. We then expect decay rates $m_n > m_1$. Let P_n denote the orthogonal projection onto the subspace spanned by Euclidean vectors

$$|X_n> \equiv (I - \sum_{i=0}^{n-1} P_i)\phi(x_1)\ldots\phi(x_n)\Omega .$$

The projection P_n can be written in terms of a kernel $P_n(X,Y)$, and in Dirac notation

(3) $$P_n = \int dXdY|X_n>P_n(X,Y)<Y_n| ,$$

where the integration extends over $R^{nd} \times R^{nd}$ and where the kernel $P_n(X,Y)$ is the inverse (in the sense of integral operators) to the kernel $<X_n|Y_n>$.

The kernel

$$P_1(X,Y) = -K_1(X,Y) = -\Gamma^{(2)}(x-y)$$

is the familiar inverse propagator, while the kernel $P_2(X,Y)$ is related to the Bethe-Salpeter kernel $K_2(X,Y)$, namely

$$P_2(X,Y) = -K_2(X,Y) + \frac{1}{2} K_1 \otimes K_1 .$$

In terms of Feynman graphs, K_2 is the connected part of $-P_2$. More generally, we can define an n-body Bethe-Salpeter kernel $K_n(X,Y)$ as the connected part of $-P_n$. In fact, P_n can be written as a sum of tensor products of Bethe-Salpeter kernels, generalizing the expressions for P_1, P_2 above:

$$P_n(X,Y) = \sum_{r=1}^{n} \sum_{|\alpha_1|+\ldots+|\alpha_r|=n} (-1)^r \binom{n}{\alpha}^{-2} K_{|\alpha_1|} \otimes \ldots \otimes K_{|\alpha_r|} , \tag{4}$$

where $\binom{n}{\alpha} \equiv n! \left(\prod_{i=1}^{r} |\alpha_i|! \right)^{-1}$ and the sum extends over all cluster decompositions into elastic channels. For instance, the term with $r=1$ is just $-K_n$, while the term with $r=n$ equals $(-1)^n (n!)^{-1} \Gamma \otimes \ldots \otimes \Gamma$.

The existence of the Bethe-Salpeter kernel K_n, for $n=1,2,3$ is proved in [19,45,23]. We will publish an analysis of (4). A much deeper question than existence of K_n is the magnitude of the exponential decay rates associated with these kernels. Let

$$|X_{n,r}\rangle \equiv (I - \sum_{i=0}^{n-1} P_i) \phi(x_1) \ldots \phi(x_r) \Omega .$$

We seek estimates generalizing (2) of the form

$$\langle X_{n,r} | Y_{n,r} \rangle = O(e^{-m_n d}), \tag{5}$$

i.e. Euclidean cluster properties, or decay estimates

$$K_n(X,Y) = O(e^{-\overline{m}_n d}) . \tag{6}$$

In order to analyze the decay of K_n, Spencer has derived a new cluster expansion [45] which generalizes [26] by giving higher particle subtractions. Explicitly in the case $n=2$, for weak coupling even $P(\phi)_2$ models, Spencer proves that $\overline{m}_2 \geq 4m(1-\varepsilon)$, where $\varepsilon \to 0$ as the coupling tends to zero. He and Zirilli expect that the decay rate for the two body Bethe-Salpeter kernel will provide information related to asymptotic completeness, up to the threshold $\overline{m}_2$, i.e. for $M \leq 4m(1-\varepsilon)$ [46]. Similar methods give the decay rate for the part of $S_T^{(4)}$ that is two particle irreducible in each channel [2]; this amplitude is obtained by a second Legendre transformation.

If one imagines an extension of this structure analysis to arbitrary n, it appears that the allowed size of the weak coupling region would be n-dependent, and tend to zero as n tends to infinity. The question of dealing with the particle structure away from weak coupling is also of great importance. In this case we only have results for the ϕ^4 interaction, which is repulsive, as described in the following section.

Bound States. For $d = 2,3$, bound states should occur for weak as well as strong attractive forces; they should be missing for repulsive forces. For single phase ϕ^4 models, it is known that no even bound states can occur [44,7,26]. We expect that odd bound states are also missing, cf [24]. For weak $(\phi^6-\phi^4)_2$ models, it is known that mass spectrum occurs in the bound state interval $(m,2m)$ [26]. The Bethe-Salpeter equation combined with improved decay estimates above should allow a complete analysis of the bound state problem for weak coupling $P(\phi)_2$ models; partial results have been obtained [46]. An interesting question is whether bound states occur in ϕ^4 models with symmetry.

Fermions and Many Body Systems. The original construction of the infinite volume limit for the d=2 Yukawa model was given by Schrader [41]; see also [14]. Some portions of this construction have been derived in a Euclidean formalism [43,1]. Aside from the increased simplicity which may accompany a covariant Euclidean construction, the Euclidean formalism is important as a natural framework for a cluster expansion and a study of particles.

Federbush [5,6] has simplified the Dyson-Lenard proof of the stability of matter. The methods were suggested in part by constructive field theory techniques, including techniques he previously employed for the Yukawa_2 model. [6] contains a cluster expansion, which should have a number of applications.

Phase Transitions. Fröhlich [10] has shown that the Euclidean decomposition of the measure $d\phi$ into time translation invariant components coincides with the direct integral decomposition of the associated quantum fields into pure phases.

Three Dimensions. The original semiboundedness proof for the ϕ_3^4 Hamiltonian [16] and related Schwinger function bounds [8,36] were given in a finite volume. A cluster expansion for ϕ_3^4 has been established by Feldman and Osterwalder [9], yielding the first nontrivial d=3 model of the Osterwalder-Schrader axioms. We can hope that further progress will soon bring ϕ_3^4 to the level of understanding we have for d=2 .

REFERENCES

1. D. Brydges, Boundedness below for fermion model theories. Preprint.

2. C. Burnap, private communication.

3. J. Dimock, The $P(\phi)_2$ Green's functions: smoothness in the coupling constant.

4. J.-P. Eckmann, J. Magnon and R. Seneor, Decay properties and Borel summability for Schwinger functions in $P(\phi)_2$ theories. Commun. Math. Phys. To appear.

5. P. Federbush, A new approach to the stability of matter I,II. J. Math Phys. to appear.

6. ________, The semi-Euclidean approach in statistical mechanics I. Basic expansion steps and estimates II. The custer expansion, a special example. Preprint.

7. J. Feldman, On the absence of bound states in the $\lambda\phi_2^4$ quantum field model without symmetry breaking. Canadian J. Phys. 52, 1583-1587 (1974).

8. ________, The $\lambda\phi_3^4$ field theory in a finite volume, Commun. Math. Phys. 37, 93-120 (1974).

9. J. Feldman and K. Osterwalder, The Wightman axioms and mass gap for ϕ_3^4, these proceedings.

10. J. Fröhlich, Schwinger functions and their generating functionals, II. Adv. Math, to appear.

11. ________, The quantized "Sine-Gordon" equation with a nonvanishing mass term in two space-time dimensions. Preprint.

12. J. Glimm, The mathematics of quantum field theory. Adv. Math. To appear.

13. ________, Analysis over infinite dimensional spaces and applications to quantum field theory. Proceedings Int. Congress Math., 1974.

14. J. Glimm and A. Jaffe, Quantum field models, in: Statistical mechanics and quantum field theory, ed. by C. de Witt and R. Stora, Gordon and Breach, New York, 1971.

15. ________, Boson quantum field models, in: Mathematics of contemporary physics, ed. by R. Streater, Academic Press, New York, 1972.

16. ________, Positivity of the ϕ_3^4 Hamiltonian, Fort. d. Physik, 21, 327-376 (1973).

17. ________, ϕ^4 quantum field model in the single phase region: Differentiability of the mass and bounds on critical exponents, Phys. Rev. D10, 536-539 (1974).

18. ________, A remark on the existence of ϕ_4^4. Phys. Rev. Lett. 33, 440-442 (1974).

19. ________, The entropy principle for vertex functions in quantum field models, Ann. l'Inst. H. Poincaré, 21, 1-26 (1974).

20. ________, Critical point dominance in quantum field models, Ann. l'Inst. H. Poincaré, 21, 27-41 (1974).

21. J. Glimm and A. Jaffe, Absolute bounds on vertices and couplings, Ann. l'Inst. H. Poincaré, 22, to appear.

22. ____________________, On the approach to the critical point, Ann. l'Inst. H. Poincaré, 22, to appear.

23. ____________________, Two and three body equations in quantum field models, Preprint.

24. ____________________, On three-particle structure of ϕ^4 and the infinite scaling limit. Preprint.

25. J. Glimm, A. Jaffe and T. Spencer, The Wightman axioms and particle structure in the $P(\phi)_2$ quantum field model. Ann. Math. 100, p. 585-632 (1974).

26. ________________________________, The particle structure of the weakly coupled $P(\phi)_2$ model and other applications of high temperature expansions, in: Constructive quantum field theory, Ed. by G. Velo and A. Wightman, Springer-Verlag, Berlin, 1973.

27. F. Guerra, L. Rosen and B. Simon, Correlation inequalities and the mass gap in $P(\phi)_2$ III. Mass gap for a class of strongly coupled theories with nonzero external field. Preprint

28. D. Isaacson, Private communication.

29. A. Jaffe, States of constructive field theory. Proceedings of 17th International Conference on high energy physics, London, 1974. J.R. Smith, editor, pp. I-243 to I-250.

30. J. Lebowitz, GHS and other inequalities. Commun. Math. Phys. 35, 87-92 (1974).

31. B. McCoy and T. Wu, The two dimensional Ising model. Harvard University Press, Cambridge, 1973.

32. C. Newman, Inequalities for Ising models and field theories which obey the Lee-Yang theorem. Commun. Math. Phys. To appear.

33. ________, Moment inequalities for ferromagnetic Gibbs distributions. Preprint.

34. K. Osterwalder and R. Schrader, Axioms for Euclidean Green's functions, I. Commun. Math. Phys. 31, 83-112 (1973).

35. ________________________________, Axioms for Euclidean Green's functions, II. Preprint.

36. Y. Park, Lattice approximation of the $(\lambda\phi^4-\mu\phi)_3$ field theory in a finite volume. Preprint.

37. G. Parisi, Field theory approach to second order phase transitions in three and two dimensional systems. Cargèse Summer School, 1973.

38. J. Percus, Correlation inequalities for Ising spin lattices. Preprint.

39. J. Rosen, Private communication.

40. J. Rosen and B. Simon, Fluctuations in $P(\phi)_1$ processes. Preprint.

41. R. Schrader, Yukawa quantum field theory in two space time dimensions without cutoff. Ann. Phys. 70, 412-457 (1972).

42. B. Simon, The $P(\phi)_2$ Euclidean quantum field theory. Princeton University Press, Princeton, 1974.

43. E. Seiler, Schwinger functions for the Yukawa model in two dimensions with space-time cutoff.

44. T. Spencer, The absence of even bound states in ϕ_2^4. Commun. Math. Phys., 39, 77-79 (1974).

45. _________, The decay of the Bethe Salpeter kernel in $P(\phi)_2$ quantum field models. Preprint.

46. T. Spencer and F. Zirilli, private communication.

47. G. Sylvester, Representations and inequalities for Ising model Ursell functions, Commun. Math. Phys., to appear.

48. ___________, private communication.

49. C. Tracey and B. McCoy, Neutron scattering and the correlation functions of the Ising model near T_c. Phys. Rev. Lett. 31, 1500-1504 (1973).

50. A. Wightman, Introduction to some aspects of the relativistic dynamics of quantized fields, in: 1964 Cargèse Summer School Lectures, Ed. by M. Lévy, Gordon and Breach, New York (1967), p. 171-291.

DISCUSSION

Masuo Suzuki (comment): I hope that by using your inequalities you can obtain such qualitative results as the dependence of critical exponents upon the dimensionality and potential-range parameter in your ϕ^4 model. In the ferromagnetic Ising model, I proved inequalities such as $\gamma(d) \geq \gamma(d+1) \geq \ldots$ and $\nu(d) \geq \nu(d+1) \geq \ldots$, using Griffiths' inequalities. (Physics Letters 38A (1972) 23.)

VIII Critical Problems in Quantum Fields

CRITICAL PROBLEMS IN QUANTUM FIELDS*

James Glimm[1]
Rockefeller University
New York, New York 10021

Arthur Jaffe[2]
Harvard University
Cambridge, Mass. 02138

RESUME Les liaisons entre le problème de la construction des champs quantiques non triviaux à quatre dimensions et le problème du comportement au point critique à quatre dimensions sont expliqués.

ABSTRACT The connections between the problem of constructing non trivial quantum fields in four dimensions and the problem of critical point behaviour in four dimensions are explained.

* Presented at the International Colloquium on Mathematical Methods of Quantum Field Theory, Marseille, June 1975.

1. Supported in part by the National Science Foundation under Grant MPS 74-13252.

2. Supported in part by the National Science Foundation under Grant MPS 73-05037.

The last two years have seen considerable progress in our understanding of the mathematical structure of quantum fields. In two areas, the progress has been close to definitive, and the problems may be largely resolved in the near future. These areas are (a) the construction of more singular superrenormalizable models: Yukawa_2 and ϕ_3^4 and (b) the detailed structure of $P(\phi)_2$ models which are close to free theories, namely particles, bound states, analyticity, unitarity in subspaces of bounded energy, and phase transitions. In two other areas there has been progress, but the progress is far from being definitive. These areas are (c) the structure away from the neighborhood of free theories, and especially near a critical point and (d) results which pertain indirectly to the construction of four dimensional models. For the results in areas (a) and (b), we merely list recent references, and we then turn to the open problems, including (c) and (d).

Yukawa_2-Euclidean methods [Br 1, Sei, McB 1,2, Br 2, Sei-Si]

ϕ_3^4 - weak coupling expansions [Fe-Os, Ma-Sen]

$P(\phi)_2$ - scattering [Sp 1, Sp-Zi]

$P(\phi)_2$ - analyticity [EMS]

$P(\phi)_2$ - phase transitions [GJS 2,3]

The central problem of constructive quantum field theory has not changed over many years (cf. [St-Wi; p. 168]): the construction of nontrivial quantum fields in four dimensions. We explain how this problem is related to critical point theory in four dimensions, and how a number of simpler problems (of independent interest, and involving two or three dimensional quantum fields) are related to this central problem.

The simplest four dimensional interactions, ϕ_4^4 and Yukawa_4 are renormalizable, but not superrenormalizable. This means that the bare and physical coupling constants are dimensionless. In addition to this dimensionless constant, the field theory is parametrized by two or more parameters with dimension of $(\text{length})^{-1}$. Namely, there are one or more masses and an ultraviolet cutoff κ. To make the exposition explicit, we choose the ultraviolet cutoff as a lattice, and then $\kappa^{-1} = \varepsilon$ is the lattice spacing.

The goal of the construction is to take the limit $\kappa \to \infty$, i.e. $\varepsilon \to 0$. Because scaling is a unitary transformation, and because scaling multiplies all lengths by an arbitrary parameter s, the theory with ε small and mass $m = 1$ is equivalent to the theory with $\varepsilon = 1$ and mass small. In this equivalence, the test functions also scale, and so if we choose $\varepsilon = 1$, a typical test function will have support on a set of large diameter $O(m^{-1})$. Thus if we choose $\varepsilon = 1$, we must focus on the long distance behavior, i.e. on the distance scale $O(m^{-1})$ in a theory with small mass. It follows that the limit $\kappa \to \infty$, $\varepsilon \to 0$ which removes the ultraviolet cutoff is equivalent to the limit $m \to 0$ with $\varepsilon = 1$, if in this latter limit we consider the behavior on the distance scale $O(m^{-1})$. This latter limit (correlation length $= m^{-1} \to \infty$) and distance scale is traditionally considered in critical point theories, namely the "scaling limit" in statistical mechanics. Thus we see that the critical point limit, with fixed lattice spacing $\varepsilon = 1$, is equivalent to the removal of the ultraviolet cutoff and to the construction of a (continuum) quantum field ($\varepsilon = 0$). Since the long distance (infrared) singularities are worse in two and three dimensions, we see that critical point theories in two and three dimensions provide a very realistic test for the mathematical difficulties presented by four dimensions. Indeed the two and three dimensional infrared behavior is typical of nonrenormalizable field theories. A simplification of the two and three dimensional problem (and one which we hope will prove to be minor) is that the critical point can be approached by Lorentz covariant fields satisfying Wightman axioms, in place of the lattice theories introduced above, see [GJ5]. For this reason, in two and three dimensions, the spectral representation of the two point function and (presumably) the particle structure and S-matrix theory can be used as tools to study the theories which are approaching the critical point.

To construct the critical point limit, there are four essential steps:

(i) mass renormalization

(ii) field strength renormalization

(iii) uniform estimates up to the critical point

(iv) nontriviality of the limit.

The first three steps concern existence for this question we would be happy to allow a compactness principle and selection of a convergent subsequence, while hoping that the full sequence converged also. This follows principles well accepted in other branches of mathematics (e.g. partial differential equations) where questions of existence and uniqueness are often studied by separate methods. The last step (nontriviality) depends upon the correct choice of charge renormalization. We will see below that for the ϕ^4 interaction each step can be studied independently of the others.

We now examine each of these four steps in turn. We will see which portions have been solved, which portions seem feasible for study at the present time, which steps are highly interesting in their own right, independently of their role in a possible construction of ϕ_4^4, and which portions seem to present essential difficulties and whose resolution will presumably require essentially new ideas.

The first step, mass renormalization, is the step nearest to completion. For a $\lambda\phi^4 + \frac{1}{2} m_0^2\phi^2$ theory (or more generally for an even $P(\phi)$ theory) the physical mass m is a monotonic function of m_0 for a single phase theory [GRS]. This statement also pertains to a lattice theory (as required for the four dimensional program) if the mass is defined as the exponential decay rate of the two point function. For a ϕ_2^4 theory at least, the mass $m(m_0^2)$ is differentiable [GJ 2] for $m > 0$. The analysis of [Ba] suggests that $m(m_0^2)$ is continuous for $m = 0$; in the lattice case this has been rigorously established [J. Ro 2].

Assuming that the ϕ_2^4 mass is continuous for $m \geq 0$ in a single phase $\lambda\phi^4 + \frac{1}{2} m_0^2\phi^2$ theory, then the mass renormalization is defined as the inverse function

$$m_0^2 = m_0^2(m, \lambda).$$

To see that m may take on all values, $0 \leq m < \infty$, we argue by continuity. For $m_0^2 \to \infty$, $m \to \infty$ also [GJS 1], and so we require a critical theory $(m = 0)$ at the end $m_0^2 = m_{0,c}^2$ of the single phase region, with

$$m(m_0^2) \searrow 0 \quad \text{as} \quad m_0^2 \searrow m_{0,c}^2 .$$

To be explicit, we define

$$M(m_0^2) = \lim_{|x-y|\to\infty} \langle\phi(x)\phi(y)\rangle^{1/2}$$

and

$$m_{0,c}^2 = \sup_{m_0^2}\{m_0^2 \mid M(m_0^2) > 0 \quad \text{or} \quad m(m_0^2) = 0\} .$$

The existence of a phase transition for a ϕ^4 lattice theory [Nel] and for ϕ_2^4 [GJS 2,3] shows that $m_{0,c}^2$ is finite. Combining this fact with the method of [GJ 2, Ba], it can be shown that $m \searrow 0$ as $m_0^2 \searrow m_{0,c}^2$ [J.Ro 2], at least in the lattice case. We summarize the problems of this section under the name: existence of the critical point.

For ϕ_3^4, one expects a similar structure for phase transitions. Assuming this conjecture and using the decay at infinity of the zero mass free field, it follows that $M(m_{0,c}^2) = 0$, but the question of whether $m(m_{0,c}^2) = 0$ remains open. For ϕ_2^4 and for a lattice theory the reasoning concerning M does not apply. In two dimensions the zero mass free field two point function does not decay at infinity, and in a lattice theory, the absence of a Lehmann spectral formula means that the free field is not known to bound the lattice two point function.

For the Yukawa interaction, none of the above results have been obtained. Major steps for the ϕ^4 interaction depend on correlation inequalities, which are presumably not valid for the Yukawa interaction. For the pseudoscalar Yukawa interaction, a phase transition associated with a breaking of the $\phi \to -\phi$ symmetry may be expected on formal grounds. For the scalar Yukawa theory, should one expect an absence of phase transitions and $m_{0,c}^2 = -\infty$? What about cases closer to strong interaction physics, such as one charged and one neutral fermion coupled to three mesons (charged $\pm 1, 0$) ? In general the problem here is: to locate the critical point. This problem is important because renormalizable fields (e.g. ϕ_4^4, Y_4) are equivalent to lattice or ultraviolet cutoff fields studied in the critical point limit. From this point of view, one reason for studying phase transitions in field theory is as an aid in locating the critical points.

The second step is to introduce the renormalized field $\phi_{ren} = Z^{-\frac{1}{2}}\phi$, where Z is defined in terms of the spectral representation for the two point function:

$$<\phi(x)\phi(y)>^{\sim} = \frac{Z}{p^2+m^2} + \int_{a>m^2} \frac{d\rho(a)}{p^2+a} .$$

The essential problem is to show that $Z \neq 0$ for $m > 0$, or in other words to show that for each field in a noncritical theory, there is a corresponding elementary particle. Furthermore we expect only delta function contributions to $d\rho(a)$ (bound states) below the two particle threshold $a = (2m)^2$, and for an even $P(\phi)$ theory, the same should be true below the three particle threshold, because of the $\phi \to -\phi$ symmetry. In this more general form, the problem could be called Hunziker's theorem for field theory.

We now split the discussion of Hunziker's theorem into two independent paths, which we call the repulsive route and the general route. The repulsive route seeks to make maximum use of the special features of the ϕ^4 interaction, in particular of the presumably repulsive forces in this field theory. Since the ϕ^4 terms should dominate for a $P(\phi)$ critical point which is not near a tricritical point, we expect that the main results obtained in the repulsive route should be valid for general $P(\phi)$ theories near critical points which are not tri, or multicritical.

The repulsive route makes essential use of correlation inequalities. An example is the absence of even bound states for single phase ϕ^4 theories [GJS 1, Fe1, Sp 2]. However correlation inequalities by themselves cannot control the critical point behavior, because the ferromagnetic spin ½ Ising model on the Caley tree lattice has anomalous approach to the critical point [Zi]. Thus correlation inequalities must be used in conjunction with the lattice structure, as reflected in the Euclidean invariance and Hamiltonian structure of ϕ_2^4 and ϕ_3^4. A proposed correlation inequality, $\Gamma^{(6)} \leq 0$, implies that $d\rho(a)$ is supported above the three particle threshold, in the interval $[(3m)^2, \infty)$, and that $Z \gtrneqq 0$ [GJ 5]. Thus $\Gamma^{(6)} \leq 0$ would completely settle step two, following the repulsive route. Lowest order perturbation theory suggests that $\Gamma^{(6)} \leq 0$ for weak coupling, and $\Gamma^{(6)} \leq 0$ has been checked for the one dimensional Ising model [J. Ro 1, Ro-Sy]. In view of its importance here and in step three below, further investigation of this inequality would be very desirable. Numerical calculations in some simple cases support the conjecture $\Gamma^{(6)} \leq 0$ [Is - Mar].

The general route is contained in and is substantially equivalent to the problem of asymptotic completeness. In fact the problem of step two -- the existence of a (discrete mass) particle at the bottom of the energy spectrum -- is equivalent at higher energies to the absence of continuous mass spectrum beyond that associated with multiparticle states. For weak coupling and bounded energies, the problem has been solved using cluster expansions. (In [Sp - Zi], energies up to the three particle threshold are allowed for even $P(\phi)$ interactions). For other regions of convergence of the cluster expansion, large external field [Sp 1] or low temperature [GJS 2,3], the situation is expected to be the same. For weak coupling but arbitrary energies, the present methods do not apply.

Outside of the region of convergence of the cluster expansion, the problem seems to involve all major elements of structure of the field theory, including the structure of the vacuum, bound states and superselection sectors [DHR]. The relation of bound states and superselection sectors to asymptotic completeness is well known, since extra bound states as well as extra elementary particles in some extra charge one superselection sector give rise to extra multiparticle continuous spectrum. The relation of the vacuum structure of phase transitions to solitons and superselection rules in two dimensions is contained in [Go-Ja, DHN, Fr 3]. It is an old question to ask whether the Goldstone picture provides a qualitatively correct picture of phase transitions, and now we ask whether the ideas of [Go-Ja, DHN, Fr 3] are sufficient to describe all superselection sectors for $P(\phi)_2$ fields. Can the reasoning be reversed in the sense that $Z=0$ and suppt $\rho(a) = [m^2, \infty)$ would imply existence of a new superselection sector?

On a less etherial level, we ask whether each pure phase for a $P(\phi)$ interaction can be obtained by an appropriate choice of boundary conditions, as is the case in statistical mechanics. Can the

FKG inequalities be used to prove convergence and Euclidean invariance of the infinite volume limit, for general $P(\phi)$ theories? See the introduction to [GJS 3] for a further discussion of the Goldstone picture of phase transitions.

The third step is a bound, uniform as the critical point is approached, on the renormalized n-point Schwinger functions. We follow the repulsive route in our reliance on correlation inequalities. For the ϕ^4 interaction, a correlation inequality reduces this problem to a bound on the renormalized two point Schwinger function

$$S^{(2)} = \langle\phi_{ren}(x)\phi_{ren}(y)\rangle = Z^{-1}\langle\phi(x)\phi(y)\rangle ,$$

[GJ 1]. The reduction applies to all cases (ϕ_2^4, ϕ_3^4 and lattice ϕ_4^4) considered here. It is convenient to choose the scale parameters so that $m=1$ and $\varepsilon\to 0$, and then a sufficient bound on $S^{(2)}$ is e.g.

$$\int S^{(2)}(x)dx \leq \text{const.}$$

with a constant independent of ε, or more generally, $|S^{(2)}|_{\mathscr{S}'} \leq$ const. for some $\mathscr{S}'$-norm$|\cdot|_{\mathscr{S}'}$ independent of ε. In the absence of level crossings, the required bound on $S^{(2)}$ is equivalent to a bound on CDD zeros [GJ 5]. The conjectured inequality, $\Gamma^{(6)} \leq 0$, would imply an absence of bound states (and thus of bound state level crossings), and of CDD zeros below the three particle threshold [GJ 5]. Thus $\Gamma^{(6)} \leq 0$ would bound $S^{(2)}$ and complete the third step. This application of the $\Gamma^{(6)} \leq 0$ inequality was derived in the context of the ϕ_2^4 interaction. The methods extend without change to the ϕ_3^4 interaction. The adaptations of these methods to the lattice ϕ_4^4 interaction is an open problem.

As a concluding remark on the repulsive route, we mention that considerable progress has been made in deriving new correlation inequalities and in finding interrelations between, and simplified proofs of, old ones. See [Sy, New 1,2, El-New, Du-New, El-Mo].

Unfortunately, we have little to say about a possible general route for the third step. In particular there is no argument for believing (or disbelieving) that the phenomena considered above -- absence of bound states and of CDD zeros below the three particle threshold -- occur in the scaling limit approach to a ϕ^6 tricritical point. An absence of level crossings between bound states and the elementary particle and a bound on CDD zeros away from the elementary particle mass might be a general picture, and would yield a bound on the two point function, but not on the general n point functions. In the approach to a general $P(\phi)$ critical point which is not near a tricritical point, the ϕ^4 (repulsive) critical behavior should dominate. Is there any argument (even heuristic) which can be used to discuss bound states, CDD zeros and/or bounds on the renormalized two point function in this region, other than the proposed $\Gamma^{(6)}$ inequality?

The fourth step is the nontriviality of the limit. We hold m fixed (for example m=1) throughout the discussion and consider first $d = 2,3$ dimensions, then $d = 1$ and finally $d = 4$. Starting with a lattice field theory with $\lambda < \infty$ and $\varepsilon > 0$, we have the definition $(d \leq 3)$

$$\phi^4\text{-field theory} = \lim_{\varepsilon \to 0} .$$

We believe that

$$\text{Ising model} = \lim_{\lambda \to \infty} \quad \text{(lattice spacing } \varepsilon)$$

$$\text{scaling limit } \phi^4 \text{ field theory} = \lim_{\lambda \to \infty} \lim_{\varepsilon \to 0}$$

$$\text{scaling limit Ising model} = \lim_{\varepsilon \to 0} \lim_{\lambda \to \infty} .$$

It is reasonable to conjecture that the ε and λ limits above can be interchanged and thus that the scaling limits of the ϕ^4 field theory and the Ising model coincide. This conjecture is a variant of the universality principle for critical exponents in statistical mechanics. Because the Ising critical exponents are known to be nontrivial for $d = 2,3$, we can expect the scaling limit for ϕ_2^4, ϕ_3^4 to be nontrivial.

For $d=1$, all steps one-four have been completed [Is], including interchange of the ε-λ limits. The one dimensional Ising model is already scale invariant, and so the $\varepsilon \to 0$ limit has a trivial form. Control over the $\lambda \to \infty$ limit is obtained from an analysis of anharmonic oscillator eigenvalues and eigenfunctions in a neighborhood of the critical point $m_0^2 = -\infty$.

In $d=4$ dimensions the situation is somewhat different from $d<4$. In terms of the Callan-Symanzik equations, the sign of the crucial function $\beta(\lambda)$ is reversed. This change in the sign of β has its origin in the fact that λ is dimensionless (and thus scale invariant). In terms of the above constructions, the scale invariance means that λ is not taken to infinity by an infinite scale transformation. Rather λ, the bare charge, must be chosen (renormalized) to yield some desired value λ_{phys} of the physical charge. We define the physical charge by

$$\lambda_{\text{phys}} = \lambda_{\text{phys}}(\lambda,\varepsilon m) = -\varepsilon^3 Z^{-2} \chi^{-4} \sum_{x_1,x_2,x_3} \langle \phi(x_1)\cdots\phi(x_4)\rangle^T ,$$

where $\langle \ \ \rangle^T$ denotes the connected, Euclidean Green's function (Ursell function). By Lebowitz' inequality,

$$0 \leq \lambda_{\text{phys}} .$$

We have taken advantage of the scale invariance of λ_{phys} to write it as a function of the scale invariant parameters λ and εm. Recalling that $\lambda = \infty$ is the Ising model, we define

$$\lambda_I(\varepsilon m) = \lambda_{\text{phys}}(\infty,\varepsilon m) .$$

Also note that $\lambda = 0$ is a free lattice field, and

$$0 = \lambda_{\text{phys}}(0,\varepsilon m) .$$

To simplify the discussion of renormalization, we suppose that $\lambda_{phys}(\lambda,\varepsilon m)$ is monotone increasing as a function of λ for fixed εm. (However, we have no argument to support such an hypothesis.)

We claim that λ_{phys} should be continuous in λ and εm. We assume that $G^{(4)} \equiv \langle\phi(x_1)\cdots\phi(x_4)\rangle^T Z^{-2}$ is continuous in λ and εm. Then upper bounds on the two point function, suggested by perturbation theory, substituted in the inequalities of [GJ 3] yield an integrable upper bound for $|G^{(4)}| = -G^{(4)}$, independent of λ and εm, for m fixed, $m > 0$. Continuity of λ_{phys} follows from the Lebesgue bounded convergence theorem.

By definition, charge renormalization is the inverse function,

$$\lambda = \lambda(\lambda_{phys},\varepsilon m),$$

and by continuity, we can choose $\lambda = \lambda(\varepsilon m)$ so that $\lambda_{phys} = \lambda_{phys}(\lambda(\varepsilon m),\varepsilon m))$ approaches any desired value in the interval

$$[0,\lambda_I(0)],$$

as $\varepsilon m \to 0$, $m \neq 0$. Nontriviality of the Ising model (in its critical point limit) is the statement that $\lambda_I(0) \neq 0$. We conclude that the ϕ_4^4 fields constructed here should be nontrivial if and only if the critical behavior of the Ising model is.

According to conventional ideas, $\lambda(\lambda_{phys},\varepsilon m) \to \infty$ as $\varepsilon m \to 0$ in order to ensure $\lambda_{phys} \neq 0$ (infinite charge renormalization). In order to discuss the long and short distance scaling limits of the ϕ_4^4 field, we also suppose

$$\lambda(\lambda_{phys},\varepsilon m) \nearrow \infty$$

as $\varepsilon m \searrow 0$.

In the context of the Callan-Symanzik equations, one changes m_0^2, followed by a scale transformation to keep m fixed. The decrease of

m_0^2 is called long distance scaling; the increase of m_0^2 is called short distance scaling. According to conventional ideas, there are two fixed points to this transformation, the points $\lambda_{phys} = 0$, $\lambda_{phys} = \lambda_I(0)$. The zero mass theories associated with these fixed points are scale invariant.

At the endpoint $\lambda_{phys} = 0$ (assuming χ is finite), the field is Gaussian [Newman]. Presumably it is the free field, invariant under the above transformation group (the renormalization group). At the endpoint $\lambda = \lambda_I(0)$, we expect the field theory to coincide with the long distance scaling limit of the Ising model.

We now consider λ_{phys} lying in the interval $(0,\lambda_I(0))$. For such a theory, according to conventional ideas, the short distance behavior is governed by the fixed point $\lambda_{phys} = \lambda_I(0)$, while the long distance behavior is governed by the fixed point $\lambda_{phys} = 0$. We show that λ_{phys} is monotone increasing in its dependence on m_0^2. Since λ_{phys} is dimensionless, and hence unchanged under scale transformations, this also shows that λ_{phys} decreases under long distance renormalization group transformations and increases under short distance transformations, i.e. $\beta \geq 0$.

Consider two values of the bare mass, m_0, m_0^* satisfying $(m_0^*)^2 < m_0^2$. Let $m^* < m$ be the corresponding masses. By definition

$$\lambda_{phys} = \lim_{\varepsilon \to 0} \lambda_{phys}(\lambda(\lambda_{phys}, \varepsilon m), \varepsilon m)$$

$$\lambda_{phys}{}^* = \lim_{\varepsilon \to 0} \lambda_{phys}(\lambda(\lambda_{phys}, \varepsilon m), \varepsilon m^*) .$$

Since ε is a dummy variable, we replace it in $\lambda_{phys}{}^*$ by $\varepsilon m/m^*$

$$\lambda_{phys}{}^* = \lim_{\varepsilon \to 0} \lambda_{phys}(\lambda(\lambda_{phys}, \varepsilon m^2/m^*), \varepsilon m) .$$

Since $m/m^* > 1$, we have by monotonicity of λ_{phys} in λ and monotonicity of λ in εm that

$$\lambda_{phys}^{*} \leq \lambda_{phys} .$$

This completes the proof.

The statement that the charge renormalization is infinite is equivalent to the statement that the lattice ϕ_4^4 field is free in its critical point behavior (e.g. $Z \to 1$ as $\varepsilon \to 0$, with $\lambda = \text{const.} < \infty$). This is, of course, an open problem.

The existence of ϕ_3^4 and $P(\phi)_2$ fields suggests that the critical point scaling limit exists for the corresponding lattice fields; in the $P(\phi)_2$ case, tri- and multi-critical point limits should also exist. More generally, we summarize the discussion up to this point by asserting that a Euclidean quantum field is the critical point scaling limit of a corresponding lattice field. In the limit of strong physical coupling, the lattice field is replaced by an Ising model.

An alternate approach to nontriviality of the ϕ_4^4 field theory could be based on existence of the classical limit $\hbar \to 0$. Scattering for $\hbar = 0$ is known to be nontrivial. We thank Raczka for this comment.

Turning away from the construction of ϕ_4^4 via critical point theory, we note that recent work [Co, Fr 1] solves the (renormalizable but not superrenormalizable) massive Thirring model in two dimensions. Does this solution provide insight into the problems of charge and wavefunction renormalization? Can other solvable two dimensional models be used as a starting point to prove existence of fields, for interactions of the form (explicitly solvable) + (superrenormalizable)?

Most thinking in contemporary particle physics uses nonabelian gauge fields and a Higgs mechanism as an ingredient. There are many problems here, including a proof of the existence of a Higgs mechanism, even in the lattice case.

We conclude by mentioning two other problems in mathematical physics which may be related to the critical and nonrenormalizable infrared problems considered above. First, the approach of Kolmogoroff to a statistical theory of turbulence uses scaling arguments to deduce exponents governing (short distance) asymptotic behavior. The second problem is the divergence of the virial expansion for the transport coefficients. Here the problem is infrared (slow decay of large time correlations) and infrared. In some cases, the leading divergences can be resummed, and the leading nonanalytic dependence on the density explicitly determined, on a formal level, cf. [Ha - Co].

Bibliography

[Ba] G. Baker, Self-interacting boson quantum field theory and the thermodynamic limit in d dimensions, J. Math. Phys. 16 (1975) 1324-1346.

[Bi-Wi] J. Bisognano and E. Wichmann, On the duality condition for a Hermitian scalar field, J. Math. Phys. 16 (1975) 985-1007.

[Br 1] D. Brydges, Boundedness below for fermion model theories, I and II.

[Br 2] __________, Cluster expansions for fermion fields by the time dependent Hamiltonian approach.

[Co] J. Coleman, The quantum sine-Gordon equation as the massive Thirring model.

[DHN] R. Dashen, B. Hasslacher and A. Niveau, Nonperturbative methods and extended-hadron models in field theory I, II, III, Phys. Rev. D 10 (1974) 4114-4142.

[DHR] S. Doplicher, R. Haag and J. Roberts, Fields, observables and gauge transformations, Commun. Math. Phys. 13 (1969) 1-23.

[Du-New] F. Dunlop and C. Newman, Multicomponent field theories and classical rotators.

[EMS] J.-P. Eckmann, J. Magnen and R. Seneor, Decay properties and Borel summability for Schwinger functions in $P(\phi)_2$ theories, Commun. Math. Phys. 39 (1975) 251-272.

[El-Mo] R. Ellis and J. Monroe, A simple proof of GHS and further inequalities, Commun. Math. Phys. 41 (1975) 33-38.

[El-New] R. Ellis and C. Newman, A tale of two inequalities: Concave force implies concave magnetization.

[Fel] J. Feldman, On the absence of bound states in the $\lambda\phi_2^4$ quantum field model without symmetry breaking, Canad. J. Phys. 52 (1974) 1583-1587.

[Fel-Os] J. Feldman and K. Osterwalder, The Wightman axioms and the mass gap for weakly coupled $(\phi^4)_3$ quantum field theories.

[Fr 1] J. Fröhlich, The quantized Sine-Gordon equation with a nonvanishing mass term in two space-time dimensions.

[Fr 2] J. Fröhlich, The pure phases, the irreducible quantum fields and dynamical symmetry breaking in the Symanzik-Nelson positive quantum field theories.

[Fr 3] ________, New super-selection sectors ('solition states') in two dimensional Bose quantum field models.

[GJ 1] J. Glimm and A. Jaffe, A remark on the existence of ϕ_4^4. Phys. Rev. Lett. 33 (1974) 440-441.

[GJ 2] ________, ϕ_2^4 quantum field model in the single phase region: Differentiability of the mass and bounds on critical exponents. Phys. Rev. D 10 (1974) 536-539.

[GJ 3] ________, Absolute bounds on vertices and couplings, Ann. Inst. H. Poincaré, 22 (1975).

[GJ 4] ________, On the approach to the critical point, Ann. Inst. H. Poincaré, 22 (1975).

[GJ 5] ________, Three particle structure of ϕ^4 interactions and the scaling limit. Phys. Rev. D 11 (1975).

[GJS 1] J. Glimm, A. Jaffe, and T. Spencer, The particle structure of the weakly coupled $P(\phi)_2$ models and other applications of high temperature expansions. In: Constructive Quantum Field Theory, G. Velo and A. Wightman (eds.), Springer-Verlag, Berlin, 1973.

[GJS 2] ________, Phase transitions for ϕ_2^4 quantum fields. To appear.

[GJS 3] ________, A cluster expansion for the ϕ_2^4 quantum field theory in the two phase region. To appear.

[Go-Ja] J. Goldstone and R. Jackiw, Quantization of nonlinear waves. Phys. Rev. D 11 (1975) 1486-1498.

[GRS] F. Guerra, L. Rosen and B. Simon, The $P(\phi)_2$ Euclidean quantum field theory as classical statistical mechanics, Ann. Math. 101 (1975) 111-259.

[Ha-Co] E. Hauge and E. Cohen, Divergences in non-equilibrium statistical mechanics and Ehrenfest's wind tree model. Arkiv for det Fysiske Seminar i Trondheim, No. 7 - 1968.

[Is] D. Isaacson, Private communication.

[Is-Mar] D. Isaacson and D. Marchesin, Private communication.

[Kl-La] A. Klein and L. Landau, An upper bound on the energy gap in the $(\lambda\phi^4+\sigma\phi^2)_2$ model.

[McB 1] O. McBryan, Finite mass renormalizations in the Euclidean $Yukawa_2$ field theory, Commun. Math. Phys. To appear.

[McB 2] ________, Volume dependence of Schwinger functions in the $Yukawa_2$ quantum field theory.

[Ma-Sen] J. Magnen and R. Seneor, The infinite volume limit of the ϕ_3^4 model.

[Nel] E. Nelson, Probability and Euclidean field theory. In: Constructive quantum field theory, G. Velo and A. Wightman (eds.) Springer-Verlag, Berlin, 1973.

[New 1] C. Newman, Inequalities for Ising models and field theories which obey the Lee-Yang theorem, Commun. Math. Phys. 41 (1975) 1-10.

[New 2] ________, Gaussian correlation inequalities for ferromagnets.

[J. Ro 1] J. Rosen, The number of product-weighted lead codes for ballots and its relation to the Ursell functions of the Ising model, J. Comb. Anal. To appear.

[J. Ro 2] ________, Private communication.

[Ro-Sy] J. Rosen and G. Sylvester, Private communication.

[Sei] E. Seiler, Schwinger functions for the Yukawa model in two dimensions with space-time cutoff, Commun. Math. Phys.

[Sei-Si] E. Seiler and B. Simon, On finite mass renormalization in the two dimensional Yukawa model.

[Sp 1] T. Spencer, The mass gap for the $P(\phi)_2$ quantum field model with a strong external field, Commun. Math. Phys. 39 (1974) 63-76.

[Sp 2] T. Spencer, The absence of even bound states in ϕ_2^4, Commun. Math. Phys. 39 (1974) 77-79.

[Sp 3] ________, The decay of the Bethe-Salpeter kernel in $P(\phi)_2$ quantum field models, Commun. Math. Phys. To appear.

[Sp-Zi] T. Spencer and F. Zirilli, Private communication.

[St-Wi] R. Streater and A. Wightman, PCT, Spin and Statistics and all that, Benjamin, New York, 1964.

[Sy] G. Sylvester, Continuous-spin inequalities for Ising ferromagnets.

[Zi] J. Zittartz, New type of phase transition, International Symposium on Mathematical Problems in Theoretical Physics, Kyoto, January 1975.

IX Existence of Phase Transitions for φ_2^4 Quantum Fields

Existence of Phase Transitions for φ_2^4 Quantum Fields

James Glimm [1]
Rockefeller University
New York, N.Y. 10021

Arthur Jaffe [2]
Harvard University
Cambridge, Mass. 02138

Thomas Spencer [2,3]
Harvard University
Cambridge, Mass. 02138

RESUME L'existence des transitions de phase pour les champs quantiques $\lambda\varphi_2^4$ dans la région $\lambda \gg 1$ de couplage nu est établie. La brisure de symétrie pour l'interaction $\lim \mu \to \pm 0$ $(\varphi^4 - \mu\varphi)$ est aussi démontrée. On fait la distinction entre les transitions de phase et la brisure de symétrie.

ABSTRACT The existence of phase transitions for $\lambda\varphi_2^4$ quamtum fields in the region $\lambda \gg 1$ of bare coupling is established. Symmetry breaking for the interaction $\mu \to \pm 0$ $(\varphi^4 - \mu\varphi)$ is alsoproved and the distinction between phase transitions and symmetry breaking is emphasized.

1 - New Results

We prove the existence of phase transitions for $\lambda\varphi_2^4$ quantum fields in the region $\lambda \gg 1$ of bare coupling. The same methods apply in principle to even $\lambda P(\varphi)_2$ models. We demonstrate the existence of long range order in the (even) $P(\varphi)_2$ theory defined with zero Dirichlet boundary data. (However, we restrict attention in this talk to φ_2^4.) We also prove the existence of symmetry breaking for the interaction

$$\lim_{\mu \to \pm 0} (\varphi^4 - \mu\varphi) .$$

As in statistical mechanics, where phase transitions may occur without symmetry breaking [4], we expect phase transitions in certain quantum field models which do not possess a symmetry group, such as the interaction

$$(\varphi^2 - \sigma^2)^4 + \epsilon\varphi^3 - \mu\varphi ,$$

with $\sigma >> 1$, $\epsilon << 1$, $\mu = \mu(\epsilon, \sigma)$. Thus we emphasize this distinction between phase transitions and symmetry breaking.

In a separate article [5], we give a cluster expansion for strong (bare) coupling of even φ_2^4 models. This expansion allows us to construct two pure phases, each satisfying the Wightman and Osterwalder-Schrader axioms, with a unique vacuum and with a mass gap.

In contrast to our detailed study based on the cluster expansion [5], we present at this conference a simple, direct proof that phase transitions occur. The details of this talk will be published separately [6]. An alternative approach to the problem of phase transitions has been announced in [7], but the proof has not appeared.

<u>Theorem 1.</u> Consider the $\lambda :\varphi_2^4:_{m_0^2} + \frac{1}{2} m_0^2 :\varphi^2:_{m_0^2}$ theory with Wick ordering mass m_0, bare mass m_0, and zero Dirichlet boundary conditions. For λ/m_0^2 sufficiently large, there is long range order (lack of clustering).

<u>Theorem 2.</u> Consider the model

$$\lim_{\mu\to 0+}(\lambda :\varphi_2^4:_{m_0^2} + \tfrac{1}{2} m_0^2 :\varphi^2:_{m_0^2} - \mu\varphi)$$

with Wick ordering mass m_0 and bare mass m_0. For λ/m_0^2 sufficiently large, there is symmetry breaking, i.e.

$$\lim_{\mu\searrow 0} \langle\varphi\rangle > 0 ,$$

where $\langle \cdot \rangle$ denotes the vacuum expectation value. Likewise the model defined by $\mu \to 0-$ has $\langle\varphi\rangle < 0$.

Our proof of these theorems is based on a Peierls argument, similar to the proof of phase transitions in statistical mechanics. The basic idea is to study the average field

$$\varphi(\Delta) = \int_{\Delta} \varphi(x)dx$$

where the average is taken over a unit square Δ in Euclidean space-time. The average (low momentum) field dominates the description of phase transitions, while the error

$$\delta\varphi(x) = \varphi(x) - \varphi(\Delta), \qquad x \in \Delta,$$

the "fluctuating field" is estimated in terms of the kinetic part of the action, $\frac{1}{2}(\nabla\varphi)^2$. Technically, we use φ^j bounds to establish the estimates which give the convergent Peierls expansion, and show the probability of "flipping" values of $\varphi(\Delta)$ is small.

In place of repeating the material in [6], we explain the classical (mean field) approximation to the φ^4 theory. This classical picture is the basis for our convergent expansions about the mean field.

2. Classical Approximation

Consider a quantum field defined by the Euclidean action density

$$\tfrac{1}{2}:(\nabla\varphi)^2:_{a^2} + :\mathcal{V}(\varphi):_{a^2} = :\tfrac{1}{2}(\nabla\varphi)^2 + \tfrac{1}{2}a^2\varphi^2:_{a^2} + :P(\varphi):_{a^2}$$

Here $:\ :_{a^2}$ denotes Wick ordering with respect to mass a, and by convention we include a bare mass a in the free part of the action, $\frac{1}{2}:(\nabla\varphi)^2 + a^2\varphi^2:$. The classical approximation for the ground state of the field φ is obtained by regarding $\frac{1}{2}(\nabla\varphi)^2$ as a kinetic term and $\mathcal{V} \equiv \frac{1}{2}a^2\varphi^2 + P(\varphi)$ as a potential term. Then in the classical approximation the vacuum expectation (mean) $\langle\varphi\rangle$ of φ equals φ_c, a value of φ which minimizes $\mathcal{V}$. The classical mass m_c is given by

$$m_c^2 = \mathcal{V}''(\varphi_c) = a^2 + P''(\varphi_c) .$$

In other words the classical low mass states of φ are those of a free field with action density

$$\mathcal{V}_c = :\tfrac{1}{2}(\nabla\varphi)^2 + \tfrac{1}{2}m_c^2(\varphi - \varphi_c)^2:_{m_c^2} .$$

For convenience, we choose the constant in P so that $P(0) = 0$. (The same then holds for $\mathcal{V}$.)

We expect the classical approximation to be accurate (up to higher order quantum corrections) for those interaction polynomials $\mathcal{V}$ such that

(i) $\mathcal{V} - \mathcal{V}_c$ is small for $\varphi - \varphi_c$ small,

and

(ii) $a^2 = m_c^2$.

We say that an interaction $:P:_{a^2}$ satisfying (i) and (ii) is classical.

To understand the conditions (i) and (ii) concretely, we write $\mathcal{V}(\varphi)$ in terms of its Taylor series about $\varphi = \varphi_c$, namely

$$\mathcal{V}(\varphi) = \mathcal{V}(\varphi_c) + \tfrac{1}{2}m_c^2(\varphi - \varphi_c)^2 + \sum_{i\geq 3} \lambda_i(\varphi - \varphi_c)^i ,$$

where

$$\lambda_i = (i!)^{-1}\mathcal{V}^{(i)}(\varphi_c)$$

$$= (i!)^{-1}P^{(i)}(\varphi_c) , \qquad i \geq 3 .$$

In particular, condition (i) is satisfied if

(1) $$|\lambda_i/m_c^2| \ll 1 ,$$

where $i \geq 3$.

To achieve (ii) will normally require Wick reordering, and in preparation, we calculate the a dependence of the Wick constant

$$c(a^2) = \frac{1}{(2\pi)^2}\int\frac{d^2p}{p^2 + a^2} .$$

Then

$$-\frac{d}{da^2}c(a^2) = \frac{1}{(2\pi)^2}\int\frac{d^2p}{(p^2 + a^2)^2} = (4\pi a^2)^{-1} ,$$

where we interpret this formula as a $\kappa \to \infty$ limit of cutoff equations in which $p^2 \leq \kappa^2$. We expect that (ii) will be satisfied after Wick reordering if

(2) $$(\lambda_i/m_c^2)\ln^{i/2}(m_c^2/a^2) \ll 1\,, \quad i \geq 3\,.$$

In the following section we carry out this choice for the φ_2^4 model. The classical approximation is also referred to as the Goldstone approximation or the mean field approximation.

3. The φ^4 Interaction

The conventional definition of the φ^4 interaction is

(3) $$:\mathcal{V}:_{m_0^2} = \lambda:\varphi^4:_{m_0^2} + \tfrac{1}{2}m_0^2:\varphi^2:_{m_0^2}$$

$$= :P(\varphi):_{m_0^2} + \tfrac{1}{2}m_0^2:\varphi^2:_{m_0^2}\,.$$

The weak coupling region $\lambda/m_0^2 \ll 1$ satisfies (1) and (2), and hence is also a classical region. In this region $\varphi_c = 0$, $m_c = m_0$. Thus the classical picture of weakly coupled φ^4 is a field with mean zero, and particles of mass $m_c = m_0$. The $\varphi \to -\varphi$ symmetry preserves $\langle\varphi\rangle = 0$ as an exact identity, but we expect quantum corrections to give a physical mass

(4) $$m = m_c(1 + O(\lambda/m_c^2)), \quad \lambda/m_c^2 \to 0\,.$$

In fact the weak coupling region is well understood from the cluster expansion [8], which yields a Wightman-Osterwalder-Schrader theory for $\lambda/m_0^2 \ll 1$, and $m - m_c = o(\lambda/m_c^2)$.

We now turn our attention to the region $\lambda/m_0^2 \gg 1$. In this region $:\mathcal{V}:_{m_0^2}$ given by (3) is clearly not classical, since both (1) and (2) fail. In order to obtain a classical interpretation, we rewrite (3) in terms of a new Wick ordering mass a satisfying

$$a^2 \gg \lambda \gg m_0^2. \tag{5}$$

Then we write (3) as a new polynomial $:\mathcal{V}_1(\varphi):_{a^2}$ satisfying $\mathcal{V}_1(0) = 0$. Thus

$$\begin{aligned} :\mathcal{V}(\varphi):_{m_0^2} + \text{const.} &= :\mathcal{V}_1(\varphi):_{a^2} \\ &= :\lambda\varphi^4 - \left(\frac{6\lambda}{4\pi} \ln \frac{a^2}{m_0^2} - \tfrac{1}{2}m_0^2\right)\varphi^2:_{a^2} \end{aligned} \tag{6}$$

Here

$$\frac{1}{4\pi} \ln \frac{a^2}{m_0^2} = c(m_0^2) - c(a^2) .$$

Likewise

$$:P_1(\varphi):_{a^2} = :\lambda\varphi^4 - \left(\frac{6\lambda}{4\pi} \ln \frac{a^2}{m_0^2} + \tfrac{1}{2}a^2 - \tfrac{1}{2}m_0^2\right)\varphi^2:_{a^2} \tag{7}$$

and

$$\mathcal{V}_1 = P_1 + \tfrac{1}{2}a^2\varphi^2 .$$

By (5), the coefficient of φ^2 in $\mathcal{V}_1$ is negative, so $\mathcal{V}_1$ has a double minimum at $\varphi = \varphi_c = \pm\sigma$. Here m_c and σ are related by

$$m_c^2 = 8\lambda\sigma^2 = \frac{6\lambda}{\pi} \ln \frac{a^2}{m_0^2} - 2m_0^2 . \tag{8}$$

We now choose a so that $m = a$, as can be achieved by letting x solve the equation

$$x = \frac{\lambda}{m_0^2} \frac{6}{\pi} \ln x - 2 .$$

For λ/m_0^2 sufficiently large, this equation has exactly two solutions. The larger solution determines a by the relation $x = (a/m_0)^2$. The smaller solution is spurious in the sense that is gives an interaction satisfying (ii) but not (i).

Next we perform a scale transformation so the classical mass becomes one. Since the Wick ordering mass transforms similarly, it also becomes 1. Thus after the scale transformation. we obtain an interaction polynomial $:\mathcal{V}_2:_1$ given by

$$:\mathcal{V}_2:_1 = :\frac{1}{8\sigma^2}\varphi^4 - \frac{3}{4}\varphi^2:_1 + \tfrac{1}{2}:\varphi^2:_1 \tag{9}$$

$$= :P_2:_1 + \tfrac{1}{2}:\varphi^2:_1 .$$

By (8), we see that $\sigma \gg 1$. Thus the interaction (3), in the strong coupling region $\lambda/m_0^2 \gg 1$, is equivalent to the weakly coupled φ^4 interaction (9), with a negative quadratic term, with bare mass 1 and with Wick mass 1. For the interaction (9), we find that

$$\varphi_c = \pm\sigma, \qquad m_c = 1, \tag{10}$$

$$\lambda_3 = \pm(2\sigma)^{-1}, \qquad \lambda_4 = (8\sigma^2)^{-1}.$$

Thus for σ large, both (1) and (2) are satisfied and (9) is classical. It exhibits the two phase classical approximation to strongly coupled φ_2^4, since φ_c has two possible mean field values. In our second paper [5], we present a systematic expansion about the classical field φ_c, combined with a Peierls argument to select a given phase. We find that in each of two pure phases, the physical mass is positive.

References and Footnotes

1. Supported in part by the National Science Foundation under Grant MPS 74-13252.

2. Supported in part by the National Science Foundation under Grant MPS 73-05037.

3. On leave from Rockefeller University, New York, NY 10021.

4. S. A. Priogov and Ya. G. Sinai, Phase transitions of the first kind for small perturbations of the Ising model, Funct. Anal. and its Appl. 8, 21-25 (1974). (English Trans.)

5. J. Glimm, A. Jaffe and T. Spencer, A cluster expansion in the wo phase region, in preparation.

6. J. Glimm, A. Jaffe and T. Spencer, Phase transitions for φ_2^4 quantum fields, Commun. Math. Phys., to appear.

7. R. Dobrushyn and R. Minlos, Construction of a one-dimensional quantum field via a continuous Markov field, Funct. Anal. and its Appl. 7, 324-325 (1973). (English Trans.)

8. J. Glimm, A. Jaffe and T. Spencer, The Wightman Axioms and particle structure in the $P(\varphi)_2$ quantum field model, Ann. Math. 100, 585-632 (1974).

X Critical Exponents and Renormalization in the φ^4 Scaling Limit

Critical Exponents and Renormalization in the ϕ^4 Scaling Limit

J. Glimm[+]
Rockefeller University
New York, New York 10021

A. Jaffe[++]
Harvard University
Cambridge, Mass. 02138

Abstract

For dimensions $d \leq 3$, the ϕ^4 scaling limit defines a nonrenormalizable field theory. The standard relations between critical exponents and renormalization are presented. Arguments supporting the existence of the scaling limit are based on correlation inequalities and the numerical values of Ising model exponents, $2\eta \lneq \eta_E$ for d=2,3.

Contents

[+] Supported in part by the National Science Foundation under Grant 74 - 13252.

[++] Supported in part by the National Science Foundation under Grant 75 - 21212.

1. Formulation of the Problem

The Euclidean action density

$$:(\nabla\phi^2 + \lambda\phi^4 + \sigma\phi^2 + \mu\phi): \tag{1.1}$$

determines a quantum field theory for $d=1,2,3$ [7,21,22,4,20,26, 6,23]. The Wick order is defined with a bare mass 1. Dirichlet boundary conditions and (for d=3) additional mass counterterms are implied in (1.1). For large positive values of σ, the field has a unique state Ω and a positive mass

$$m = m(\lambda,\sigma,\mu) > 0, \tag{1.2}$$

[13,4,20]. For d=2 (and presumably for d=3), for $\mu=0$, and for large negative value of σ, the vacuum is degenerate [14]. We define the critical value of σ, $\sigma_c=\sigma_c(\lambda)$ as the supremum of the values of σ for which either the vacuum is degenerate, or for which (1.2) fails.

On the interval $(\sigma_c(\lambda),\infty)$, $m(\lambda,\sigma,0)$ is monotone increasing and Lipschitz continuous [16,9] in σ, for d=2 at least, and guided by theorems concerning lattice fields [1,24] (and arbitrary $d\geq 2$), we expect that $m\searrow 0$ as $\sigma\searrow\sigma_c$. This is the only mathematically rigorous statement which can be made about the critical point: for ϕ_2^4 fields (and Ising models), $\sigma_c<\infty$ while for lattice ϕ^4 fields (and Ising models) $m(\sigma)\searrow 0$ as $\sigma\searrow\sigma_c$. In addition to the corresponding continuum problem, as mentioned above, the uniqueness of the vacuum at $\sigma=\sigma_c$ and the conjecture

$$m(\lambda,\sigma_c,0) = 0$$

are open problems, both for Ising models ($d\geq 3$), and lattice and continuum fields ($d\geq 2$). The d=2 Ising model, I_2, is of course a special case, because the existence of a closed form solution makes the detailed critical structure accessible [2].

Because of the absence of a mathematical theory of critical behavior, the remainder of our discussion will be mainly on a heuristic level. Let $\langle\ldots\rangle = \int \ldots d\phi$ be the Euclidean vacuum expectation associated with the quantum field defined by (1.1). The two point Schwinger function

$$S^{(2)}(x,y) = \int \phi(x)\phi(y)d\phi$$

can be represented, according to the Lehmann spectral theorem, in the form

$$S^{(2)}(x,y) = \int C_a(x-y)d\rho_a, \qquad (x \neq y)$$

where C_a is the convolution inverse to $(-\Delta+a)$. Necessarily,

$$m^2 = \inf \text{suppt } d\rho_a$$

and we make the further assumption (absence of bound states in $S^{(2)}$) that

$$S^{(2)}(x,y) = Z_3 C_{m^2}(x-y) + \int_{a \geq (3m)^2} C_a(x-y)d\rho_a \tag{1.3}$$

The evidence in favor of this assumption will be presented below. In particular, we take $\mu=0$, $\sigma>\sigma_c$, since otherwise bound states may be expected. Here the wave function renormalization constant Z_3 is defined by (1.3).

The problem of critical exponents is to understand the leading singularity of the long distance behavior of the field ϕ, as $\sigma \to \sigma_c$, in particular as expressed in such formulae as

$$m \sim \text{const} \left(\frac{\sigma-\sigma_c}{\sigma_c}\right)^{\nu} \tag{1.4}$$

$$\chi \equiv S^{(2)}(x-y)dx \sim \text{const} \left(\frac{\sigma-\sigma_c}{\sigma_c}\right)^{-\gamma} \tag{1.5}$$

$$Z_3 \sim \text{const} \left(\frac{\sigma - \sigma_c}{\sigma_c} \right)^{\zeta_3} \tag{1.6}$$

$$S^{(2)}(x-y) \Big|_{\sigma=\sigma_c} \sim \text{const} \, |x-y|^{2-d-\eta} \quad \text{as } |x-y| \to \infty \tag{1.7}$$

2. The Scaling and Critical Point Limits

The ϕ^4 field theory, $d\leq 3$, has two intrinsic length scales. The longer length scale is the correlation length,

$$\xi = m^{-1}$$

and this length governs the long distance decay of the correlation functions

$$S^{(n)} = \int \phi(x_1)\dots\phi(x_n)d\phi$$

In fact for n=2,

$$S^{(2)}(x-y) \sim Z_3 m^{d-2}(mr)^{(1-d)/2}e^{-mr}, \quad mr\to\infty$$

assuming (1.3). The short distance behavior of a ϕ^4 field is canonical (free) for $d\leq 3$, cf. [8,10]. The short length scale is the distance scale on which this canonical behavior becomes dominant. For a lattice field or Ising model, the short length scale could be the lattice spacing. The lattice spacing, of course, functions as an ultraviolet cutoff, but we will see below that the short distance scale is always an ultraviolet cutoff. In the case of a Lorentz covariant ϕ^4 continuum field theory, the short distance scale cuts off the nonrenormalizable infrared singularities, and substitutes a Lorentz covariant short distance canonical behavior, for d=2,3. It could happen (for example in the continuum limit of lattice theories) that the lattice spacing is very short relative to the above Gaussian short distance scale. This case, in which there are two short distance scales and one long distance scale, will not be discussed further.

In the case of two length scales, one can always be eliminated by a scale transformation, and only the dimensionless ratio

$$\varepsilon = \frac{\text{short distance scale}}{\text{long distance scale}} \tag{2.1}$$

has an intrinsic signifiance. The scale transformation U_s is a unitary operator connecting the Hilbert spaces of the fields (1.1) having the same value of dimensionless bare charge

$$g = \lambda\sigma^{(d-4)/2} \tag{2.2}$$

and differing values of σ. U_s is defined to multiply all lengths by the factor s, so that

$$U_s\phi(x)U_s^{-1} = s^{(2-d)/2}\phi(s^{-1}x) \equiv \psi_s(x) \tag{2.3}$$

defines a field with the new parameter values

$$\lambda_s = s^{d-4}\lambda, \qquad \sigma_s = s^{-2}\sigma$$

and Wick ordering mass s^{-1}. (A reWick ordering then gives the action of the field ψ_s the form (1.1).)

There are two distinguished choices for the parameter s, namely

$$s = 1 \qquad \text{or} \qquad s = \varepsilon$$

In the first choice, s=1, the short distance scale is fixed at one; this choice defines the <u>unscaled</u> theory. In the second choice, s=m, the long distance scale ξ = 1/mass, is fixed at one; this choice defines the <u>scaled</u> theory. We note ε is the mass of the unscaled theory and the lattice spacing or short distance cutoff of the scaled theory.

It follows that there are three distinguished distance intervals:

$$r \le \text{short distance scale} \tag{2.4a}$$

$$\text{short distance scale} \le r \le \text{long distance scale} \tag{2.4b}$$

$$\text{long distance scale} \le r \tag{2.4c}$$

We call these intervals (a) the canonical (or free, or Gaussian) interval, (b) the scaling (or critical) interval, and (c) the one particle interval. In the two extreme intervals, i.e. in the canonical and one particle intervals, the behavior of ϕ is well understood. In cases where these two intervals dominate, (high or low temperature) there is a convergent cluster expansion, and ϕ is well understood on all distance scales [13,14,15].

Near the critical point, ε becomes small, and the critical, or scaling interval (2.4b) dominates. As $\varepsilon \to 0$, we may construct three limiting theories.

i) The critical point is defined by $\varepsilon \to 0$, $\sigma \searrow \sigma_c$ in the unscaled theories. This limit is characterized by

$0 = m$

$0 \neq$ short distance scale.

ii) The scaling limit is defined by $\varepsilon \to 0$ in the scaled theories, and is characterized by

$1 = m$

$0 =$ short distance scale

In this limit, we use the renormalized field $\phi_{ren} = Z_3^{-1/2}\phi$ in place of ϕ.

iii) The scaled critical point is defined by a long distance scaling of i) or a short distance scaling, $m \to 0$, of ii), and is characterized by

$m = 0 =$ short distance scale

These two definitions of iii) are related by an interchange of order of limits.

Alternately, we may say that i) eliminates (2.4c), ii) eliminates (2.4a) and iii) eliminates both, in either order. To check this interchange in the order of limits, we analyze its influence on η, using Ising model exponents. In the limit iii),

$$S^{(2)} = r^{2-d-\eta} \qquad (2.5)$$

and η as defined in (1.7), refers to the long distance scaling of i).

We now assert, as a scaling hypothesis, that the decay rate (2,5) holds for $S^{(2)}$ for the entire interval (2.4b). As a check on this assertion, we use it to compute (in the unscaled theory)

$$\chi = \int S^{(2)}(r)dr \sim \int_{|r|\leq\varepsilon^{-1}} r^{2-d-\eta}dr \sim \varepsilon^{-2+\eta}$$

In (1.4-6), the unscaled theory is understood, and so

$$\varepsilon \sim \left(\frac{\sigma-\sigma_c}{\sigma_c}\right)^{\nu} ; \quad Z_3 \sim \varepsilon^{\zeta_3/\nu} ; \quad \chi \sim \varepsilon^{-\gamma/\nu}$$

Thus the consistency check follows from the identity

$$\gamma = (2-\eta)\nu \tag{2.6}$$

valid for Ising model exponents, d=1,2,3.

We next consider the scaled theory. At the short distance end of (2.4b),

$$S^{(2)}_{ren}(\varepsilon) = Z_3^{-1}S^{(2)}_{free}(\varepsilon) = Z_3^{-1}\varepsilon^{2-d} = \varepsilon^{2-d-\zeta_3/\nu} \tag{2.7}$$

neglecting logarithms in d=2 dimensions. Thus

$$\begin{aligned} S^{(2)}_{ren}(r) &= \varepsilon^{2-d-\zeta_3/\nu}(r/\varepsilon)^{2-d-\eta} \\ &= \varepsilon^{\eta-\zeta_3/\nu} r^{2-d-\eta} \end{aligned} \tag{2.8}$$

on (2.4b) by the scaling hypothesis. We note that the (rigorous) inequalities

$$2\nu - \zeta_3 \leq \gamma \leq 2\nu - \eta\nu$$

[5,9] imply that

$$\eta - \zeta_3/\nu \leq 0 \tag{2.9}$$

so that (2.8) does not vanish as $\varepsilon \to 0$.

We further require equality in (2.9), as is known for I_2, and as follows from the existence of the limit ii), in the special case (2.8). Then in the scaling limit ii), we have

$$S^{(2)}_{ren}(r) \cong r^{2-d-\eta}, \quad r << 1 \qquad (2.10)$$

and scaling from short distances gives $S^{(2)}_{ren}(r) = r^{2-d-\eta}$ for all r. Thus the two definitions of the limit iii) agree, at least in their two point function. In particular the two definitions of η agree. In summary we can say that interchange of limits in iii) follows from a scale relation, such as (2.5), valid over the entire scaling interval (2.4b), and that this hypothesis can be checked against known values of the exponents.

A comprehensive exposition of the scaling behavior and the theory of the renormalization group is given in [28].

3. Renormalization of the $\phi^2(x)$ Field

The field $\phi^2(x)$ requires both an additive and a multiplicative renormalization. The additive renormalization is Wick ordering in the physical vacuum, defined by

$$:\phi^2(x): \; = \phi^2(x) - \langle\phi^2(x)\rangle \tag{3.1}$$

(These Wick dots do not coincide with those of (1.1), but in the term $\sigma\int:\phi^2:$ the difference is a constant, and of no consequence). The multiplicative renormalization

$$\begin{aligned}(:\phi^2(x):)_{ren} &= Z_E^{-1/2}:\phi^2(x): \\ &\neq \; :(\phi_{ren})^2(x): \end{aligned} \tag{3.2}$$

is not obtained by renormalizing each factor ϕ in $\phi^2(x)$. Rather we define the exponents

$$\langle:\phi^2(x)::\phi^2(y):\rangle\Big|_{\sigma=\sigma_c} \sim \text{const}\,|x-y|^{4-2d-\eta_E} \quad \text{as} \quad |x-y|\to\infty \tag{3.3}$$

$$C_H = \int\langle:\phi^2(x)::\phi^2(y):\rangle d(x-y) \sim \left(\frac{\sigma-\sigma_c}{\sigma_c}\right)^{-\alpha} = \varepsilon^{-\alpha/\nu} \tag{3.4}$$

Here C_H is the specific heat, and if $Z(\sigma,\mu)$ is the partition function for (1.1), then

$$C_H = \frac{\partial^2}{\partial\sigma^2}\frac{\ln Z}{V}$$

$$\chi = \frac{\partial^2}{\partial\mu^2}\frac{\ln Z}{V}$$

As in §2, we make a scaling hypothesis, that (3.3) is valid over the scaling interval (2.4b), and then the identity in the unscaled theory

$$\varepsilon^{-\alpha/\nu} = \int_{1\le|r|\le\varepsilon^{-1}} r^{4-2d-\eta_E}dr + \int_{|r|\le 1} r^{4-2d}dr$$

$$= \varepsilon^{-(4-d-\eta_E)}$$

implies

$$\alpha/\nu = 4-d-\eta_E = \begin{matrix} 0, & d=2 & \text{Ising} \\ 2 & d=3 & \text{Ising} \end{matrix} \tag{3.5}$$

This identity is analogous to the identity (2.6) for the ϕ-two point function. As in §2, a matching of the canonical form of the ϕ^2 two point function on (2.4a) with its scaling form on (2.4b) implies $\zeta_E/\nu = \eta_E$. In fact we assume in the unscaled theory that

$$<:\phi^2(x)::\phi^2(y):> \sim c_a r^{4-2d} \qquad \text{on (2.4a)}$$

$$\sim c_b(\varepsilon) r^{4-2d-\eta_E} \qquad \text{on (2.5b)}$$

Because $\sigma=\sigma_c$ is a regular point for the canonical part of ϕ, we may take $c_a=c_a(\varepsilon)\big|_{\varepsilon=0}$ to be independent of ε. Equality of the two asymptotic forms at r=1 implies $c_b=c_a$ is also regular in ε. It follows that in the scaled theory

$$<:\phi^2(x)::\phi^2(y):> \sim c_b \varepsilon^{4-2d}(r/\varepsilon)^{4-2d-\eta_E} \tag{3.6a}$$

and

$$<:\phi^2:_{ren}:\phi^2(y):_{ren}> \sim c_b \varepsilon^{\eta_E-\zeta_E/\nu} r^{4-2d-\eta_E} \tag{3.6b}$$

on (2.4b). We now define $Z_E = \varepsilon^{\zeta_E/\nu}$ and

$$\zeta_E/\nu = \eta_E \tag{3.7}$$

The fact that

$$\eta_E = \begin{cases} 2, & d=2 \\ .8, & d=3 \end{cases} > 2\eta = \begin{cases} 1/2, & d=2 \\ .082, & d=3 \end{cases} \tag{3.8}$$

implies the following important result:

$$:(\phi_{ren})(x)^2: = 0 \tag{3.9}$$

in the scale limit. Here we obtain the values for η_E from (3.5).

Theorem 3.1. For the scaled theories,

$$\left\|\int :(\phi_{ren})(x)^2:dx\right\|_{L_2}^2 \sim \varepsilon^{\eta_E-2\eta}$$

Proof. We use (3.6)-(3.7) as follows:

$$\left\|\int_\Lambda :(\phi_{ren})(x)^2:dx\right\|_{L_2}^2 = \varepsilon^{\eta_E-2\eta}\left\|\int_\Lambda :\phi(x)^2:_{ren}dx\right\|_{L_2}^2$$

$$= O(1)\varepsilon^{\eta_E-2\eta}\int_{\varepsilon\le|r|\le 1} r^{4-2d-\eta_E}dr$$

$$= O(1)\varepsilon^{\eta_E-2\eta}$$

In more intuitive language, $:(\phi_{ren})^2:$ and $:\phi^2:_{ren}$ differ by an infinite multiple. Since the constant Z_E is defined so that the larger, $:\phi^2:_{ren}$, is finite, the smaller, $:(\phi_{ren})^2:$, must vanish.

4. Existence of the Scaling Limit

Existence of the scaling limit, in the weakest sense, means bounds uniform in ε on the renormalized Schwinger functions

$$S_{\text{ren}}^{(n)} = Z_3^{-n/2} S^{(n)}$$

so that by compactness, a convergent subsequence may be selected. By the explicit introduction of an invariant mean, the limit may be taken to be covariant under the translation subgroup. The Osterwalder-Schrader reconstruction theorem [21,22] then guarantees the existence of a scaling limit field theory which is at least translation covariant.

The Lebowitz correlation inequalities bound $S_{\text{ren}}^{(n)}$ by a sum of products of two point functions [11]. Thus existence follows from a uniform bound on $S_{\text{ren}}^{(2)}$. The required bounds on $S_{\text{ren}}^{(2)}$ are implied [12] by a conjectured correlation inequality

$$\Gamma^{(6)}(\text{xxx yyy}) \leq 0 \tag{4.1}$$

Here $\Gamma^{(6)}(x_1,\ldots,x_6)$ is the six point vertex, or direct correlation function. In graphical language, $\Gamma^{(6)}$ is one particle irreducible, and in (4.1) we may take $\Gamma^{(6)}$ in either its unamputated or its amputated form. Let $G^{(n)}$ denote the n-point truncated (Ursell) function. Choosing the unamputated form for the Γ's, and defining

$$\Gamma^{(2)} = -(G^{(2)})^{-1}$$

(convolution inverse), we have the explicit formulae

$$\Gamma^{(4)} = G^{(4)}$$

$$\Gamma^{(6)} = G^{(6)} + \frac{1}{2} \sum_{\text{Permutations}} (3!)^{-2} G^{(4)}(x_{i_1},\ldots x_{i_3},z) \times$$

$$\Gamma^{(2)}(z,z')G^{(4)}(a',x_{i_4},\ldots x_{i_6})$$

The $G^{(n)}$ are connected parts of the $S^{(n)}$.

The identity $\phi^2 \equiv$ const. in the scaling limit, allows an explicit calculation of $\Gamma^{(6)}$(xxx yyy). We first consider the more elementary calculation

$$G^{(4)}(xxyy) = S^{(4)}(xxyy) - S^{(2)}(xx)S^{(2)}(yy) - 2S^{(2)}(xy)^2$$

$$= \int(\phi^2(x) - \langle\phi^2(x)\rangle)(\phi^2(y) - \langle\phi^2(y)\rangle)d\phi - 2S^{(2)}(xy)^2$$

$$= \int:\phi^2(x)::\phi^2(y):d\phi - 2S^{(2)}(xy)^2 = -2S^{(2)}(xy)^2$$

For $\Gamma^{(6)}$ we have

$$\Gamma^{(6)}(xxxyyy) = -24\ G^{(2)}(xy)^3 \tag{4.2}$$

(4.2) holds only for $\varepsilon=0$, and thus does not establish (4.1) as $\varepsilon\to 0$, but it certainly makes the inequality (4.1) highly plausible, at least for small ε. Other tests of (4.1) include the one-dimensional Ising model, in which

$$\Gamma^6(x_1,\ldots,x_6) = -24\ G^{(2)}(x_1x_6)G^{(2)}(x_2x_5)G^{(2)}(x_3x_4)$$

for

$$x_1 \le x_2 \le \cdots \le x_6$$

[25], and numerical studies of the d=1 ϕ^4 field theory -- i.e., the anharmonic oscillator [19].

We remark that (4.1) has one other consequence: an absence of bound states in the two point function [12]. In particular if (4.1) holds, then $S^{(2)}$ has the form (1.3), with $0<Z_3\le 1$. Moreover in case (4.1) holds, there are no CDD zeros, which means that $\Gamma^{(2)} = -(G^{(2)})^{-1}$ has a 3m decay rate [12]. In particular the scaling limit has neither three particle bound states nor CDD zeros. Even bound states are also excluded by a general argument [3,27], which also applies for $\sigma>\sigma_c$.

The absence of bound states may be special to $\sigma>\sigma_c$, $\mu=0$. In fact for the two dimensional Ising model with $T<T_c$, it is suggested [15] that as $\mu \searrow 0$, the number of bound states becomes infinite, and fills in to form a continuum threshold. Here we appeal to the explicit solution of the Ising model for $T<T_c$, $\mu=0$. If the Ising and ϕ^4 theories have identical scaling limits, then a similar phenomena should occur in ϕ^4 field theory for $\sigma \lesssim \sigma_c, \mu \searrow 0$.

For d=1, existence of the scaling limit was established [17] on the basis of asymptotic estimates on the eigenvalues and eigenvectors of the anharmonic oscillator. This scale limit coincides with the one dimensional Ising model, proving universality in this case.

5. The Josephson Inequality

The Josephson inequality states

$$2-\alpha \leq d\nu \qquad (5.1)$$

[18]. For I_2, (5.1) is an equality, and for I_3, (5.1) is correct as an inequality, and nearly (or possibly) correct as an equality. For the Gaussian (free field) case, the exponents are

$$\nu = 1/2, \quad \alpha = 2-d/2, \quad 0 = \eta = \eta_E = \zeta_3 = \zeta_E, \quad \gamma = 1$$

and (5.1) is again an equality.

Following a similar calculation in [9] based on Lebowitz's inequality, we have

$$\varepsilon^{-\alpha/\nu} \sim C_H \leq \int \langle\phi(x)\phi(y)\rangle^2 dx$$

$$\leq O(1) \int_{|r|\leq\varepsilon^{-1}} r^{4-2d-2\eta} dr$$

$$\leq O(\varepsilon^{-(r-d-2\eta)}) \qquad \text{for } d = 2,3,4.$$

Thus

$$\alpha \leq (4-d-2\eta)\nu \qquad d = 2,3,4. \qquad (5.2)$$

Here we have used the scaling hypothesis for the two point function. Without this hypothesis, the same reasoning implies $\alpha \leq (4-d)\nu$ as a rigorous inequality, d=2,3.

Combined with (5.1), (5.2) reproduces the inequality $\nu \geq (2-\eta)^{-1} \geq \frac{1}{2}$ Furthermore of α/ν is strictly less than its Gaussian value, 4-d, then ν is strictly greater than its canonical value, 1/2. From Lebowitz's inequality, we have the (rigorous) inequality

$$\eta_E \geq 2\eta \geq 0 = \eta_{E\ \text{Gaussian}} = \eta_{\text{Gaussian}} \tag{5.3}$$

for d=1,2,3.

We complete this section with a formal derivation of (5.1), based on field theory ideas. The Hamiltonian and particle structure occur as hypotheses, and use is made of Lebowitz inequality. From (1.4), we have

$$\frac{dm}{d\sigma} \sim \left(\frac{\sigma-\sigma_c}{\sigma_c}\right)^{\nu-1} = \varepsilon^{1-1/\nu}$$

Let $\langle 1|$ be the one particle zero momentum state. Then

$$m = \langle 1|H|1\rangle$$

and

$$\frac{dm}{d\sigma} = \langle 1|\frac{dH}{d\sigma}|1\rangle$$

Equating these two expressions gives

$$\varepsilon^{1-1/\nu} \sim \langle 1|\int :\phi^2(x):d\vec{x}|1\rangle$$

and because $m\int_0^{1/m} e^{-mt}dt = 1-e^{-1}$ and $m = \varepsilon$,

$$\varepsilon^{-1/\nu} \sim \langle 1|\int_{|x_o|\leq m^{-1}} :\phi^2(x):dx|1\rangle \tag{5.4}$$

The right side of (5.4) will be bounded by a Schwarz inequality. To justify this step, we use two approximations

$$\langle 1| \to \text{const}\int_{|x|\leq m^{-1}} \phi(x)\,d\vec{x}\Omega \tag{5.5}$$

$$\text{const} = \Big(\int_{\substack{|\vec{x}|\leq m^{-1}\\ |\vec{y}|\leq m^{-1}}} S^{(2)}(x-y)\,d\vec{x}\;d\vec{y}\Big)^{-1/2}$$

and

$$\int_{|x_o|\leq m^{-1}} :^2(x):dx \to \int_{|x|\leq m^{-1}} :\phi^2(x):dx \tag{5.6}$$

With these approximations, $\langle 1|\varepsilon L_4$, with an L_4 norm bounded independently of ε. In fact $\|\langle 1|\|_{L_4}^4$ is a four point function divided by a product of two two point functions. The four point function can be expanded as

$$S^{(4)} = G^{(4)} - \sum G^{(2)} G^{(2)}$$

and by Lebowitz inequality

$$0 \leq -G^{(4)} \leq G^{(2)} G^{(2)}$$

Here the two point functions $G^{(2)}$ are bounded by the two point functions occurring in the denominator, and so $\langle 1|\varepsilon L_4$ as asserted.

Applying the Schwarz inequality to (5.4) yields

$$\varepsilon^{-1/\nu} \leq O\left(\int_{|x|,|y|\leq m} \langle :\phi^2(x)::\phi^2(y):\rangle dx\, dy\right)^{1/2}$$

$$\sim (\varepsilon^{-d-\alpha/\nu})^{1/2}$$

Thus

$$1/\nu \leq \tfrac{1}{2}(d + \alpha/\nu)$$

and (5.1) follows.

Because of the wave function renormalization in the scaling limit, we include a discussion of the substitution (5.5). Let $\langle 1_{appx}|$ denote the right side of (5.5). From the scaling hypothesis, we find that

$$\text{const} = O(\varepsilon^{(d+2-\eta)/2}) = Z_3^{-1/2} O(\varepsilon^{d/2})$$

in (5.5). The factor $Z_3^{-1/2}$ replaces ϕ by ϕ_{ren}, and ensures that the one particle contribution to ϕ_{ren} has unit strength. The factor is proportional to

$$\left\| \int_{|\vec{x}|\leq m^{-1}} \phi_{free}(x)\, d\vec{x}\, \Omega_{free} \right\|^{-1}$$

and guarantees that the one particle portion of $\langle 1_{appx}|$ has strength $O(1)$, uniformly as $\varepsilon \to 0$.

References

1. G. BAKER, Self-interacting boson quantum field theory and the thermodynamic limit in d dimensions, J. Math. Phys. 16, 1324 - 1346 (1975).
2. E. BAROUCH, B. MCCOY, C. TRAY and T. T. WU, The spin-spin correlation functions for the two dimensional Ising-model Exact theory in the scaling limit, to appear.
3. J. FELDMAN, On the absence of bound states in the ϕ^4 quantum field model without symmetry breaking, Cand. J. Phys. 52, 1583 - 1587 (1974).
4. J. FELDMAN and K. OSTERWALDER, The Wightman axioms and the mass gap for weakly coupled $(\phi)^4_3$ quantum field theories, to appear.
5. M. FISHER, Rigorous inequalities for critical point correlation exponents, Phys. Rev. 180, 594 - 600 (1969).
6. J. FRÖHLICH, Existence and analyticity in the bare parameters of the $|\lambda(\vec{\phi}\,\vec{\phi})^2 - \sigma\phi_1^2 - \mu\phi_1|$ quantum field models, I. Manuscript.
7. J. GLIMM and A. JAFFE, The $(\lambda\phi^4)_2$ quantum field theory without cutoffs III. The physical vacuum, Acta Math. 125, 203 - 261 (1970).
8. ———————, The $(\lambda\phi^4)_2$ quantum field theory without cutoffs IV. Perturbation of the Hamiltonian, J. Math. Phys. 13, 1558 - 1584 (1972).
9. ———————, ϕ^4_2 quantum field theory in the single phase region: Differentiability of the mass and bounds on critical exponents, Phys. Rev. D10, 536 - 539 (1974).
10. ———————, Two and three body equations in quantum field models, Commun. Math. Phys. 44, 293 - 320 (1974).
11. ———————, A remark on the existence of ϕ^4_4. Phys. Rev. Lett. 33, 440 - 442 (1974).
12. ———————, Three particle structure of ϕ^4 interactions and the scaling limit, Phys. Rev. D11, 2816 - 2827 (1975).
13. J. GLIMM, A. JAFFE, and T. SPENCER, The particle structure of the weakly coupled $P(\phi)_2$ models and other applications of high temperature expansions. In: Constructive quantum field theory, G. Velo and A. Wightman (eds.) Springer Verlag, Berlin, 1973.

14. J. GLIMM, A. JAFFE, and T. SPENCER, Existence of phase transitions for ϕ_2^4 quantum fields, Commun. Math. Phys. To appear.
15. ———, A cluster expansion for the ϕ_2^4 quantum field theory in the two phase region. In preparation.
16. F. GUERRA, L. ROSEN, and B. SIMON, In: Constructive quantum field theory, G. Velo and A. Wightman (eds.) Springer Verlag, Berlin, 1973.
17. D. ISAACSON, The critical behavior of the autoharmonic oscillator, NYU Thesis.
18. B. JOSEPHSON, Inequality for the specific heat, I Derivation, II Applications, Proc. Phil. Soc. 92, 269 - 284 (1967).
19. D. MARCHESIN, Work in progress.
20. J. MAGNEN and R. SENEOR, The infinite volume limit of the ϕ_3^4 model, Ann. Inst. H. Poincaré, to appear.
21. K. OSTERWALDER and R. SCHRADER, Axioms for Euclidean Green's functions, Commun. Math. Phys. 31, 83 - 112 (1974).
22. ———, Axioms for Euclidean Green's functions II, Commun. Math. Phys. 42, 281 - 305 (1975).
23. Y. PARK, Uniform bounds of the pressure of the $\lambda\phi_3^4$ field model. Preprint.
24. J. ROSEN, Mass renormalization for $\lambda\phi_2^4$ Euclidean lattice field theory.
25. J. ROSEN, Private communication.
26. E. SEILER and B. SIMON, Nelson's symmetry and all that in the Yukawa_2 and ϕ_3^4 field theories. Preprint.
27. T. SPENCER, The absence of even bound states in ϕ_2^4. Commun. Math. Phys. 39, 77 - 79 (1974).
28. E. BREZIN, J. C. LEGUILLORE and J. ZINN-JUSTIN, Field theoretical approach to critical phenomena. In: Phase transitions and critical phenomena, Vol. VI., Ed. by Domb and Green, Academic Press, New York, to appear.

XI A Tutorial Course in Constructive Field Theory

A TUTORIAL COURSE IN CONSTRUCTIVE FIELD THEORY

James Glimm[1]
The Rockefeller University
New York, N.Y. 10021

Arthur Jaffe[2]
Harvard University
Cambridge, MA 02138

---

1. Supported in part by the National Science Foundation under Grant PHY76-17191.
2. Supported in part by the National Science Foundation under Grant PHY75-21212.

1.1 INTRODUCTION

Prior to 1970, the major focus of constructive field theory was the mathematical framework required to establish the existence of quantum fields [1,2]. Since 1970, the emphasis has gradually shifted, first toward verifying physical properties of the known models, and more recently toward bringing constructive field theory closer to the mainstream of physics [3]. In fact, by 1973 it had become more or less clear that the mathematical framework developed to give the first examples in $d = 2, 3$ space-time dimensions would be adequate to study $d = 4$. However, it was also clear that to solve the ultraviolet problem in $d = 4$, it would be necessary to incorporate into mathematical physics a deeper physical understanding of the questions being studied. In particular ideas of scaling and of critical behavior may be useful to select and analyze a suitable nontrivial critical point as the first step in dealing with the $d = 4$ ultraviolet problem. An infinite scaling transformation connects the problem of removing the ultraviolet cutoff with the problem of existence of scaling behavior at the critical point.

In these lectures we describe some results for boson (generally φ^4) fields. For $d = 4$, scaling arguments (e.g. the renormalization group) indicate that the φ^4 theory is not free at high energy. Thus the study of the φ_4^4 theory is a strong coupling problem which must be approached independent of perturbation theory. See, e.g. [4, 5] for a discussion of the mathematical aspects of this program. On the other hand, asymptotic freedom suggests that $d = 4$, nonabelian gauge field theories may be free at high energy and thus amenable to perturbation theory. For this reason gauge fields should play an important role in the future of the mathematical study of field theory, and the lecture of Osterwalder provides an introduction to that topic.

1.2 e^{-tH} AS A FUNCTIONAL INTEGRAL

It is now easy to formulate the connection between Euclidean field theory (functional integrals) and quantum mechanics (Hilbert space and Hamiltonians). We give a complete mathematical presentation of this connection for boson fields in the accompanying article, "Functional Integral Methods in Quantum Field Theory" [6]. Here we give a brief summary.

Assume we are given a functional integral $d\mu(\varphi)$, i.e. a measure $d\mu(\varphi)$ on generalized functions $\varphi \in \mathscr{S}'(R^d)$. Here φ is a classical field, and given a functional $F(\varphi)$, we define the expectation $\langle F \rangle$ by

$$(1.1) \qquad \langle F \rangle = \int F(\varphi) d\mu(\varphi) .$$

We require the normalization condition

$$(1.2) \qquad \langle 1 \rangle = \int d\mu(\varphi) = 1 ,$$

i.e. $d\mu(\varphi)$ is a probability measure. Of particular interest is the choice $F(\varphi) = \exp(i\varphi(f))$, which leads to the Fourier transform of $d\mu(\varphi)$, which we denote

$$(1.3) \qquad S\{f\} = \int e^{i\varphi(f)} d\mu(\varphi) .$$

$S\{f\}$ is also known as the generating functional.

We now wish to define a Hilbert space $\mathcal{H}$ of quantum mechanics in $d - 1$ space dimensions, associated with $d\mu(\varphi)$. In order to define $\mathcal{H}$, the measure $d\mu$ must satisfy three simple requirements, which we expect from every such functional integral which arises in physics. The vectors in $\mathcal{H}$ (at least a dense set of vectors) will be functionals $A = A(\varphi)$, for example $A = \varphi(f)$ or $A = \exp(i\varphi(f))$, and for certain $f \in \mathscr{S}$. The fundamental formula which relates the Hamiltonian H to functional integrals is

$$(1.4) \qquad \boxed{\langle A, e^{-tH} A \rangle_{\mathcal{H}} = \int \overline{\vartheta A}\, A_t \, d\mu(\varphi)} ,$$

where ϑ denotes time inversion, $\overline{A}$ denotes complex conjugation and A_t denotes time translation. In (1.4) A plays a dual role. On the left side, A is a vector in the Hilbert space of quantum mechanics $\mathcal{H}$, while on the right side $A(\varphi)$ is a functional of the classical field. The importance of (1.4) is that any question about H can be reduced to a question about a functional integral. For example: Does H have a unique vacuum? Does H have a mass gap? Does H have bound states and what are their masses? Do the asymptotic states span $\mathcal{H}$ (asymptotic completeness)? etc. We start with a brief explanation of (1.4), see [6] for details.

Let us start by mentioning the three requirements on $S\{f\}$ or $d\mu$ needed to obtain $\mathcal{H}$. (The generalization to include spinor or tensor fields poses no fundamental difficulty.) The conditions are

(1) Invariance
(2) Reflection Positivity
(3) Regularity

(1) We require invariance of $S\{f\}$ under a one parameter group of time translations $f(\vec{x}, s) \to f_\tau(\vec{x}, t-\tau)$ where $\tau \in R$ (or $\tau = n\delta$, $n = 0, \pm 1, \pm 2, \cdots$, i, e, lattice translations) and also invariance under time inversion $f(\vec{x}, t) \to (\vartheta f)(\vec{x}, t) = f(\vec{x}, -t)$. Then $S\{f_\tau\} = S\{\vartheta f\} = S\{f\}$.

(2) Let $f^{(j)}$, $j = 1, 2, \cdots, r$ be any sequence of real functions which vanish unless $t > 0$. Then reflection positivity is the requirement $S\{f_i - \vartheta f_j\} = M_{ij}$ is a positive matrix. In other words, if $A = \sum_{j=1}^{r} c_j \exp(i\varphi(f^{(j)}))$, then

$$\int \vartheta \bar{A}\, A d\mu \geq 0 . \tag{1.5}$$

This condition was discovered by Osterwalder and Schrader [7].

(3) The regularity assumption is important for a continuous time translation group, in which case $S\{f_\tau\} \to S\{f\}$ as $\tau \to 0$.

From (1-3) we obtain $\mathcal{H}$ and e^{-tH} (in the case of a continuous time translation group). In the case of discrete time translations we obtain $\mathcal{H}$ and a transfer matrix K, with K^t playing the role of e^{-tH}. In [6] we assume invariance under the full Euclidean group on R^d (rotations, translations and reflections in hyperplanes) and we obtain a stronger result, namely the construction of a full relativistic field theory.

In fact, assuming (2), we may define a scalar product on functionals $A(\varphi)$ at positive time, e.g. functionals of the form

$A(\varphi)$ in (1.5). We let

$$\langle A, B\rangle_{\mathcal{H}} \equiv \int \overline{\vartheta A}\, B\, d\mu(\varphi) \; . \tag{1.6}$$

The Hilbert space $\mathcal{H}$ is the completion of these positive time $A(\varphi)$ in the scalar product (1.6), and it is these positive time functionals $A(\varphi)$ which are allowed in (1.4). Time translation is defined by

$$A_t = \sum c_j \exp(i\varphi(f^{(j)}))_t = \sum c_j \exp(i\varphi(f_t^{(j)})) \; , \tag{1.7}$$

so for $t \geq 0$, A_t is also a positive time functional. Thus the formula

$$\langle A, R(t)A\rangle_{\mathcal{H}} = \int \overline{\vartheta A}\, A_t d\mu \tag{1.8}$$

defines an operator $R(t)$ on $\mathcal{H}$. As a consequence of (1) - (2), it follows that

$$0 \leq R(t) = R(t)^* \leq I, \;\; R(t + s) = R(t)R(s) \; , \tag{1.9}$$

see for example [6]. For $t = 1$, $R(1) = K$ is the self adjoint transfer matrix, and $R(t) = K^t$. In case that time is continuous, then (3) ensures that $R(t) = e^{-tH}$, where H is a positive, self adjoint Hamiltonian, $0 \leq H = H^*$. Furthermore, since the functional $A(\varphi) = 1$ satisfies $A_t = A = 1$, it follows that $R(t)1 = 1$ and 1 is a ground state for H, $H1 = 0$. The vector 1 in $\mathcal{H}$ is generally denoted by Ω, so $H\Omega = 0$.

1.3 EXAMPLES

A. Gaussian Examples

Consider the case of full Euclidean symmetry. The simplest example is the Gaussian measure $d\mu_0$ corresponding to the free field of mass $\sigma^{\frac{1}{2}}$

$$S_0\{f\} = \int e^{i\varphi(f)} d\mu_0(\varphi) = e^{-\frac{1}{2}\langle f, (-\Delta+\sigma)^{-1} f\rangle} . \tag{1.10}$$

In other words, $d\mu_0(\varphi)$ is the unique Gaussian measure with mean zero and covariance $(-\Delta + \sigma)^{-1}$. The only question to verify is whether (1.10) satisfies reflection positivity. This fact follows from

$$(1.11)\qquad \langle \vartheta f, (-\Delta + \sigma)^{-1} f\rangle = \left(\int_0^\infty \| e^{-t\mu}(2\mu)^{-\frac{1}{2}} f(\cdot, t)\|_{L_2} dt\right)^2$$

$$\geq 0 ,$$

where f vanishes unless $t \geq 0$ and where $\mu^2 = -(\vec{\nabla})^2 + \sigma$, where $\vec{\nabla}$ is the gradient in the $\vec{x}$ directions.

We can approximate R^d by a lattice Z^d, in which case we obtain

$$(1.12)\qquad d\mu_{0,\delta} = \lim_{\Lambda \nearrow Z^d_\delta} \frac{1}{N(\delta, \Lambda)} e^{-\frac{1}{2}\sum_{(n,n)} \delta^{d-2}(\varphi(i)-\varphi(i'))^2 - \frac{1}{2}\sigma\sum_i \delta^d \varphi(i)^2} \prod_i d\varphi(i).$$

In (1.12), Λ denotes a finite subset of the lattice Z^d_δ with lattice spacing δ. The sum and product over i extend over lattice sites in Λ, while the sum $\sum_{(n,n)}$ extends over nearest neighbor lattice sites in Λ. The measure $d\varphi(i)$ is Lebesgue measure, and $N(\delta, \Lambda)$ is chosen so $\int d\mu_{0,\delta} = 1$. The limit $\Lambda \nearrow Z^d$ exists in the sense of convergence of $S_{0,\delta,\Lambda}\{f\}$.

We can rewrite (1.12) as

$$(1.13)\qquad d\mu_{0,\delta} = \lim_{\Lambda \nearrow Z^d_\delta} e^{\beta \sum_{(n,n)} \varphi(i)\varphi(i')} \prod_i d\nu_0(\varphi(i)) ,$$

where $\beta = \delta^{d-2}$ and $d\nu_0(\varphi)$ is a Gaussian. Thus $d\mu_{0,\delta}$ is a ferromagnetic, nearest neighbor spin system with inverse temperature β and a single spin distribution function $d\nu_0(\varphi)$. The ferromagnetic coupling arises from the off-diagonal part of the gradient term in (1.12).

The limit $\delta \to 0$ for (1.12) - (1.13) exists and is (1.10). (This limit, however, must be taken in the form $S_{0,\delta}\{f\} \to S\{f\}$, since the measures $d\mu_{0,\delta}$ do not converge in the usual sense of convergence of measures on a finite dimensional space.) Thus formally

(1.14) $$d\mu_0 = N^{-1}\int e^{-\frac{1}{2}\int[(\nabla\varphi(x))^2+\sigma\varphi(x)^2]dx} \prod_{x\in R^d} d\varphi(x) ,$$

but the mathematically meaningful form of (1.14) is (1.10).

B. Non-Gaussian Examples

In the case of invariance of $d\mu$ under continuous time translations, we obtain non-Gaussian examples from non-relativistic quantum mechanics ($d = 1$) and nonlinear quantum fields ($d = 2, 3$). In the case of nonrelativistic quantum mechanics with one degree of freedom, the measure

(1.15) $$d\mu(q) = \frac{1}{N} e^{-\int_{-\infty}^{\infty} V(q(t))dt} d\mu_0(q)$$

formally satisfies (1-3). The quantum mechanics Hamiltonian is just

(1.16) $$H = -\frac{1}{2}\frac{d^2}{dq^2} + \frac{1}{2}q^2 + V(q) - E$$

$$= H_0 + V - E ,$$

i.e., the perturbation of the harmonic oscillator Hamiltonian H_0 by the potential V. Here E is a constant chosen so $H \geq 0$, $H\Omega = 0$. The formula (1.15) is exactly the Feynman-Kac functional integral representation of the ground state of H. In fact integration of $A(q(0))$ over $d\mu(q)$ is expectation of A in the state Ω:

(1.17) $$\int A(q_0)d\mu(q) = \lim_{T\to\infty} \frac{1}{N_T}\int e^{-\int_{-T}^{T} V(q(t))dt} A(q(0))d\mu_0(q)$$

$$= \lim_{T\to\infty} \frac{\langle e^{-TH}\Omega_0, Ae^{-TH}\Omega_0\rangle}{\|e^{-TH}\Omega_0\|^2}$$

$$= \langle \Omega, A\Omega\rangle .$$

Here Ω_0 is the ground state of H_0 and Ω is the ground state of H.

In quantum field theory, the known examples for $d = 2, 3$ arise from interaction energy densities $P(\varphi(x)) = V(x)$ for polynomials P. For instance the $\lambda\varphi_2^4$ model has the measure

$$(1.18) \qquad d\mu = \lim_{\Lambda \nearrow R^2} \frac{1}{N_\Lambda} e^{-\int_\Lambda V(x)dx} d\mu_0$$

$$= \lim_{\Lambda \nearrow R^2} \frac{1}{N_\Lambda} e^{-\int_\Lambda :\varphi^4(x):dx} d\mu_0 .$$

The Wick ordering of V is necessary since for $d > 1$, the measure $d\mu_0$ is concentrated on certain distributions, rather than on continuous functions as in the case of $d = 1$ (e.g. Wiener measure). The main work in this approach to proving existence of nonlinear quantum fields lies in establishing the existence of measures such as (1.18).

On a lattice, the local interaction V(x) in (1.18) does not effect the nearest neighbor term of the measure (1.12)-(1.13). Rather, V(x) contributes to the distribution of a single spin $d\nu(\varphi)$. Thus a lattice $P(\varphi)_d$ model has a measure

$$(1.19) \qquad d\mu_\delta = \lim_{\Lambda \nearrow Z_\delta^d} e^{\delta^{d-2} \sum_{(n,n)} \varphi(i)\varphi(i')} \prod_i d\nu(\varphi(i)) ,$$

with

$$d\nu(\varphi) = \frac{1}{N} e^{-\delta^d V(\varphi)} d\nu_0(\varphi) .$$

Here N is an appropriate normalization constant. For example, the lattice $\lambda\varphi_d^4$ model has

$$(1.20) \qquad d\nu(\varphi) = \frac{1}{N} e^{-\delta^d \lambda (\varphi^2 - c)^2} d\nu_0(\varphi) ,$$

where

$$c = 3\int \varphi^2 d\mu_0(\varphi) = \begin{cases} O(\ln \delta^{-1}) & d = 2 \\ O(\delta^{-(d-2)}) & d > 2 \end{cases}$$

is the Wick ordering constant.

1.4 APPLICATIONS OF THE FUNCTIONAL INTEGRAL REPRESENTATION

The usefulness of formula (1.4) is that it provides a computational framework to answer questions concerning the spectrum of H. For instance consider the question: Is the ground state Ω unique? In other words, we ask whether $e^{-tH}\theta - P_\Omega\theta$ converges to zero as $t \to \infty$. Here P_Ω is the projection onto Ω,

$$(1.21) \qquad P_\Omega\theta = |\Omega\rangle\langle\Omega|\theta\rangle = \int \theta(\varphi)d\mu(\varphi) .$$

Equivalently, is there a dense set of positive time functionals $\theta(\varphi)$ such that

$$(1.22) \quad F_\theta(t) \equiv \langle\theta, e^{-tH}(1 - P_\Omega)\theta\rangle = \int \overline{\vartheta\theta}\, \theta_t d\mu - \left|\int \theta d\mu\right|^2$$

converges to zero as $t \to +\infty$? If so, then Ω is the unique ground state of H. On the other hand if there is a positive time functional θ for which $F_\theta(t) \not\to 0$, then Ω is degenerate. In terms of the measure $d\mu(\varphi)$, the uniqueness of the vacuum Ω is equivalent to ergodicity of $d\mu(\varphi)$ under the group of time translations.

If, more generally, there are exactly r normalized, orthogonal vacuum states Ω_i, $i = 1, 2, \cdots, r$, then the function

$$F(t) = \int \overline{\vartheta\theta}\; \theta_t d\mu - \sum_{i=1}^{r} \left|\int \Omega_i \theta d\mu\right|^2$$

would converge to zero as $t \to +\infty$, rather than (1.22).

We see later that in the case of $\lambda\varphi_2^4$ or $\lambda\varphi_3^4$ quantum field models, the vacuum Ω is unique for $\lambda \ll 1$. However, Ω is degenerate for $\lambda \gg 1$. The construction of $d\mu(\varphi)$ outlined above, yields for $\lambda \gg 1$ an even mixture of vacuum states and $\langle\varphi\rangle = 0$. By introducing boundary conditions to select a particular vacuum, we may obtain solutions breaking the $\varphi \to -\varphi$ symmetry, $\langle\varphi\rangle \neq 0$, and with a unique vacuum. See the lectures of Fröhlich and Spencer for further discussion of phase transitions, and in particular a discussion of continuous symmetry breaking.

A second question concerning H is the existence of a mass gap, i.e. a gap in the spectrum corresponding to massive particles. The occurrence of a gap (0, m) in the spectrum is equivalent to

$$|F_\theta(t)| \le O(1)e^{-mt}, \tag{1.23}$$

where the constant O(1) depends on θ. Thus again the spectral properties of H are reduced to asymptotic decay rates of certain functional integrals. The proof of such decay rates in models has been established by expansion methods or by using correlation inequalities as described below.

In a theory which is even (e.g. a φ^4 model in which the symmetry $\varphi \to -\varphi$ is not broken) we can decompose the Hilbert space $\mathcal{H} = \mathcal{H}_e + \mathcal{H}_o$ into subspaces even or odd under the transformation $\varphi \to -\varphi$. The vacuum lies in $\mathcal{H}_e$, while the one particle states lie in $\mathcal{H}_o$. Let m denote the bottom of the spectrum on $\mathcal{H}_o$. On $\mathcal{H}_e$, we thus expect a mass gap of magnitude m', where $m < m' \le 2m$. (Since two particle scattering states occur in $\mathcal{H}_e$, the Hamiltonian will always have spectrum throughout the interval $[2m, \infty)$.) The statement $m' = 2m$ is the statement that two particle bound states do not occur in $\mathcal{H}_e$. This is equivalent to

$$|F_\theta(t)| \le O(1)e^{-2mt} \tag{1.24}$$

as θ ranges over a dense set of $\mathcal{H}_e$. We discuss this further in the next section.

Finally, in order to analyze the bound states or scattering of several particles, it is useful to study kernels (e.g. exact Bethe-Salpeter kernels) which characterize the Hamiltonian for n-body processes. Such kernels have a functional integral representation, and a detailed study has been made by Spencer and Zirilli [7, 8] in the case n = 2. (See also [2, 3, 9]

1.5 ISING, GAUSSIAN AND SCALING LIMITS

We briefly mention the qualitative structure of the φ_d^4 lattice quantum field model of §1.3, in its dependence on the parameters δ, λ, σ. In particular, we discuss the measures $d\mu_\delta$ defined in (1.19)-(1.20); an analogous discussion could be given for $P(\varphi)$ models. See [4, 5, 10].

To begin with, consider the (λ, δ) parameter space with fixed σ. For δ fixed, we study $\lambda \to 0$ and $\lambda \to \infty$, the minimum and maximum coupling. It is clear that the $\lambda \to 0$ limit of (1.19)-(1.20) is Gaussian, in fact $d\mu_\delta \to d\mu_{0,\delta}$. (In every case we define convergence as convergence

(1.25) $$S_\delta\{f\} = \int e^{i\varphi(f)} d\mu_\delta(\varphi) \to S_{0,\delta}\{f\} = \int e^{i\varphi(f)} d\mu_{0,\delta}(\varphi)$$

of generating functionals.) On the other hand, for $\lambda \to \infty$, with δ fixed, the measure $d\nu(\varphi)$ becomes concentrated at the points where $|\varphi| = c^{\frac{1}{2}}$, i.e. $\varphi = \pm c^{\frac{1}{2}}$. Since the integral of $d\mu_\delta$ is normalized to one, in this limit

(1.26) $$d\mu_\delta = \lim_{\Lambda \nearrow Z^d_\delta} e^{\beta \sum_{(n,n)} \varphi(i)\varphi(i')} \prod_i d\nu(\varphi_i)$$

where $d\nu(\varphi) = \frac{1}{N}(\delta(\varphi - c^{\frac{1}{2}}) + \delta(\varphi + c^{\frac{1}{2}}))$. In other words, $d\mu_\delta$ is an Ising model with lattice spacing δ and spin φ normalized to take the values $\pm c^{\frac{1}{2}}$. The mathematical existence of this Ising limit was established [11]. Furthermore, for $d = 2$ (or with $d = 3$ and the proper choice $\sigma = \sigma(\delta) = O(\ln \delta^{-1})$ to ensure massrrenormalization, the $\delta \to 0$ limit can be taken with λ fixed. This continuum limit yields the Euclidean φ^4 model.

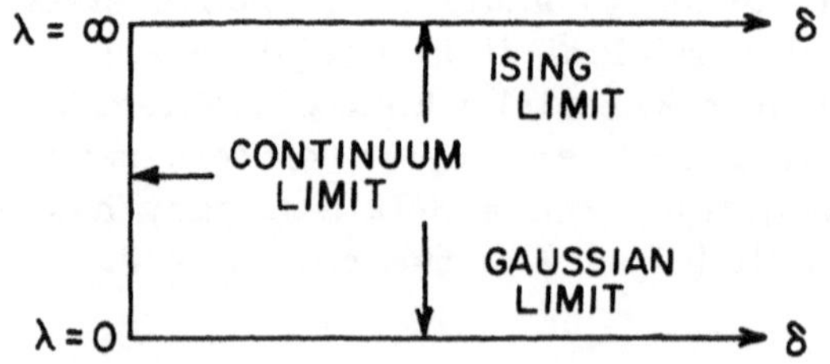

We next modify this picture slightly by fixing the mass gap m. On the lattice, m is defined as the gap in the spectrum of $-\ln K$ (K is the transfer matrix); if $\delta = 0$, m is defined as the gap in the spectrum of H. Since $m = m(\lambda, \delta, \sigma)$, we achieve this by choosing σ in such a way that we remain in the single phase region and such that $m = \text{const.}$ The required continuity of m follows by [12, 13]. We now plot the projection of such a $\sigma = \sigma(\lambda, \delta)$ surface in the (λ, δ) plane.

One can now ask whether the $\delta \to 0$ limit of Ising models exist, and whether the $\lambda \to \infty$ limit of continuum models exists. The first is a scaling limit of the Ising model, the second is a scaling limit of the $\lambda\varphi^4$ model. In the continuum theory $(d = 2, 3)$, increasing λ with σ fixed would result in a phase transition and $m \to 0$. Thus in the scaling limit, with m fixed, it follows that $\sigma(\lambda, \delta = 0) \to \infty$ as $\lambda \to \infty$, i.e. there is an infinite change of scale. Also the scaling limit is formally an infinite scaling of a $m = 0$ (critical) theory. We conjecture that both the $\lambda \to \infty$ and the $\delta \to 0$ scaling limits exist, and that they agree. (See §1.5.)

In studying this limit, it is also useful to consider curves in the (λ, δ) plane with constant unrenormalized, dimensionless charge $g_0 \doteq \lambda \delta^{4-d}$. For $d < 4$ (the superrenormalizable case) these curves lead to the scaling limit $(\lambda = \infty, \delta = 0)$ discussed above, i.e. strong coupling. For $d > 4$ (the nonrenormalizable case) these curves lead to $\lambda = \delta = 0$, i.e. weak coupling.

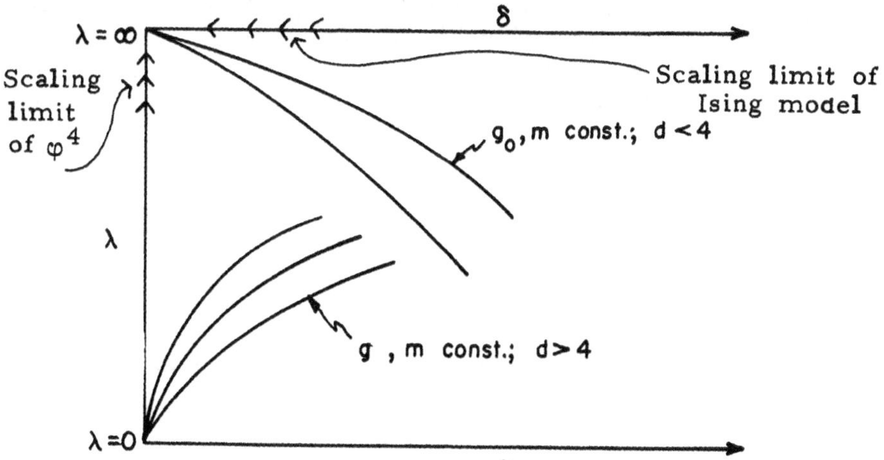

This picture leads us to conjecture that for $d > 4$, the $\delta \to 0$ limit for g_0 bounded is Gaussian (trivial), but a nontrivial theory could result with $g_0 \to \infty$ as $\delta \to 0$ (charge renormalization). See §2.5 for further discussion, and also [20, 10].

1.6 MAIN RESULTS

We sketch the main results for the φ^4 model, φ^4 lattice fields and Ising models, some of which we discuss in detail in the next two chapters. There are two main methods to derive these results: correlation inequalities (discussed in Chapter 2) and expansion techniques (discussed in Chapter 3). The correlation inequalities for the φ^4 model express in part the repulsive character of the forces in these models (in the single phase region). We obtain mathematical proofs for portions of the critical behavior in these models, as well as an initial analysis of the elementary particle and bound state problems.

A. The critical point has a conventional φ^4 structure. For $\sigma >> 0$, there is a unique phase, independent of boundary conditions, and for $\sigma << 0$ and $d \geq 2$, there are at least two pure phases, depending on the boundary conditions [14, 15, 16] The critical value $\sigma_c = \sigma_c(\lambda, \delta)$ is defined as the largest (and presumably the only) value of σ for which $m = m(\sigma) \to 0$ as $\sigma \searrow \sigma_c$. This critical σ_c exists [13] and $m(\sigma)$ is monotonic increasing for $\sigma \geq \sigma_c$ [17]. For $d \geq 3$ (and presumably for $d = 2$ also) $m(\sigma_c) = 0$, and for $\sigma = \sigma_c$ there is a unique phase and zero magnetization [13]. For $\sigma \geq \sigma_c$, the physical charge g (defined as the amputated connected four point function evaluated at zero momentum) is finite and bounded, uniformly as $\sigma \searrow \sigma_c$, and $\lambda \to \infty$. (See §2.3.) Furthermore the n-point Schwinger functions $S^{(n)}$ satisfy a Gaussian upper bound

$$0 \leq S^{(n)}(x_1 \cdots x_n) \leq \sum_{\text{pairing}} \prod_{\text{pairs}} S^{(2)}(x_i x_{i'}) ,$$

see [18, 19]. Closely related are critical exponent bounds of the form

$$\text{Gaussian exponent} \leq \varphi\text{-exponent} \leq \varphi^2\text{-exponent} .$$

For example with η the anomalous dimension of the field φ and η_E the anomalous dimension of the field $:\varphi^2(x):$,

$$0 \leq 2\eta \leq \eta_E .$$

See [20, 21, 22] for other recent exponent inequalities.

B. Particles do not form even bound states, for $\sigma \geq \sigma_c$. In the single phase region, particles exist for a.e. $m > 0$ [21]. (Presumably they exist for all $m > 0$, but at least for $d = 2, 3$, the particles should not exist for $m = 0$, i.e. $\sigma = \sigma_c$.) Here particles are poles in the two point function at Minkowsky momenta, or δ-functions in a Källen-Lehmann representation for the two point function or (at least in the Euclidean invariant case of a continuum field) an Ornstein-Zernike decay rate

$$\langle\varphi(x)\varphi(y)\rangle \sim Zr^{(1-d)/2}e^{-mr}$$

as $r = |x - y| \to \infty$. These particles do not form even bound states (with energies below the two particle continuum) [23, 24] and there are indications that they do not form odd bound states (with energies below the three particle continuum) [23, 24].

C. A heuristic interchange of the $\delta \to 0$ (continuum) limit and the $\lambda \to \infty$ (Ising) limit "shows" the identical critical point behavior [25,19]. Combining this idea with the known spectral properties of the $d = 2$ Ising suggests that for φ_2^4 continuum or lattice fields, with $\sigma < \sigma_c$, the elementary particle is actually a two soliton bound state. Furthermore in this picture the binding energy goes to zero (relative to the soliton mass) as $\sigma \nearrow \sigma_c$ and the field strength renormalization $Z \to 0$, and it probably vanishes faster than the strength of the two soliton continuum, thus suggesting that because of the solitons, the intermediate renormalization is correct, but the mass shell renormalization is incorrect, for $\sigma \nearrow \sigma_c$, $d = 2$, unless the mass shell renormalization includes all sectors, with the soliton as elementary particle. Introducing a small external field $-\mu\varphi(x)$ in the action, we have argued [26] that for $d = 2$ and $\sigma < \sigma_c$, but $\sigma \approx \sigma_c$, the limit $\mu \to 0$ introduces many bound states which coalesce to form a two soliton continuum at $\mu = 0$. For $\sigma << \sigma_c$ the same reasoning suggests many resonances coalescing to form continuous spectrum.

D. Within its region of convergence, the cluster expansion allows a nearly complete analysis of the field theory. To begin with, one can prove convergence of the infinite volume limit and uniqueness of the vacuum. (In a two phase region, suitable boundary conditions are required to select a pure phase.) The convergence is valid for λ complex in a sector about $\lambda = 0$ and it follows that the correlation functions are also analytic in λ and other parameters, for $\lambda \neq 0$. For the case of a $\lambda\varphi_2^4$ interaction in the region $\lambda << 1$, the perturbation series about $\lambda = 0$ is Borel summable to the exact solution [27]. For general $P(\varphi)_2$ interactions in the region (5.1), the perturbation series about $\lambda = 0$ for the Euclidean and Minkowsky correlation functions and for the S-matrix is also asymptotic [28, 29, 30]. The particles (whose scattering is described by the S-matrix) are also constructed from the cluster expansion [15]. Criteria (in terms of P) for existence or nonexistence of weak coupling $P(\varphi)_2$ bound states are given in [31] following earlier work of [8]. Here the cluster expansion permits the study of the exact Bethe-Salpeter equation, and shows that the low order terms (ladder approximation + $\cdots$) give the dominant effects. Up to some energy level, this analysis of the Bethe-Salpeter equation also shows asymptotic completeness [8]. Unfortunately, the energies allowed by present techniques apparently do not reach up to the two soliton threshold, in the two phase region.

2.1 CORRELATION INEQUALITIES

In this section we derive some basic correlation inequalities and state some others. In the following section we derive some consequences of these inequalities, and finally we consider the conjectured inequality $\Gamma^{(6)} \leq 0$.

For positive integers $a_1, \cdots, a_n$ define

(2.1) $$\varphi_A = \varphi(x_1)^{a_1} \cdots \cdot \varphi(x_n)^{a_n} .$$

Theorem 2.1: For a lattice $(\lambda\varphi^4 - \mu\varphi)_d$ quantum field with $\lambda, \mu \geq 0$,

(GKS 1) $$0 \leq \langle\varphi_A\rangle$$

(GKS 2) $$0 \leq \langle\varphi_A\varphi_B\rangle - \langle\varphi_A\rangle\langle\varphi_B\rangle .$$

These are the first and second Griffiths (Griffiths, Kelly, Sherman) inequalities. Since they say that certain quantities are positive, they are preserved under limits, e.g. $\delta \to 0$ or

$\lambda \to \infty$, whenever such limits exist. The solution of the $P(\varphi)_2$ ultraviolet problem [1] was extended by Guerra, Rosen and Simon to lattice cutoffs [17] in order to prove GKS and related inequalities for these models. See [32] for φ_3^4.

Proof of GKS 1: We write on a finite lattice

$$(2.4)\qquad \langle\varphi_A\rangle = \int\varphi_A d\mu_\delta = \int\varphi_A e^{\beta\Sigma\varphi(i)\varphi(i')+\mu\Sigma\varphi(i)} \prod_i d\nu(\varphi(i)) .$$

Note that $d\nu(\varphi) = d\nu(-\varphi)$, so

$$(2.5)\qquad \int\varphi(i)^{a_i} d\nu(\varphi(i)) = \begin{cases} 0 \text{ if } a_i \text{ odd} \\ \text{positive if } a_i \text{ even} \end{cases}$$

The basic idea of the proof is to expand the exponentials in (2.4) in power series and then factor the resulting integrals over lattice sites. Using (2.5), we obtain a sum of products of positive terms, and hence (2.4) is positive. We then take the limit as the finite lattice increases to Z^d.

Proof of GKS 2: The basic idea is to use the technique of duplicate variables. Let φ, ψ be independent, lattice fields. Define an expectation of functionals $A = A(\varphi, \psi)$ by

$$(2.6)\qquad \langle A\rangle = \int A(\varphi, \psi)d\mu(\varphi)_\delta d\mu(\psi)_\delta .$$

Here

$$d\mu(\varphi)_\delta = e^{\beta\Sigma\varphi(i)\varphi(i')}\prod_i d\nu(\varphi(i))$$

$$d\mu(\psi)_\delta = e^{\beta\Sigma\psi(i)\psi(i')}\prod_i d\nu(\psi(i)) ,$$

and for simplicity we let $\mu = 0$. Define

$$(2.7)\qquad t = \varphi + \psi, \quad q = \varphi - \psi$$

as the even and odd combinations of φ, ψ, under interchange of φ and ψ. Note

(2.8) $$\langle t\rangle = \langle\varphi\rangle + \langle\psi\rangle = 2\langle\varphi\rangle ,$$

$$\langle q\rangle = 0 ,$$

$$d\mu(\varphi)_\delta d\mu(\psi)_\delta = e^{\frac{1}{2}\beta\Sigma(t(i)t(i')+q(i)q(i'))}\prod_i d\nu\left(\frac{t(i)+q(i)}{2}\right)d\nu\left(\frac{t(i)-q(i)}{2}\right) .$$

First we remark that $d\nu(t+q)d\nu(t-q)$ is even under the transformation $t \to -t$ and also under $q \to -q$. Thus

(2.9) $$\int q^\alpha t^\beta d\nu(t+q)d\nu(t-q) = \begin{cases} 0 \text{ if } \alpha \text{ or } \beta \text{ odd} \\ \text{positive if } \alpha, \beta \text{ both even} . \end{cases}$$

Following the proof of GKS 1, and using (2.9), we find that for all A, B,

(2.10) $$0 \le \langle q_A t_B\rangle .$$

To complete the proof of GKS 2, we write

(2.11) $$\langle\varphi_A\varphi_B\rangle - \langle\varphi_A\rangle\langle\varphi_B\rangle = \langle\varphi_A(\varphi_B - \psi_B)\rangle$$

$$= 2^{-|A|-|B|}\langle(t+q)_A[(t+q)_B - (t-q)_B]\rangle .$$

But $(t+q)_B(t-q)_B$ is a polynomial in t, q with positive coefficients. Hence (2.10) shows that (2.11) is positive.

The proof above follows the presentation of Sylvester [33], which we recommend for proofs of other correlation inequalities. We now state three inequalities:

Theorem 2.2: For a lattice $(\lambda\varphi^4 - \mu\varphi)_d$ quantum field theories with $\lambda, \mu \ge 0$,

(2.12) $$\langle t_A t_B\rangle - \langle t_A\rangle\langle t_B\rangle \ge 0 ,$$

(2.13) $$\langle q_A q_B\rangle - \langle q_A\rangle\langle q_B\rangle \ge 0 ,$$

$$\langle q_A t_B \rangle - \langle q_A \rangle \langle t_B \rangle \le 0 . \tag{2.14}$$

The inequalities (2.14) were first proved for the Ising model by Lebowitz [34], and have a number of interesting consequences. We remark that two special cases of (2.14) are

$$\langle \varphi(x_1)\varphi(x_2)\varphi(x_3) \rangle_T \le 0 , \quad \mu \ge 0 , \tag{2.15}$$

$$\langle \varphi(x_1)\varphi(x_2)\varphi(x_3)\varphi(x_4) \rangle_T \le 0 , \quad \mu = 0 . \tag{2.16}$$

Here $\langle \ \rangle_T$ denotes the truncated (connected) expectation values, defined by

$$\langle \varphi(f)^n \rangle_T = \frac{d^n}{d\alpha^n} \ln \langle e^{\alpha\varphi(f)} \rangle \Big|_{\alpha=0} , \tag{2.17}$$

and extended to $\langle \varphi(f_1)\cdots\varphi(f_n) \rangle_T$ by multilinearity. We obtain (2.15), the Griffiths-Hearst-Sherman inequality, by expanding $\langle t_1 q_2 q_3 \rangle - \langle t_1 \rangle \langle q_2 q_3 \rangle \le 0$. The inequality (2.16) follows from evaluating $\langle t_1 t_2 q_3 q_4 \rangle - \langle t_1 t_2 \rangle \langle q_3 q_4 \rangle \le 0$ in case $\mu = 0$.

2.2 ABSENCE OF EVEN BOUND STATES

In a single phase, even φ^4 model, i.e. for $\sigma > \sigma_c$, we now show that the Hamiltonian, restricted to $\mathcal{H}_{even}$, has no spectrum in the interval $(0, 2m)$, i.e. two particle bound states do not exist. We remark that $\mathcal{H}_{even}$ is spanned by vectors Ω, $\varphi(f_1)\cdots\varphi(f_n)\Omega$, $n = 2, 4, \cdots$, where suppt f_j is contained in $t > 0$.

Theorem 2.3: Consider a φ^4 field or Ising model with zero external field and $\sigma > \sigma_c$, and let A and B have an even number of elements. Then

$$\langle \varphi_A \varphi_B \rangle - \langle \varphi_A \rangle \langle \varphi_B \rangle \le \sum_{\substack{A_1 \subset A,\ A_1 \text{ odd} \\ B_1 \subset B,\ B_1 \text{ odd}}} \langle \varphi_{A_1} \varphi_{B_1} \rangle \langle \varphi_{A-A_1} \varphi_{B-B_1} \rangle$$

Corollary 2.4: Under the hypothesis of Theorem 3.3, there are no even bound states with energy below the two particle threshold.

Proof: Let Ω be the vacuum state, unique since it is assumed that $\sigma > \sigma_c$. We write $x = x_1, \cdots, x_d$ as

$$x = (t, \vec{x})$$

with $\vec{x} \in R^{d-1}$. In particular if

$$A + s = \{(t + s, \vec{x}) : t, \vec{x} \in A\} ,$$

then the equation (1.4), namely

$$\langle \varphi_A e^{-sH} \varphi_B \rangle = \langle \vartheta \varphi_A \varphi_{B+s} \rangle ,$$

is valid when the times in A precede the times in B. In particular we choose A to have only negative times $t \le 0$, and B, chosen as

$$B = \{(-t, \vec{x}) : (t, \vec{x}) \in A\}$$

then has only positive times. With this choice of A and B, and with P_Ω the projection onto the vacuum state, we recognize

$$\langle \varphi_{A+s} \varphi_{B+s} \rangle - \langle \varphi_{A+s} \rangle \langle \varphi_{B+s} \rangle = \| e^{-sH} (I - P_\Omega) \varphi_A \Omega \|^2$$

so that the Theorem 2.3 gives a bound on the decay rates which occur in

$$e^{-sH}(I - P_\Omega).$$

For A_1 odd, $\varphi_{A_1} \Omega$ is perpendicular to the vacuum $((\Omega, \varphi_{A_1} \Omega) = \langle \varphi_{A_1} \rangle = 0)$, and so $\langle \varphi_{A_1 - s} \varphi_{B_1 + s} \rangle$ has as its slowest exponential decay rate, m, by definition the mass of the theory. Thus by definition of m,

$$\langle \varphi_{A_1 - s} \varphi_{B_1 + s} \rangle \le C_{A_1, B_1} e^{-ms}$$

for some constant C_{A_1, B_1} depending on A_1 and B_1. The same bound holds for $\langle \varphi_{(A - A_1) - s} \varphi_{(B - B_1) + s} \rangle$, and so by Theorem 2.3,

$$\| e^{-sH}(I - P_\Omega)\varphi_A \Omega \|^2 \le \text{const.}\, e^{-2ms} .$$

Thus there are no even states, except Ω, with energy below 2m, hence in particular no even bound states in this energy range.

Proof of Theorem 2.3:

$$\langle t_A q_B \rangle = \sum_{\substack{A=A_1 \cup A_2 \\ B=B_1 \cup B_2}} (-1)^{|B_2|} \langle \varphi_{A_1} \psi_{A_2} \varphi_{B_1} \psi_{B_2} \rangle$$

$$= \sum (-1)^{|B_2|} \langle \varphi_{A_1} \varphi_{B_1} \rangle \langle \varphi_{A_2} \varphi_{B_2} \rangle$$

$$\le \langle t_A \rangle \langle t_B \rangle = \sum (-1)^{|B_2|} \langle \varphi_{A_1} \rangle \langle \varphi_{A_1} \rangle \langle \varphi_{B_1} \rangle \langle \varphi_{B_2} \rangle .$$

We drop zero terms from the right hand side (A_2 or B_2 odd), and from the left hand side (only one of A_2, B_2 odd). For the terms with B_2 even and the partition nontrivial, we combine right and left sides and eliminate from the inequality, using Theorem 2.2. The terms remaining yield Theorem 2.3.

2.3 BOUND ON g

Define the dimensionless φ^4 coupling constant by

$$g = -m^{d-4} \chi^{-4} \int \langle \varphi(x_1) \cdots \varphi(x_4) \rangle_T dx_1 dx_2 dx_3$$

where

$$\chi = \int \langle \varphi(x_1) \varphi(x_2) \rangle dx_1 = \int_0^\infty \frac{d\rho(a)}{a} .$$

By GKS 1, $\chi \ge 0$. For a massive, single phase, even φ^4 interaction $g \ge 0$ by (2.16). We now assume in addition that the proper field strength renormalization has been performed;

in the case of an isolated particle of mass m, this means $d\rho(a) = \delta(a - m^2)da + d\sigma(a)$, where inf suppt $d\sigma > m^2$. We then prove an upper bound on g.

<u>Theorem 2.5</u> [35]: Under the above assumptions,

$$0 \le g \le \text{const.} ,$$

where the dimensionless constant is independent of all parameters (e.g. λ, σ).

We outline the proof. For details, see the original paper. We use the basic inequality GKS 2 to derive (writing 1 for $\varphi(x_1)$, etc.)

$$(2.18)\ 0 \le \langle 1234\rangle - \langle 12\rangle\langle 34\rangle = \langle 1234\rangle_T + \langle 13\rangle\langle 24\rangle + \langle 14\rangle\langle 23\rangle .$$

By (2.16), $\langle 1234\rangle_T \le 0$ and

$$0 \le -\langle 1234\rangle_T \le \langle 13\rangle\langle 24\rangle + \langle 14\rangle\langle 23\rangle .$$

After symmetrization over the choices of variables,

$$(2.19)\ -\langle 1234\rangle_T \le (\langle 13\rangle\langle 24\rangle + \langle 14\rangle\langle 23\rangle)^{\frac{1}{3}}(\langle 12\rangle\langle 34\rangle + \langle 13\rangle\langle 24\rangle)^{\frac{1}{3}}$$

$$\times (\langle 14\rangle\langle 23\rangle + \langle 12\rangle\langle 34\rangle)^{\frac{1}{3}} .$$

From elementary properties of the Green's function for the Poisson operator (i.e. $\ker(-\Delta + a)^{-1}(x, y)$) we find

$$\langle xy\rangle = \int_0^\infty \ker(-\Delta + a)^{-1}(x, y)d\rho(a)$$

$$\le \text{const.}\chi |x - y|^{-d}\exp(-m|x - y|/2) .$$

Inserting this in our bound (2.19) for $-\langle 1234\rangle_T$ gives

$$g \le \text{const.}\, m^{-4}\chi^2 .$$

Since

$$\chi = \int_{m^2}^{\infty} \frac{d\rho(a)}{a} \geq m^{-2},$$

we obtain $g \leq$ const. as claimed.

Observe that the final bound does not depend on m, and hence also holds in the limit $m \to 0$. Hence the critical point (which for $d < 4$ should be an infrared stable fixed point of the renormalization group) occurs for finite g.

2.4 BOUND ON $dm^2/d\sigma$ AND PARTICLES

Here we consider a canonical, single phase φ^4 model (i.e. without field strength renormalization). We establish

$$\frac{dm^2(\sigma)}{d\sigma} \leq Z(\sigma) \tag{2.20}$$

from which our next result follows by approximation methods:

Theorem 2.6: (See [21].) For almost every value of m, particles exist, i.e. $Z \neq 0$.

Proof of (2.20): Consider $\Gamma(p) = -S(p)^{-1}$, where $S(p)$ is the Fourier transform of $\langle\varphi(x)\varphi(0)\rangle$. Note

$$S(p) = \frac{Z}{p^2 + m^2} + \int \frac{d\sigma(a)}{p^2 + a},$$

and

$$Z^{-1} = -(d\Gamma/dp^2)_{p^2=-m^2} \,. \tag{2.21}$$

Since $\Gamma = 0$ on the one particle curve $p^2 = -m^2(\sigma)$, $\nabla\Gamma$ must be orthogonal to the vector $(dm^2/d\sigma, 1)$ in the $-p^2, \sigma$ space. Thus for $p^2 = -m^2$,

$$0 = -\frac{\partial\Gamma}{\partial p^2}\frac{dm^2}{d\sigma} + \frac{\partial\Gamma}{\partial\sigma} = Z^{-1}\frac{dm^2}{d\sigma} + \frac{\partial\Gamma}{\partial\sigma}.$$

The desired inequality follows from

Theorem 2.7: Under the above assumptions,

$$\left(\frac{\partial\Gamma}{\partial\sigma}\right)_{p^2=-m^2} \geq -1 . \qquad (2.22)$$

Proof: Let $\chi(p) = \int\langle\varphi(x)\varphi(0)\rangle e^{-px}dx$. Then

$$-\frac{d\chi(p)}{d\sigma} = \frac{1}{2}\int\int[\langle xozz\rangle - \langle xo\rangle\langle zz\rangle]dze^{-px}dx$$

by (2.16), (2.18),

$$\leq \int\int\langle xz\rangle\langle yz\rangle e^{-p(x-z)}e^{-pz}dxdz$$

$$= \chi(p)^2 .$$

Thus

$$0 \leq \frac{d\chi(p)^{-1}}{d\sigma} \leq 1 .$$

However $\chi(p) = -\Gamma(p)|_{p^2=-m^2}$, so (2.22) is proved.

2.5 THE CONJECTURE $\Gamma^{(6)} \leq 0$

The unamputated six point vertex function is defined by

$$\Gamma^{(6)}(xxxyyy) = \langle xxxyyy\rangle_T + \int\langle xxxz\rangle_T\Gamma(zz')\langle z'yyy\rangle_T dzdz' + 9\int\langle xxyz\rangle_T\Gamma(zz')\langle z'xyy\rangle_T dzdz' . \qquad (2.23)$$

The conjecture

$$\Gamma^{(6)}(xxxyyy) \leq 0 \qquad (2.24)$$

has a number of interesting consequences: e.g., the absence of three particle bound states in the propagator, the existence of the scaling limit, and certain bounds on critical exponents, see [25, 5, 20].

There is some evidence for (2.24) in single phase, even φ^4 models. For example it is true in perturbation theory (i.e. for $\sigma >> 0$ or high temperature). It holds in the one dimensional Ising model [36] and numerical studies indicate that it holds for the anharmonic oscillator [37]. There is a heuristic argument that it holds near σ_c. However, some good new idea is needed to prove (2.24).

In this section we illustrate some uses of (2.24). For example, we have

Theorem 2.8: If (2.24) holds, then

$$0 \le \Gamma(x) \le e^{-3m|x|}, \quad |x| \to \infty . \tag{2.25}$$

Remark: The bound (2.25) excludes spectrum in $\Gamma(x)$ in the interval $(0, 3m)$, and hence spectrum in $d\sigma(a)$ in the interval $(m, 3m)$. Thus no three particle bound states occur in the propagator, i.e. in the states spanned by $\varphi(x)\Omega$.

Outline of Proof: We use the integration by parts formula [25]

$$\int \varphi(x)A(\varphi)d\mu(\varphi) = \langle \varphi(x)A \rangle \tag{2.26}$$

$$= \int dyS(x - y)[\langle \frac{\delta A}{\delta\varphi(y)} \rangle - \langle \mathcal{V}'(y)(I - P_1)A \rangle] .$$

Here $\mathcal{V} = \lambda :\varphi^4:$ is the interaction, and $P_1A = \int \varphi(z)\Gamma(z - z')\langle\varphi(z')A\rangle dzdz'$. From (2.26), it follows that for $x \neq 0$,

$$\Gamma(x - y) = \langle \mathcal{V}'(x)(I - P_1)\mathcal{V}'(y) = \lambda^2 \langle \varphi^3(x)(I - P_1)\varphi^3(y) \rangle , \tag{2.27}$$

see [25]. Expanding (2.27), and using (2.23),

$$\lambda^{-2}\Gamma(x - y) = 6\langle xy \rangle^3 + 9\langle xxyy \rangle_T \langle xy \rangle \tag{2.28}$$

$$- 9\int \langle xxyz \rangle_T \Gamma(zz')\langle z'xyy \rangle_T dzdz' + \Gamma^{(6)}(xxxyyy).$$

The first term in (2.28) is $o(e^{-m|x-y|})^3$, for $|x - y| \to \infty$. The second term is negative. The third term, also has a three particle decay, which can be established using the absence of two particle bound states in $\langle xxyz\rangle_T$, see [25]. Thus (2.24) results in

$$\Gamma(x - y) \le e^{-3m|x-y|}, \quad |x - y| \to \infty .$$

The positivity of $\Gamma(x - y)$ follows from the fact that it is the Fourier transform of a Herglotz function. This completes the outline of the proof.

We finish this section with the statement of another consequence of (2.24), and an Ornstein-Zernicke upper bound

$$(2.29) \qquad \langle\varphi(x)\varphi(0)\rangle \le K \min(|x|, m^{-1})^{-(d-2+\eta)} e^{-m|x|} .$$

Theorem 2.8 [20]: Assume (2.24), (2.29) and $\lambda < \infty$. Then

$$\eta \le \begin{cases} .4 & d = 4 \\ .8 & d = 3 \\ 1.2 & d = 2 \end{cases} .$$

Also $\eta = 0$, $Z^{-1} < \infty$ for $d \ge 5$.

Corollary 2.9 [20]: Assume (2.24). If the $\delta \to 0$ limit of the $\lambda\varphi_d^4$ lattice field theory is Euclidean invariant for $g_0(\delta) = \lambda\delta^{4-d} \le$ const. (finite charge renormalization), then the limit is a free field for $d \ge 6$.

3. CLUSTER EXPANSIONS

3.1 THE REGION OF CONVERGENCE

The cluster expansion, in field theory as in statistical mechanics, provides almost complete information for parameter values away from critical, and it provides only limited information for parameter values near critical. In statistical mechanics, this expansion is a variant of the virial, high temperature and low temperature (Peierls' contour) expansions. These names distinguish various regions of the coupling constants, and expansion parameters. In field theory, convergence of the cluster expansion is known for the corresponding parameter values. In particular, for a two dimensional $P(\varphi)$ field theory, the cluster expansion is convergent in the following asymptotic regions [14, 15, 38, 26, 39].

(3.1) $$P(\varphi) = \lambda :P_0(\varphi): + \sigma\varphi^2, \quad \lambda \to 0$$

(3.2) $$P(\varphi) = \lambda :P_0(\varphi): - \mu\varphi, \quad \mu \to \infty$$

(3.3) $$P(\varphi) = \lambda :\varphi^4: + \sigma\varphi^2, \quad \sigma \to -\infty \text{ or } \lambda \to +\infty,$$

or more generally, whenever P, expanded about a suitable global minimum φ_c of $P(\varphi)$ has a dominant quadratic term. For the Yukawa$_2$ and φ_3^4 interactions, convergence of the expansion is known in the high temperature region ($\lambda \to 0$ as in (3.1)) [40-43].

3.2 THE ZEROTH ORDER EXPANSION

The expansion is adapted from the virial expansion of statistical mechanics. In the zeroth order, all couplings are removed. We divide Euclidean space time into cells (lattice cubes) and then remove the coupling between distinct cells. In the zeroth approximation, all correlations factor,

$$\langle \varphi(x_1)\cdots\varphi(x_n)\rangle_0 = \prod_{\Delta=\text{cell}} \langle \prod_{x_j\in\Delta} \varphi(x_j)\rangle_0 .$$

Consequently the long distance behavior is trivial and all states have infinite energy in the zeroth approximation.

To define the zeroth approximation, let $\partial\Delta$ be the boundary of the cell Δ.. Then formally

(3.4) $$\mathcal{a}_0(\varphi) = \infty \sum_{\Delta} \int_{\partial\Delta} \varphi^2(x)dx + \mathcal{a}$$

is the action defining the zeroth approximation, if $\mathcal{a}$ is the action of the full theory. To rewrite this expression in mathematical language, we combine the $\partial\Delta$ term in $\mathcal{a}_0$ with the gradient term in $\mathcal{G}$,

$$(3.5) \qquad \infty \sum_{\Delta} \int_{\partial\Delta} \varphi^2 dx + \int \nabla\varphi^2(x) dx = \langle \varphi - \Delta_D \varphi \rangle ,$$

where $-\Delta_D$ is the Laplace operator with zero Dirichlet boundary conditions on all cube faces $\partial\Delta$.

3.3 THE PRIMITIVE EXPANSION

Graphical expansions in statistical mechanics are generated by the identity

$$(3.6) \qquad \prod_{i<j} e^{-V(r_i - r_j)} = \prod_{i<j} [1 - e^{-V(r_i - r_j)} - 1)]$$

$$= \sum_{\text{sets of pairs}} \quad \prod_{\text{pairs in set}} (e^{-V(r_i - r_j)} - 1) .$$

Now a pair $\{r_i, r_j\}$ is represented by the line segment connecting r_i to r_j, so that a set of pairs is a graph. Thus (3.6) is a graph expansion. Before adapting (3.6) to field theory, we modify it slightly. Let $V^{(s)}$ be a one parameter family of potentials, $0 \le s \le 1$, with

$$V^{(0)} = 0 , \quad V^{(1)} = V .$$

Then

$$e^{-V} - 1 = \int_0^1 \frac{d}{ds} e^{-V^{(s)}} ds ,$$

and introducing a parameter s_{ij} for each pair $\{r_i, r_j\}$, we have

$$(3.7) \qquad \prod_{i<j} e^{-V(r_i - r_j)} = \sum_{\Gamma} \int \prod_{\{ij\} \in \Gamma} \frac{d}{ds_{ij}} e^{-V^{(s_{ij})}} ds_{ij}$$

where Γ is a graph constructed from pairs $\{i, j\}$. In (3.7), the s_{ij} for $\{i, j\}$ not in Γ are evaluated at $s = 0$. Writing

$$s^{\Gamma} = \{s_{ij} : \{i,j\} \in \Gamma\}$$

$$s^{\Gamma^c} = \{s_{ij} : \{i,j\} \notin \Gamma\} ,$$

this formula can be expressed in the compact notation

$$\prod_{i<j} e^{-V(r_i - r_j)} = \sum_{\Gamma} \int \partial s^{\Gamma} \prod_{\{i,j\} \in \Gamma} e^{-V(s_{ij})} ds^{\Gamma} \Big|_{s^{\Gamma^c}=0} . \tag{3.8}$$

To adapt this formula to field theory, we only need to choose a multiparameter family $s^{\Gamma} = \{s^{\gamma} : \gamma \in \Gamma\}$ interpolating between the totally decoupled Laplacian Δ_D and the ordinary Laplacian Δ. This can easily be done, and we only mention that it is convenient to choose the s-dependence to be linear in the inverse operators $(-\Delta + m_0^2)^{-1}$. In field theory, (3.8) becomes

$$\langle F(\varphi) \rangle = \sum_{\Gamma} \int \partial s^{\Gamma} \langle F(\varphi) \rangle_s ds^{\Gamma} \Big|_{s^{\Gamma^c}=0} . \tag{3.9}$$

Here $F(\varphi)$ is some function of φ and $\langle \quad \rangle_s$ is the expectation defined by the interpolating "Laplacian" $\Delta(s)$. Observe that $s^{\gamma} = 0$ corresponds to Dirichlet data on γ, and no coupling, while $s^{\gamma} > 0$ allows coupling across Γ. Thus the hypercube faces $\gamma \in \Gamma$ transmit coupling from one cube to its neighbors, while the faces $\gamma \in \Gamma^c$ do not.

3.4 FACTORIZATION AND PARTIAL RESUMMATION

In graphical language, we are concerned here with the cancellation of disconnected vacuum contributions. In (3.9), the expectation $\langle \cdot \rangle$ is normalized so that $\langle 1 \rangle = 1$, while $\langle \cdot \rangle_s$ is normalized by the same denominator, so that

$$\langle 1 \rangle_s = \int e^{-\int P(\varphi) dx} d\varphi_{0,s} \Big/ \int e^{-\int P(\varphi) dx} d\varphi_0$$

$$= Z(s)/Z$$

where $d\varphi_{0,s}$ is the interpolating Gaussian measure defined by the parameter s. It follows that the expansion (3.9) defines disconnected graphs.

Suppose $F(\varphi)$ depends on $\varphi(x)$ only for x in some set $S \subset R^d$. In (3.9), we consider the term with Γ fixed. Because of the Dirichlet data on Γ^c, this term decouples across Γ^c. More precisely suppose Γ^c divides R^d into a certain number of connected components. Let Y be the closure of the union of the components which meet S; graphically, Y is the region connected to S by Γ. Then the expectation $\langle F(\varphi)\rangle_s \big|_{s_{\Gamma^c}=0}$ factors across ∂Y, allowing us to write it as a product of an "inside" expectation associated with Y times an "outside" expectation associated with $R^d \sim Y$.

Now let us fix Y in place of Γ and sum over all Γ which yield this fixed Y. This sum factorizes into independent sums over $\Gamma \cap Y$ and $\Gamma \sim Y$. The sum over $\Gamma \sim Y$ of the outside expectations has no constraints, and sums to a partition function $Z(\sim Y)$ for the region $\sim Y$, defined with Dirichlet data on ∂Y. (This resummation is just the reverse of the expansion (3.9), but for the region $\sim Y$ in place of R^d.) In a similar fashion we define $Z(Y)$, the partition function for the region Y, with Dirichlet data on ∂Y. Let

$$\langle F(\varphi)\rangle_Y Z(Y)/Z$$

denote the sum over $\Gamma \cap Y$ of the inside expectations (including the overall normalization denominator Z). Then (3.9) becomes

$$\langle F(\varphi)\rangle = \sum_Y \langle F(\varphi)\rangle_Y \frac{Z(Y)Z(\sim Y)}{Z} . \tag{3.10}$$

In this expression, all disconnected vacuum components have been resummed into the single factor $Z(Y)Z(\sim Y)/Z$. Somewhat lengthy estimates, including a Kirkwood-Salzburg equation, prove the convergence of (3.10) for the parameters and interactions indicated in §3.1.

3.5 TYPICAL APPLICATIONS

To prove exponential decay of correlations, i.e. a positive mass, we use the method of duplicate variables. Thus we write

$$|\langle F(\varphi)G(\varphi)\rangle - \langle F(\varphi)G(\varphi)\rangle| = 2^{-1}|\langle [F(\varphi) - F(\psi)][G(\varphi) - G(\psi)]\rangle|$$

$$\leq \sum_{Y} |\langle [F(\varphi) - F(\psi)][G(\varphi) - G(\psi)]\rangle_{Y}|$$

$$\times \frac{Z(Y)Z(\sim Y)}{Z} .$$

Here we choose Y to be the closure of the union of the connected components containing the support of F alone. However Y's which do not also meet the support of G then vanish, by the $\varphi \to \psi$ symmetry. The convergence proof shows that the remaining terms have the desired exponential decay rate.

To establish the upper mass gap and the isolated one particle hyperboloid, the cluster expansion can be applied to the vertex functions $\Gamma^{(n)}$ in place of the Schwinger functions considered above. Similarly a cluster expansion in the Bethe-Salpeter kernel K defined graphically by

(3.11) 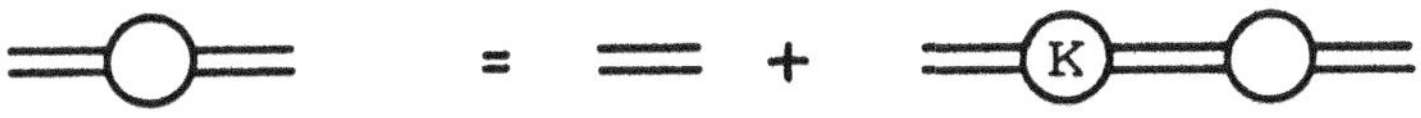

shows that K has three particle (and for even theories four particle) decay. In Minkowski momentum space, this means that K has an analytic continuation up to a neighborhood of the three or four particle threshold. Returning to (3.11), the only singularities of the four point function up to these energies come from the two particle cut. Hence asymptotic completeness up to these energies follows.

To extend these methods to the two phase region (3.3), we choose a length L >> 1 and write

$$\bar{\varphi} = L^{-d} \int_{L\text{-cube}} \varphi(x)dx$$

$$\sigma(\bar{\varphi}) = \operatorname{sgn} \bar{\varphi} .$$

Then σ is an Ising variable, defined on the lattice $(LZ)^d$. An expansion in Peierls contours isolates large regions of R^d in which σ is a constant. In these regions, the field φ is concentrated within a single well of the W-shaped potential, and a Gaussian approximation to this well is possible. This Gaussian approximation is controlled by a cluster expansion. Thus the final expansion is a double expansion, first in Peierls contours and then Dirichlet contours, as in (3.10).

REFERENCES

[1] J. Glimm and A. Jaffe, Quantum Field Models, in Statistical Mechanics, C. DeWitt and R. Stora (eds.) 1970 Les Houches Lectures, Gordon and Breach Science Publishers, New York, 1971.

[2] ________, Boson Quantum Field Theory, in Mathematics of Contemporary Physics, R. Streater (ed.), London Mathematical Society, Academic Press, 1972.

[3] J. Glimm, A. Jaffe and T. Spencer, The Particle Structure of the Weakly Coupled $P(\varphi)_2$ Model and Other Applications of High Temperature Expansions, Part I. Physics of Quantum Field Models, in Constructive Quantum Field Theory, G. Velo and A. S. Wightman (eds.), 1973 Erice Lectures, Springer Lecture Notes in Physics, Vol. 25, Springer-Verlag, 1973.

[4] J. Glimm and A. Jaffe, Critical Problems in Quantum Fields, presented at the International Colloquium on Mathematical Methods of Quantum Field Theory, Marseille, June 1975.

[5] ________, Critical Exponents and Renormalization in the φ^4 Scaling Limit, to appear in Quantum Dynamics: Models and Mathematics, L. Streit (ed.), Springer.

[6] ________, Functional Integral Methods in Quantum Field Theory, proceedings of the 1976 Cargèse Summer Institute.

[7] T. Spencer, The Decay of the Bethe-Salpeter Kernel in $P(\varphi)_2$ Quantum Field Models, Commun. Math. Phys. 44, 143-164 (1975).

[8] T. Spencer and F. Zirilli, Scattering States and Bound States in $\lambda P(\varphi)_2$, Commun. Math. Phys. 49, 1-16 (1976).

[9] J. Glimm and A. Jaffe, Particles and Bound States and Progress Toward Unitarity and Scaling, presented at the International Symposium on Mathematical Problems in Theoretical Physics, Kyoto, January 23-29, 1975.

[10] R. Schrader, A Constructive Approach to φ_4^4. I. Commun. Math. Phys. 49, 131-153 (1976); II. preprint; III. Commun. Math. Phys. 50, 97-102 (1976).

[11] J. Rosen, The Ising Model Limit of φ^4 Lattice Fields, preprint, 1976.

[12] J. Glimm and A. Jaffe, The φ_2^4 Quantum Field Model in the Single Phase Region: Differentiability of the Mass and Bounds on Critical Exponents, Phys. Rev. D10, 536-539 (1974).

[13] O. McBryan and J. Rosen, Existence of the Critical Point in φ^4 Field Theory, preprint 1976.

[14] J. Glimm, A. Jaffe and T. Spencer, The Particle Structure of the Weakly Coupled $P(\varphi)_2$ Model and Other Applications of High T mperature Expansions, Part II. The Cluster Expansion, in Constructive Quantum Field Theory, G. Velo and A. S. Wightman (eds.), 1973 Erice Lectures, Springer Lecture Notes in Physics, Vol. 25, Springer-Verlag, 1973.

[15] __________, The Wightman Axioms and Particle Structure in the $P(\varphi)_2$ Quantum Field Model, Ann. Math. 100, 585-632 (1974).

[16] J. Fröhlich, B. Simon and T. Spencer, Infrared Bounds, Phase Transitions and Continuous Symmetry Breaking, Commun. Math. Phys. 50, 79-95 (1976).

[17] F. Guerra, L. Rosen, B. Simon, The $P(\varphi)_2$ Euclidean Quantum Field Theory as Classical Statistical Mechanics, Ann. Math. 101, 111-259 (1975).

[18] J. Glimm and A. Jaffe, A Remark on the Existence of φ_4^4, Phys. Rev. Lett. 33, 440-442 (1974).

[19] C. Newman, preprint.

[20] J. Glimm and A. Jaffe, Particles and Scaling for Lattice Fields and Ising Models, Commun. Math. Phys. 51, 1-14 (1976).

[21] __________, Critical Exponents and Elementary Particles, Commun. Math. Phys., to appear.

[22] R. Schrader, preprint.

[23] J. Feldman, On the Absence of Bound States in the $\lambda\varphi_2^4$ Quantum Field Model without Symmetry Breaking, Canad. J. Phys. 52, 1583-1587 (1974).

[24] T. Spencer, The Absence of Even Bound States for $\lambda(\varphi^4)_2$, Commun. Math. Phys. 39, 77-79 (1974).

[25] J. Glimm and A. Jaffe, Three Particle Structure of φ^4 Interactions and the Scaling Limit, Phys. Rev. D11, 2816-2827 (1975).

[26] J. Glimm, A. Jaffe and T. Spencer, A Convergent Expansion about Mean Field Theory, Part I. The Expansion, Ann. Phys. 101, 610-630 (1976).

[27] J.-P. Eckmann, J. Magnen, and R. Sénéor, Decay Properties and Borel Summability for the Schwinger Functions in $P(\varphi)_2$ Theories, Commun. Math. Phys. 39, 251-271 (1975).

[28] J. Dimock, The $P(\varphi)_2$ Green's Functions: Asymptotic Perturbation Expansion, Helv. Phys. Acta 49, 199-216 (1976).

[29] K. Osterwalder and R. Sénéor, The Scattering Matrix is Non-Trivial for Weakly Coupled $P(\varphi)_2$ Models, Helv. Phys. Acta 49, 525-534 (1976).

[30] J.-P. Eckmann, H. Epstein and J. Fröhlich, Asymptotic Perturbation Expansion for the S-Matrix and the Definition of Time-Ordered Functions in Relativistic Quantum Field Models, Ann. de l'Inst. H. Poincaré 25, 1-34 (1976).

[31] J. Dimock and J.-P. Eckmann, Spectral Properties and Bound State Scattering for Weakly Coupled $P(\varphi)_2$ Models, to appear in Annals of Physics.

[32] Y. Park, Lattice Approximation of the $(\lambda\varphi^4 - \mu\varphi)_3$ Field Theory in a Finite Volume, J. Math. Phys. 16, 1065-1075 (1975).

[33] G. Sylvester, Continuous-Spin Ising Ferromagnets, Ph.D. thesis, M.I.T., 1976; J. Stat. Phys., to appear.

[34] J. Lebowitz, GHS and Other Inequalities, Commun. Math. Phys. 35, 87-92 (1974).

[35] J. Glimm and A. Jaffe, Absolute Bounds on Vertices and Couplings, Ann. de l'Inst. H. Poincaré 22, 1-11 (1975).

[36] J. Rosen, Mass Renormalization for Lattice $\lambda\varphi_2^4$ Fields, preprint, 1976.

[37] D. Marchesin, private communication.

[38] T. Spencer, The Mass Gap for the $P(\varphi)_2$ Quantum Field Model with a Strong External Field, Commun. Math. Phys. 39, 63-76 (1974).

[39] J. Glimm, A. Jaffe and T. Spencer, A Convergent Expansion about Mean Field Theory, Part II. Convergence of the Expansion, Annals of Phys. 101, 631-669 (1976).

[40] J. Magnen and R. Sénéor, The Infinite Volume Limit of the φ_3^4 Model, Ann. de l'Inst. H. Poincaré 24, 95-159 (1976).

[41] J. Feldman and K. Osterwalder, The Wightman Axioms and the Mass Gap for Weakly Coupled $(\varphi^4)_3$ Quantum Field Theories, Annals of Phys. 97, 80-135 (1976).

[42] J. Magnen and R. Sénéor, Wightman Axioms for the Weakly Coupled Yukawa Model in Two Dimensions, to appear in Commun. Math. Phys.

[43] A. Cooper and L. Rosen, The Weakly Coupled Yukawa_2 Field Theory: Cluster Expansion and Wightman Axioms, preprint.